£45.00

Design Management

A HANDBOOK OF ISSUES AND METHODS

Design Management

A HANDBOOK OF ISSUES AND METHODS

EDITED BY
MARK OAKLEY

ADVISORY EDITORS
BRIGITTE BORJA DE MOZOTA, COLIN CLIPSON

Blackwell Reference

Copyright © Basil Blackwell 1990
© Editorial organization Mark Oakley 1990

First published 1990

Basil Blackwell Ltd
108 Cowley Road, Oxford OX4 1JF, UK

Basil Blackwell Inc.
3 Cambridge Center, Cambridge, MA 02142, USA

British Library Cataloguing in Publication Data

Design management.
1. Design.   Management aspects
I. Oakley, Mark
658.5'7
ISBN 0–631–15404–3

Library of Congress Cataloging-in-Publication Data

Design management: a handbook of issues and methods/edited by Mark Oakley.
    Includes bibliographical references.
    ISBN 0–631–15404–3
    1. Design, Industrial—Management.
    I. Oakley, Mark.
TS171.4.D483   1990   89–37145   CIP
658.5'72—dc20

Printed in Great Britain by
Butler & Tanner Ltd, Frome, Somerset.

Contents

Foreword

The 1980s saw design emerge as a key weapon in the fight for industrial and commercial survival. Companies which had already updated manufacturing methods, labour relations, product distribution and customer service now began to realize that it was the design of their products that could determine whether they would prosper or not.

For many manufacturers in Britain, the United States and elsewhere, the power of product excellence was harshly brought home by the success of new foreign producers. In sector after sector, high quality, well-designed products from overseas were pushing aside – and, in some cases, entirely displacing – domestically produced goods. True, relatively low selling prices, deriving from lower overseas wage costs, sometimes had important effects. But increasingly, as the decade wore on, many foreign products dominated their markets not because of low cost but because of good design.

In Britain, the 'Corfield Report' published by the National Economic Development Office in 1979 provided a much needed stimulus to action. The report graphically underlined the importance of design by citing telling examples of products and companies. It spelt out the consequences of neglecting or misunderstanding design – and drove home that design, which they too often chose to leave to others, was a responsibility of *managers*.

Many initiatives followed from Corfield. The Thatcher government acted by supporting 'funded consultancy schemes' to get more designers into companies. The Design Council intensified its efforts to promote the effective use of design. The Council for National Academic Awards recognized that the training of young managers at business schools did not generally include reference to design management. It set out to rectify this deficiency by initiating work (undertaken by the editor of this present volume) to identify the elements of a suitable curriculum – resulting in a blueprint which has been widely adopted in the UK and abroad.

During 1984, Tony Sweeney, then a commissioning editor at Basil Blackwell, noted the interest being generated in design management and found that there was no competent reference book covering the topic. The original plan was to produce a kind of 'referenced dictionary' containing relatively short descriptions and discussions. However, it soon became clear that the subject area was still at such an embryonic stage of development as a management discipline – with little agreement about terminology, priorities and 'best practice' – that a prescriptive text would be inappropriate and potentially misleading. What was needed, it was concluded, was a set of contributions from practitioners, academics and others that would constitute an effective survey of design management issues, problems, techniques and methods. Thus began the long haul of identifying eminent contributors, agreeing topics and editing the resulting submissions into a coherent whole. The result – like all humanly designed products – is not perfect. There are omissions and overlaps; some contributors promised much and eventually produced nothing; other contributors deviated from agreed topics; the scope of the project turned out to be so vast and time-consuming that some of the contents may now be ready for further development.

Yet, at the time of writing this foreword, there is still no other volume which attempts to cover the same ground. As a whole, the book has successfully achieved the original objective of bringing together a common understanding about design management. There is an abundance of practical guidance from writers who are all experts in their subjects – guidance which will help managers to approach design decision-making with confidence and with the expectation of achieving good design results.

Many acknowledgements are due. First, to all the contributors for agreeing to put pen to paper and then tolerating the long process of development and refinement; next, to Dr Brigitte Borja de Mozota, based in France, and Professor Colin Clipson, in the USA, who agreed to join the project as Advisory Editors; they have identified many important contributors and provided much useful advice. Finally, at Basil Blackwell, to Caroline Richmond and Alyn Shipton for providing greatly appreciated continuity of support, without which the undertaking might well have faltered.

MARK OAKLEY
Aston University
February 1990

PART I

DESIGN AND DESIGN MANAGEMENT

I Design and Design Management

MARK OAKLEY

Aston Business School, UK

Introduction

This book aims to assist the achievement of good results from design. It contains a cross-section of views and advice from more than 40 writers, all experts in their respective fields. In the preparation of the book, two main concerns have predominated. The first is to identify the management issues and responsibilities judged to be most significant in influencing the progress and outcomes of design projects. The second is to provide information about appropriate management methods and approaches.

Predominantly, this is not a book about design techniques; it does not set out to influence how designers use their skills. Even less does it attempt to turn managers into designers, although it may lead to a better understanding of what designers actually *do*. Rather, the focus is firmly on the *context* in which designing takes place. As competitive pressures give rise to a steadily increasing emphasis on design, more managers are finding that they are having to plan design work, manage projects, decide between different design approaches – and generally be responsible for ensuring that the best possible design results are achieved.

The book addresses the needs of managers in both small and large companies. Whether the focus is a one-person business or the largest of enterprises, the key tasks in managing design are much the same; only the scale is different. Thus, the book presents guidelines and insights that can help all managers to arrive at good design results – even if they lack prior experience of the processes of design or are unsure of the methods of working used by designers.

Design is now an inescapable concern of every company, of every manager. All of us are much more conscious of design than we were just a few years ago. In our personal lives we have come to demand much higher standards – for example, in the way that we furnish our homes, in the experience that we expect to enjoy if we stay at a hotel, in the comfort and features we take for granted in a new motor car. These raised standards in products and services come about partly because competition between providers ensures that the search for distinctive uniqueness never ends.

It is the act of specifying and assembling the product or service which mainly determines whether any individual company will succeed or fail in first gaining, and then maintaining, a profitable presence in its chosen market. In other words, the key is how well design is commissioned and conducted.

Hence, design is an indispensable part of every business and has to be managed with at least as much skill as all other business activities. All firms exist in order to provide customers with products or services which must be designed either from scratch or by drawing on ideas which already exist. Of course, for those who own these firms or who work in them, much more is at stake than just satisfying the needs of customers. A regular salary, social interaction, intellectual stimulation and the demands of attending to a range of tasks – all are part of the mix of concerns and challenges which make up an enterprise and its culture. But sometimes, especially in very large businesses, some employees – even very senior managers – may lose sight of the basic fact that their livelihoods depend on customers continuing to buy what the firm has to offer.

If complacency sets in, design may be neglected, resulting in out-of-date products or services which customers reject in favour of better ones from other companies. If the neglect persists, markets may be lost for ever. To make matters more difficult, in many market sectors it is no longer sufficient to provide something which is merely adequate or competent. Customers do not now expect to have to search out acceptable levels of quality and performance from amongst available alternatives. Unlike during the recent past, when often the challenge was to identify the one or two manufacturers of a particular product who could be relied upon to offer something competent and durable, it is now taken for granted that certain minimum standards can be expected.

Furthermore, purchasers are increasingly prepared to pay above the cost of an 'ordinary' model in order to enjoy, for example, the latest technology or a particularly attractive appearance. Hence, any searching by customers is likely to be for attributes such as distinctiveness, variety, flexibility – and value for money. Similarly, higher than average standards of quality and reliability, which are determined to a large extent at the design stage, can attract custom not just at regular prices but at premium prices, resulting in better profitability for the successful supplier companies.

Good Design Means Business Success

Some commentators say that design is now the major determinant of success in competitive markets. They point to a steady evolution from the period of shortages after World War II, when almost any product, however badly designed and manufactured, could find a ready market. Through the 1960s and 1970s supply caught up with demand, and the emphasis shifted to competition on price, speed of delivery and basic quality levels. Today many manufacturers are able to achieve similar levels of price, quality, durability and reliability, leaving design-determined factors such as appearance, variety and specification as the major differences around which consumers can exercise a choice.

During the earlier post-war years, improved reliability and durability came about largely through greater attention to manufacturing methods, better specification and control of quality, and only to a relatively small extent through more attention to design. Today, the qualities being demanded in products are influenced more than anything else by design.

As part of this, more and more manufacturers of all kinds of products are having to come to terms with what the motor companies have already grasped: appearance – or presentation – is often the most crucial factor in customer decision making. In just the same way that car styling has become subject to frequent revisions to keep the market in a state of interest and excitement, we now find the same phenomenon applying to other products. Right across the spectrum, from hi-fi sets to hiking boots, from saucepans to sofas, from computers to carpets – styling is often the major factor influencing what people buy, how much they will pay ... and how soon they will want to make a repeat purchase.

Having said this, it would be a mistake to suggest that outward appearance should be the only concern. Technical aspects of products remain vitally important. In fact, there is much renewed interest in the functional aspects of products, not because of quality considerations as previously, but because of the many new technical possibilities becoming available. Cheap, reliable electronics, including microprocessors and computing devices, mean that many traditional (and not so traditional) products are having new life breathed into them. Examples include sewing machines, telephone answering machines, typewriters, cameras and domestic appliances. A consequential effect of this is a sharpening of appetites for more ingenuity and capability in all sorts of products (not just those that might incorporate electronics), such as baby buggies, furnishing systems, garden equipment and items of luggage.

As well as the product itself, the context in which it is sold and the manner in which it is packaged are also now extremely important. An indication of this can be seen in any high street where much refurbishment of stores is taking place; in the case of some retailers, interiors are completely replaced and redesigned every few months to coincide with the launching of new ranges of products. The object, of course, is to retain or reawaken the interest of customers and to maintain levels of sales. In addition to the packaging and display of products, all accompanying literature must be carefully prepared. In the era of laser printers, blotchy copies from a poorly serviced photocopier do not properly reinforce an image of high standards.

The same kind of arguments apply to services: competitive forces raise basic standards to uniform levels until design becomes the main differentiating factor. Within any given price range, the services offered by businesses such as hotel chains, airlines, banks, insurance companies or retail stores tend to be much the same regardless of company. But the appropriate use of design can give one firm an advantage over competitors, often by identifying and exploiting a particular segment of a market.

For example, many insurance companies are developing new types of policies for specific groups (over 50s, non-smokers, self-employed, etc.) and then, to underline and reinforce the special differences, are using graphic design to create distinctive brochures, proposal forms and supporting literature. Despite traditional conservatism and slow response to change, a few insurance companies are going even further by completely revising their corporate identities and relationships with customers – using design on a grand scale to improve business performance. This has to be done with skill. Services must be not only distinctive but also consistently presented; too many unrelated design images would confuse customers or suggest a lack of direction and control.

Unease about Design

All this underlines the need for design to be regarded as a business function on a par with production, marketing and accounting – and requiring the same order of concern and managerial skill. Yet the response by some managers is to do one of two things: either to ignore the evidence and to deny the importance of design, or to acknowledge that design is important but to leave responsibility for it to others. Managers, who would not dream of being left out of financial or personnel decision making (for example), may quite readily distance themselves from anything to do with design.

The reasons put forward for this aversion to design often stress practical and cultural explanations. One argument is that managers typically experience great difficulty and unease when making judgements about what may be good or bad, appropriate or inappropriate design because their skills tend to be derived from analytical disciplines (such as engineering or accountancy). Consequently, they may be ill-equipped to deal with projects which involve unfamiliar design concepts, predominantly visual (rather than written) information, and the need to formulate subjective assessments. In particular, they may not be able to evaluate critically the results of design projects.

An additional explanation is that interest in appearance, colour and style may be regarded as a predominantly female concern which many men are unable to appreciate (or have been brought up not to appreciate). Since management continues to be dominated by men, especially at the higher levels, it follows that a lack of design interest persists. Other cultural influences may be involved; in education, early specialization tends to direct 'non-academic' children towards creative subjects such as art, design and craft studies, while depriving the brightest pupils of the chance to develop skills in these areas. This division is eventually carried into industry and is apparent in many companies – even some which depend on innovative, well-designed products for survival yet are managed by people who exercise control through the accounts, concentrating heavily on existing activities rather than nurturing new ideas.

Whatever the substance of these explanations, there is no doubt that there is much unease when it comes to making decisions about design and managing design projects. Yet achieving good design results need be no more uncertain than other management tasks.

The keys to success lie in adequate preparation before design projects start and then effective control while projects are running. Too much unsuccessful design occurs either because managers have no clear vision of what they need to achieve, or because they fail to take the steps to make sure that projects actually produce the right results. Direction and control of design must be set by a company's most senior managers. But few firms explicitly acknowledge design by recognizing it as a board-level issue, and even fewer allocate specific responsibility to a named director; most fudge the issue by referring to design as a general concern.

Different Outlooks of Managers and Designers

Attitudes are changing but it can still be interesting to witness the clash of cultures when managers and designers attempt to understand each other's points of view. For managers who have come to terms with it, design is a resource which can add value to products and services. If used properly, just like money spent wisely on advertising or the streamlining of manufacturing systems, an investment in design can produce good returns and contribute to improving the profitability of a business. In contrast, designers may stress other priorities in their work and may resist attempts to measure the results of their efforts in financial and strategic terms.

Conversations with designers often reveal wider concerns such as desires to improve the environment, to elevate public taste in art and design – or even to help bring about social or political changes. While managers might sympathize to a greater or lesser degree with these aims, their main interest in design is almost invariably with what might be termed 'design for profit'. In turn, this raises differences about what is good or bad design. In business terms, good design can often be defined simply as that which sells well; but sometimes designers are critical of popular design and instead champion design results which, to the uninformed eye, may seem esoteric and impractical.

Such divergence of outlook is well illustrated when design magazines and award schemes feature products such as furniture. Mass selling items are rarely included even though they may be bought by millions of customers, represent excellent value for money and provide considerable employment and prosperity. In their place are likely to be pages of photographs of strange objects that seem to have little commercial potential. Such preoccupation with unusual and bizarre ideas may, of course, serve a purpose in stimulating new directions. Unfortunately, the connection with, and the transition to, everyday objects is often totally incomprehensible to non-designers – and it is rare to find designers who are able to explain the process in straightforward, non-technical language.

It is felt important to raise this issue because it is fundamental to the perspective of this book. The main standpoint taken is of design as a business resource, a crucial element in the competitive struggle for commercial advantage. This does not mean the book is not relevant to the use of design by non-profit or non-commercial organizations; the aims of their design programmes may be different but the mechanics of deciding design directions, briefing designers and running projects are generally the same.

What is Design?

Before we look more closely at the management of design, we ought to set out some idea of what design is and what designers do. The word 'design' is much used (perhaps too much). Dictionary definitions list its meaning as a plan or a scheme. The word is also used to denote the end result, particularly appearance, of a design process – as when we refer to the 'design of a car'.

Perhaps a common-sense understanding of design is the best starting point. We can think of it as the outward appearance or physical arrangement of objects involving shapes, layouts, colours, textures, patterns and so on. Sometimes design might be thought of in terms of the technology that goes into a product or its convenience in use, including ergonomic aspects. Similarly economics, as represented by purchase price or service costs, might be seen (correctly) to be in some way related to design.

A frequent reason for using design is to help turn an invention into a successful innovation – or to extend the usefulness of an existing innovation. Many examples can be cited of basic innovations whose market potentials have been multiplied by design:

Basic innovation	*Designed innovation*
Bicycle	BMX bicycle
Cassette tape system	Walkman stereo (etc.)
Hovercraft	Hovermower

Likewise, a good idea of the tasks and skills involved in designing can be drawn from our own personal experiences. At some time or other most of us have, for example, decorated a room, remodelled a garden or made some clothes. All are design projects in which we are altering the state of something which already exists and achieving a new, more useful or more attractive result.

We could describe this as fine-tuning to achieve a result that suits our needs more accurately – and this is a good way of thinking of designing.

Such fine-tuning implies that we already have a basic product concept as a starting point – and this is almost always the case. It is a myth that designing invariably starts with a blank sheet of paper. In 99 per cent of cases a 'new' product is, to a greater or lesser extent, a derivative of an existing product. This is a most important realization with all sorts of beneficial implications for those who have to manage design projects – better use of designers' time, quicker projects and so on.

Note that this description of designing also stresses the needs of the customer. Designing may well involve clarifying or identifying these needs; designers are rarely provided with all the information they require and they have to know how to find it for themselves. So, designing is a more extensive, more complex activity than may be commonly supposed and certainly should not be viewed as a black box that produces outputs in response to specific inputs.

What Do Designers Do?

To get to grips with the tasks involved in the management of design, we need to consider the nature of the work done by designers. Judging by some publications and discussions, there seems to be a degree of confusion about this even amongst designers themselves! Definitions of designing, such as the following selection from those assembled in one textbook (Jones 1980), indicate a wide range of perceptions:

1   A goal directed problem solving activity.
2   Relating product with situation to give satisfaction.
3   The imaginative jump from present facts to future possibilities.
4   The conditioning factor for those parts of the product which come into contact with people.

These descriptions are not mutually exclusive but do differ from each other, underlining the difficulties faced by managers who seek a concise understanding of designing.

In the context of this handbook, designing may be viewed as being concerned with the preparation of appropriate solutions to marketing problems (or opportunities). These problems may be explicit or implicit; in either case, a key activity in designing is to review the associated symptoms or circumstances and to accurately define or redefine the problem. Then analysis, iteration and simulation are used to derive a feasible solution.

As part of such a process of designing, many activities may be involved, including generating novel concepts, reviewing and modifying existing concepts, carrying out experiments, building samples and seeking the constructive advice and judgements of others. Consequently, those engaged in this work must possess skills of creativity, analysis, synthesis and communication, plus knowledge of technical data, of existing solutions and of current and future trends in design.

In designing products, specific skills required include the ability to understand users' needs, wants, tastes and priorities; the ability to select the right materials and manufacturing processes; the ability to create products which fully meet aesthetic, ergonomic, quality and economic expectations; and the ability to produce drawings and explanations which communicate the final design solution to others working in the enterprise or outside.

Models of the Design Process

In practice, some of the tasks undertaken by designers are similar in nature to those in non-design areas of companies. Thus, the managing of design projects uses techniques and skills already known to managers (but perhaps applied in unfamiliar ways – for example, in defining design problems) together with additional expertise unique to design work, such as judging when to freeze a design solution. To pin-point these management aspects it is useful to consider a model of the design process. As with all generalizations, there are dangers in attempting to construct a typical model of designing, not least because it might imply a common approach suited to all situations. Hence, it is stressed that any model can only be indicative of the key tasks in design work.

Figure 1.1 A simple linear model of the design process

The basis of a simple model is shown in figure 1.1; similar models can be found in many design publications. The main weakness of this model is that its linear form suggests that a perfect design result can be achieved – that the design process has a clear end point beyond which no further work is required. In real life this is not true; designs are never perfect and there is always some extra benefit to be found. Even if this were not the case, constantly changing markets would soon create a need for some further design attention; the linear model lacks any indication of market reaction.

It is more realistic to represent designing as a circular process – or, even better, as a spiral process which stresses that design is an evolving activity. The converging spiral in figure 1.2 emphasizes two points. First, as each subsequent design cycle is completed, more knowledge is gained as familiar territory is covered, leading to quicker, more efficient designing – although major discontinuities may occur sometimes (for example, a new material previously unavailable) causing a jump inwards or outwards to another part of the spiral. Secondly, the spiral form also acknowledges that new technology is forcing previously separate activities closer together.

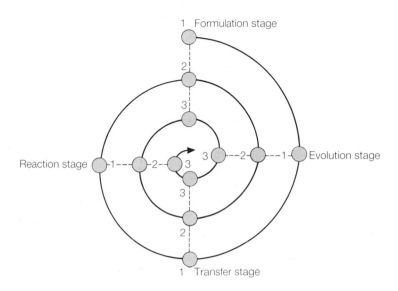

Figure 1.2 A spiral model of the design process

Computer aided design (CAD) linked with other computer controlled processes can already combine much of the evolution and transfer stages. Eventually, technology will permit very rapid design processes with far-reaching consequences; extreme product or service diversity will be possible, allowing the needs of individual consumers to be satisfied precisely and uniquely. Within the four stages in figure 1.2 there are a number of distinct activities, most of which are discussed at length in this book:

Formulation
Problem investigation
Problem definition
Design specification
Project brief.

Evolution
Idea generation
Solution refinement
Prototype development
Design freeze.

Transfer
Manufacturing drawings, data compilation
Operating system modifications

Start-up tests
Full-scale operation/production
Delivery to customers.

Reaction
Customer appraisal
After-sales service demands
Problem investigation.

Design Terminology

The working vocabulary of designers can cause problems for managers and other non-designers. In product work, the descriptions 'product design' and 'product designer' are often used, but so are a variety of other terms, usually to indicate specialisms which might be required to a greater or lesser degree in any individual project.

Package design   Most products are presented to the customer in some kind of container; often this is so important that the package is effectively a part of the product. Services, also, are increasingly packaged in the sense that they are accompanied by instructions, provided by uniformed personnel, and so on. Indeed, sales performance often depends as much, if not more, on the package than on the basic product or service. Thus package designers are concerned with choice of material (plastic, paper, glass, etc.) and manufacturing processes, together with shape, colour, decoration and lettering.

Graphic design   Specialists in graphic design may be involved in many types of projects (including the design of packages) such as the layout of brochures, company stationery, point-of-sale materials and vehicle liveries. As in all design work, an element of creative flair is necessary, but graphic designers must possess important technical skills as well. The appropriate use of upper and/or lower case type, what messages different colours and shapes can convey, how to balance the layout of print to achieve an easily legible result – all are basic graphic principles.

Building design   This is another specialism of design, often further subdivided into interior and exterior work. Buildings are only special types of products and most of the procedures are similar to those in other branches of design. Especially in service industries, the days are now long gone when the appearance and layout of premises can be left to the discretion of a local contractor. Now, the expectations of customers and users are such that the design of a building may be as critical as the business which is transacted within it. Public houses and wine bars are good examples of how building design may have far greater influence on commercial success than the basic products (beer, wine, food, etc.) offered for sale. In the retail sector, more and more companies not only pay great attention to the design of their premises but also seek to ensure that there is a consistent link with other aspects of design including product packaging and signage.

There are still further specialisms – textile design, jewellery design, exhibition design, etc. – but most are further subdivisions of specialisms already discussed. Apart from causing confusion amongst non-designers, the practical problem of this fragmentation is that companies may be involved in too many different, uncoordinated design activities resulting in a proliferation of bewildering, even conflicting, images. To avoid this, design elements may be integrated or standardized by means of a *corporate identity programme*, the aim of which is to present an appropriate, uniform image to customers, suppliers and other groups. In a company with a well-developed corporate identity, the products and services, buildings, publicity material and stationery may all share common characteristics such as the use of a standard 'house style' typeface, preferred colour schemes and distinctive shape or forms. Sometimes, details of these are compiled in a design manual which records and regulates the use of approved design forms.

Awareness of the importance of a corporate identity and all relevant specialisms of design is crucial to business success; just concentrating on one area, like product design, can be dangerous. The computer industry, for example, has long recognized this; certainly the basic product must be right, but so must be the after-sales services, supplies and information systems which support it. Customers need to be reassured that the equipment which they buy (but about which they may have little intrinsic understanding) is part of a reliable, secure operation. This reassurance can be provided to a large degree by integrated, high quality design which links the product with all other activities of the supplier company. However, there may be obstacles within a company to the implementation of a corporate design approach, possibly accompanied by an adverse impact on creativity and motivation as designers and others fear that their contributions are being devalued.

One further source of confusion in product design lies in the distinction between so-called industrial design and engineering design. The former is concerned with aspects of the product which relate to the customer or user, especially appearance and styling; the latter concentrates on the structure or function of the product and its economic manufacture. However, the skills required in the practice of industrial design frequently overlap those of engineering design and, sometimes, of other related disciplines such as architecture and ergonomics. The extent of this overlap varies but in many projects it is almost impossible to distinguish between the work of industrial and engineering designers and the concern, rightly, is with what might be called total design – or just design. Unfortunately, in too many companies, the demarcation boundaries are still in place and it can be very difficult for managers to overcome them in order to realize the maximum commercial benefits of design.

Conclusions

This chapter has provided the background and introduction to the issues which are explored in this book. Each contributor has been encouraged to provide a personal view of his or her particular topic area. In some cases this means that where several authors touch on the same issue, there may be some difference of opinion or alternative guidance; in the same way that managing in general remains predominantly an art rather than a science, managing design must always involve flexibility and judgement. However, the general message is clear: design is an important resource which must be managed with at least the same competence as other core business activities. The purpose of this book is to help achieve that competence by identifying the important issues and by providing information, ideas and advice.

Reference

Jones, J. C.: *Design Methods.* Chichester: Wiley, 1980.

2 The Future of Design and its Management

PETER GORB

London Business School, UK

The Decade of Design

For many managers in many countries the eighties was the decade in which design became a central business preoccupation.

These shifts of interest are relatively slow, with a particular specialization often dominating for up to two decades, and with the leading edge of change often coming from the USA. In the forties and early fifties, production and productivity occupied central stage as the world finished its major wars and re-equipped with capital goods in both developed and developing countries. Ten or more years of focus on the behavioural sciences gave way in the early sixties to an emphasis on marketing, with the increasing need to stimulate worldwide consumer demand for the goods and services previously only available to the wealthy in the developed world.

Not until the mid seventies did the Americans begin to worry about the competition from Japan which was eating into their home markets. The focus was shifting again, this time to design – and the leadership for the first time was coming from Europe.

Apart from a few early pioneers, the messages about the importance of design to managers hit Britain with some force in the early eighties. A main agent of change was the support given by the UK government which directed both money and manpower into promoting the design cause (Design Council 1984). Industry and education were nudged into a design courtship, and by the middle of the decade the media, television in particular, were busy celebrating the marriage. Similar activities have been evident in other countries including France, the USA and, of course, Japan.

Design is now well recognized as a powerful tool in the hands of managers who need to make and sell products or develop an effective working environment, or communicate efficiently with consumers, customers, shareholders and others. Many senior executives now expect to use design resources in the development and communication of business strategies. More and more business schools are adding design and its management to their curricula. The only remaining question is whether design as a key management preoccupation will continue to flourish.

Fashion will, of course, be a major determinant. But it needs more than fashion to keep an idea and an expertise at the commanding heights of management thinking. There is no doubt that permanent and responsible disciplines like marketing and operations management hold their place because of their abilities to embrace a range of long lasting management preoccupations; and this will also need to be true of design and design management if they are to last the course.

There have been many long running management preoccupations, but three fairly constant ones have been around since the 1940s:

1 Innovation and the place that the innovative process occupies in the search for profits.
2 Quality and its control and maintenance in both production and service organizations.
3 The development of effective line managers expressed in terms of education, training and development.

These three will be a good starting point if we are to evaluate the permanent role that design and its management may have in the coming years. We will need to take a look at how design relates and contributes to each of them.

However, before we begin to do this we ought to clear some ground. We need to be sure of what we mean by design and the scope and limitations of its relevance to the management world. We also need to clarify what we mean by design management.

A Definition

The word 'design' means many things to many people. Design includes the work of people from a wide range of disciplines: industrial designers and design engineers, architects, graphic designers, illustrators, environmental designers and all those industry- and product-related disciplines like textile design, automobile design, furniture design and many others. It also includes those people who are concerned with systems of these things. In attempting to describe design, we do need to agree about characteristics that all design and designers share, and also which of those characteristics are exclusive to design and designers.

However, we are not concerned with the legitimate extension of the word to include disciplines based on 'ideas'. Hence, in what follows, we exclude the designs of economic models, philosophical systems and so on. These ideas disciplines do share one of the three important characteristics which all designers have in common and which is exclusive to design – a methodology (of which more below). With these abstract usages excluded, a useful and simple definition of design which covers all the various uses described above is *a plan for an artefact or a system of artefacts.* More particularly it is a working definition which is highly relevant to the management world.

Managers are surrounded by artefacts. In manufacturing industry they call them products, and design is simply the plan that managers make for their products. But design is the concern not only of manufacturing industry, which makes and sells products, but also of retailers who buy products in order to sell them and of service industries which use products in order to provide a service. A consultant, for example, would be unable to operate without data; and those data come from hardware, which itself has to be designed. Furthermore, every business works in a physical environment that has to be designed. Well-designed buildings, interiors and physical distribution systems all contribute to the efficient operation of a business. Finally, all businesses need to communicate, and they do so by means of man-made things – reports, promotional and advertising materials, videos, signs and a host of other products, all of which must be designed.

All these aspects of design can be measured. For example, design input into products and their development can have an important effect on gross margins. So design pervades the manager's life and work in a measurable way. To gain its full potential benefit, therefore, a manager needs to know how best to use design and how best to understand its contribution.

The problem is that too few managers accept that artefacts dominate their world and need to dominate their thinking. Ask a manager what is of prime importance and the answer will range from profits to people, with products a long way down the list. This disregard for artefacts, and the need to make plans for them, has its origins in nineteenth century Western culture and the education system which still supports it (Weiner 1980). We are taught to value ideas above action, things spiritual above things material, the conceptual above the pragmatic and the logical above the intuitive.

As a result managers are often unable to appreciate the importance of 'things' and have a view of design which relates it to either a God-given mystery or a compensatory skill for illiterates.

Design Management

In the face of these confusions, design management has itself become the victim of some confusions, notably a series of multiple meanings covering various levels of newly emerging design-related activities. The work in all of these fields of effort is important, and each requires description.

Design office management This deals with the problems of managing a design practice, a consultancy or an in-house design department. It creates many anxieties, usually amongst designers who are promoted to design managers to deal with the specific task. It is important because, to do it properly, one needs involvement in the other four activities described below. But compared with the others it is a relatively simple task and can easily be taught in design colleges. Alas, it rarely is!

Let us move on to discuss two much more important prerequisites for a sound linking of design and management.

Educating designers for management

This means equipping designers with the languages, and at least a nodding acquaintance with the norms and values, of the world of management in which they work. This is an enormously important task if design is to take its rightful place in the management system. Yet few designers have this equipment; most are quite unable to fight their corner in the management world. For example, the performance ratio known as return on capital employed is one of the most important used by managers. Design work can affect it profoundly, yet many designers do not know how this happens. Indeed they often do not know what the ratio means. More importantly (and depressingly so) many designers do not want to know. However, education for management is about appreciating a culture as well as learning a language and even perhaps accepting a value system. Learning to love the words 'gross profit' may be difficult for some, but every successful designer must know how to value those words.

Educating managers to design

Educating managers is another side of the same bridge, an equally important starting point for a satisfactory relationship between managers and designers. It forms the core of the design management work started in the mid 1970s at the London Business School. The task is to tear down those educational and cultural barriers described above – and this is easier with the young managers than the more senior ones.

With these preliminary and important tasks described, we arrive at a key activity.

Design project management

This is about the place which design occupies in the project management process, when that process is the way in which action tasks are arranged in organizations. This is a central and vital activity, with design occupying the key role between the creative, innovative activities and the control of the preparatory stages of the operational tasks of the business.

Project management is a well-understood and well-developed expertise and the design contribution to it is generally acknowledged as an all-important one. However, it is not a universal technique. Not all organizations use project management as a way of achieving their business objectives. We need a wider frame of reference if we are to make the most practical use of design in organizations. That frame of reference in its fullest form does not yet exist, but we can describe some first steps towards it.

Design management organization

This is about the place that design occupies in the management structure of organizations and about the variations and modulations that are needed (because all organizations vary in size, shape, technology, markets, industry sectors and any other variable you care to devise) to make that relationship effective. So far, knowledge in this field is anecdotal, with the most common and most valuable contribution being the design project management method discussed above.

Some of the anecdotes are about best practice and some about disaster. Much can be learned from both. However, until norms exist and until we understand how to develop and deploy resources to meet those norms (and also the variations from norms), then the design contribution can never be effectively modulated to meet management needs. Finding out about this is part of the research which is described later in this chapter. However, first we need to look at the contribution that design makes to innovation and to quality.

Design and Innovation

We have come to accept that innovation is the lifeblood of our society. It is innovation that enriches our imagination, supports the quality of our life and determines the future of our children. Innovation contributes to the creation of wealth; it provides the bottom-line profit in science- and technology-based businesses. Indeed, innovation is now the mainspring for most managerial activities, the basis of key investments and the most highly rewarded management function.

No group has been quicker than the designers in identifying their particular skills with the innovative process. They seem right to do so. The associations are there: innovation, creativity, invention and problem solving are the words commonly found in the literature of design. Designers and innovators are perhaps the same people doing the same things – particularly in manufacturing industry.

Or are they? Are those innovation-related words quite so closely identified with each other? Is in fact innovation all good news? Can one question the value of unqualified innovation and indeed its unqualified relationship with design? Could it be that design as a function is something different from creativity?

We have defined design as a planning process for artefacts. By that definition, design becomes the key element in the planning processes of the business – the plan for the things the enterprise makes, sells, uses or communicates with. But design will never be an effective management tool if it is confused with the innovative process. Let us go back to that basket of words – innovation, invention, research and development, creativity, product development, problem solving – all used in business to describe a process of change, directed toward the new. Much has been written in an attempt to give special and related meaning to each of these words (Storr 1972). Let us use innovation to describe the whole process, recognize that it is a creative process, and argue that it is something different from design. It follows, therefore, that design is not a creative process. This is not to deny that designers are creative people. It will be argued later that they have to be. However, the design function only deserves its key role in business when we appreciate that it is another, separate and vitally important function.

The operational activities of a business enterprise – manufacturing, marketing, providing a service, for example – cannot remain relevant (and profitable) unless they continue to change. Even if a business wishes to stand still, it needs to change in order to maintain its equilibrium within a shifting environment. Many businesses with a cultural or a nostalgic element – fashion, textiles, and furniture are obvious ones – need to know how to change back to the old as well as forward to the new. But change they must. Clearly, therefore, innovative processes are often, but not always, necessary prerequisites to business operations. However, these creative and innovative activities never proceed at the same rate as the operational processes. History shows that creative flowering – the eureka principle – rarely happens as part of the calm and ordered flow of historical events. Indeed, it is often the essence of the creative act that it is out of time with society, acting as a catalyst for social change, often as a disruptive force – and sometimes as a destructive force.

For example, technical advances in twelfth century Europe, particularly in the use of water power to drive mills, were not fully exploited for nearly 200 years until the Black Death and a consequent shortage of labour brought about the necessary economic changes. Conversely, at the time of the industrial revolution, it was the marketing opportunities in the cotton textile industry which dragged forward the technical innovations which, because they had lagged so far behind, were responsible for much of the social and economic misery of the early factory systems.

Today, in information technology, trivial applications like children's games are often the first uses for innovative technology which is ahead of its markets. And in the biotechnology field, the current rate of invention far outpaces economic applications. For example, in oil extraction there is little doubt that biotechnology could more than double oil well outputs. But the implementation of this new efficiency will need to await appropriate oil prices.

The same is true in the business world. Innovation does not always fit comfortably with year-end profit statements. It can and does disrupt the commitments to stability on which the business is built – carefully planned return on capital investment, or commitment to pensions for long serving employees, for example. These considerations and many others are stabilizing and even civilizing forces. The recent collapse of some innovative information technology businesses is a case in point.

Clearly, there must be something that modulates, controls and encourages the innovative and creative inputs into the business – something that makes innovation meaningful. This something is design, which acts as a sort of thermostat for innovation, responding to the voices and views of customers, employees, capital investments and all the other factors that constrain, sustain and shape a company's culture as well as its operations.

Design is the driving force, not only of change (which is inevitable) but, more importantly, of the rate of change. It feeds innovation into systems in a way that preserves the plan for the bottom line. Planning for products (our definition of design) is in this context arguably the most important business function of all.

There are many business examples of this. In the fashion and textile industries, stock control coupled with seasonal innovation ensures that both manufacturers and retailers can adjust their end of season stocks in a way which minimizes any deterioration of gross margin. In the music industry the introduction of digital audio tape (DAT), fast on the heels of the compact disc, has created international conflicts which are only now being resolved to the mutual prosperity of the industry. The increase in the rate of introduction of plastics into motor car manufacturing will be determined by capital investment decisions as well as the availability of the new and innovative technology. All the decisions which determine these issues are, at heart, design decisions.

At this stage, it is worth reiterating that although design is not by itself a creative process, this is not to argue that designers are not creative people. The interface described above makes it necessary that they should be so. And in craft-based businesses like silversmithing and pottery, the innovative, the design and the operational activities of the enterprise are usually done by one person. Even in large organizations, the lines are not always clearly drawn. They have been separated in this discussion in order to highlight an importance for design that goes well beyond the creative act.

Once this is understood it is easy for designers and, more importantly, for design-trained line managers to deal with the key issue of innovation and its encouragement and control.

Design and Quality

The second of the important management preoccupations we undertook to relate to design is quality and its control. Once again before we can describe the relationship we need to deal with some definitions.

Quality, like innovation, is invested with many meanings. It carries an aura of exclusiveness. In the seventeenth century, people talked of 'the quality', meaning the upper classes; and as late as 1977 Fred Hirsh wrote about 'positional goods' in relating quality to exclusive ownership of goods and services by the same kinds of people. Quality also means high standards of make and finish. It carries a flavour of reliability and value for money plus a flavour of the old fashioned and the dowdy. Quality products are perhaps a little boring on the outside, but can be relied on to work. We associate quality with Heals, and Harrods, but also with Marks and Spencer. Quality is Rolls-Royce now and Rover 40 years ago (and perhaps now once again). We associate it with individual craft products, but also with mass production.

Many of these value judgements are clearly contradictory. Marks and Spencer and Harrods are not often included in one category; nor is a well-crafted product reliable in the sense that no two of them are likely to be exactly the same. A definition, therefore, which overcomes all these subjective considerations and is of relevance to management, particularly in manufacturing industry, might run as follows:

Quality is the extent to which a product meets the specification drawn up for its manufacture; and where the product is mass produced, consistently meets that specification.

Such a definition not only makes space for subjective considerations, but is also capable of measurement. How in fact is quality (defined in this way) normally measured and, given the necessary feedback procedures, controlled? Three ways are normally in use in manufacturing industry:

1   By inspection at the end of the process (either of each product, or by a sampling procedure) which compares the product with the specification.
2   By an attitude amongst the people concerned in manufacture who, through training and exhortation, place quality at the forefront of their thinking during the manufacturing process, and are constantly referring to specification. Quality circles and related organization systems fall into this category.
3   By ensuring that the specification itself is developed in such a way that it becomes very difficult for it not to be met.

All of these ways of dealing with the control of quality have their place, and they are not mutually exclusive. Nevertheless the third represents the most effective route. In effect, it shifts the problems of controlling quality to a point in the process before manufacture – at the determination of the specification.

Now, it is also the case that the last stage of the design activity is the specification (Gorb 1986); and the determination of specification constitutes the heart of all design activity. It is a difficult and complex task demanding the resolution of conflicts arising from all the management disciplines. Marketing may require product characteristics which are difficult to make, and finance may limit product options or require the cutting of production corners.

No ideal solution is ever achieved. Nevertheless it is generally recognized that designing quality into a product is better than inspecting it out. An appropriate specification is worth more than hours of exhortation or an army of inspectors. Design therefore occupies a key role in the determination of quality, as it does in the control of innovation.

Design and the Line Manager

Line management performance is perhaps the most important and permanent of the three preoccupations we have linked with design. Our final task is to explore the extent to which design can contribute to effective line management performance. That managers do in fact make use of design skills has been an interesting outcome of some recent research at the London Business School which has identified a complicated and widespread phenomenon that we have termed 'silent design' (Gorb and Dumas 1987).

What this means (in rough summary) is that, in the organizations studied, a great many people (many of them managers) who are not designers are engaged in designing; that quite often they are not aware that they are designing; and that they do not necessarily agree that what they do is designing once they are made aware of it. Furthermore, the process seems to work – though better in some cases than in others.

If this interesting observation is confirmed by further research, it will carry some far-reaching implications for the place of design in the management structure. Most are to do with the dismantling of some of the shibboleths that surround the design profession. Dismantling of this kind happened to many other business-related professions long ago. Accountancy skills are no longer the sole preserve of accountants. Line managers need to be equipped with these skills in the same way that many now use computers for themselves, which ten years ago was relatively rare. Even the most defended professions like medicine and the law have, through the advent of information technology, begun to move some of their processes into more general hands.

The Design Contribution

If designerly skills are to be available to line managers, we need to say what they are. There are three important design contributions that the manager can make good use of. These are: a care and concern for things, a set of special skills, and a special methodology. Each deserves some comment.

We have already touched on the first. A care and concern for things does not come easily to managers in our society. Their concern is based on a theory of organizations in which the manager's task is to achieve the objectives of the enterprise, expressed and measured by profit. While doing this, the manager also needs to provide optimum satisfactions for those connected with the enterprise, such as employees, shareholders and customers.

It is a praiseworthy concern, but the objectives behind it are much more easily achieved if managers have a deep interest in what they make and sell. Design training enables us to correct this imbalance – to learn, as the

Japanese learned many years ago, that without product leadership most businesses lose their competitive edge.

The second attribute of the designer is visual literacy, the ability to see and reproduce what is seen – in other words, to draw, to model, to create visual analogues. Contrary to the folklore, these are not God-given skills. Anyone can learn them, in the same way they learn to read and write. Most designers have developed these skills early and through constant reinforcement have turned them into a specialization. It is certainly the case that the lack of these skills can create great difficulties for some managers constantly faced with massive amounts of engineering drawings, architectural plans, factory layout diagrams, work flow charts and similar visually presented documents.

Finally, the methodology of a designer is action-based. In dealing with problems, the designer's concern is to find out how before finding out why. The inclination is thus to take incremental and practical steps to a solution. This way of working is a valuable corrective to the methods that have tended to dominate the education of managers. Their training motivates them towards analysis rather than synthesis and to seek general laws based on hypotheses before acting (Kuhn 1976). In other words, managers try to find out why before finding out how. Design methods are a salutary counterbalance to a view of the world that puts thinking above doing.

What therefore emerges from these three contributions by design to the management world is that while design training can help line managers to contribute to a better design of their products, it can also help them to manage better!

Conclusions

The last of these contributions is the most important one. Design is a much degraded word, used to mean a basis for the philosophical and economic life of a country as well as a label on a pair of jeans. Designers are stereotyped as back room boffins in white coats or as creative directors in bomber jackets with a two-day growth of beard. Design can make its loudest noises about trivial and ephemeral consumer products and says little about the insides of nuclear reactors or aeroplane engines in which enormous commitments to capital and to culture are involved. Designers rarely attempt to quantify their contributions, preferring to be judged by creativity, and by subjective and aesthetic considerations.

If we are to ensure that the place of design in the management hierarchy remains secure for the next decade, then we must sweep away the detritus and distortions which surround it. We have tried to do that in this chapter by exploring an appropriate role for design in the management world. We have defined design as a plan for an artefact, whether that artefact is a product made and sold (or bought and sold), a part of the business environment, or an information system that the business uses. We have sought to define a key role for design in three important management fields: innovation; quality and its control; and the development of line managers.

However, much work still needs to be done to identify and quantify how design can contribute to many more management fields than the three we have been able to cover in this chapter. Such work is needed if we are to confirm the most important of our propositions: that the design contribution to management is to enable managers to manage better.

References and Further Reading

Design Council Strategy Group: *Policies and Priorities for Design.* London: Design Council, 1984.

Gorb, P.: The business of design management. *Design Studies.* Vol. 5, no. 2, April 1986.

Gorb, P. and Dumas, A.: Silent design. *Design Studies.* Vol. 8, no. 3, July 1987.

Hirsh, F.: *The Social Limits to Growth.* London: Routledge and Kegan Paul, 1977.

Koestler, A.: *The Act of Creativity.* London: Hutchinson, 1962.

Kuhn, T. S.: *The Structure of Scientific Revolutions.* Chicago: University of Chicago Press, 1976.

Lewin, D.: On the place of design in engineering. *Design Studies.* Vol. 1, no. 2, October 1979.

Simon, H. A.: *The Science of the Artificial.* Cambridge, Mass.: MIT Press, 1982.

Storr, A.: *The Dynamics of Creation.* London: Secker and Warburg, 1972.

Weiner, M. J.: *English Culture and the Decline of the Industrial Spirit (1850–1980).* Cambridge: Cambridge University Press, 1980.

3 Design Management on London Underground

TONY M. RIDLEY
EuroTunnel, UK*

Introduction

The London Underground is widely known as a carrier of passengers and was the first underground railway in the world, starting services in 1863. Now over 800 million journeys are made each year. The Underground's aim is not just to move people, but to offer a complete package of service that encompasses all aspects of a passenger's journey. This service package represents the product we are offering to our customers and we must ensure maximum efficiency and comfort in all respects. This includes, for example, easily understood signage, comfortable trains and good publicity material.

Good design is critical to the customer's total experience from entering to leaving the system. The scale of the Underground's operations, with its many different facets and often conflicting requirements, means that management of this design effort is of paramount importance.

A Heritage of Design

Design management is not a new concept to London Underground. Examples of good design abound but, without doubt, the work which was carried out under the auspices of Frank Pick is most outstanding. There are many examples of his role as a great sponsor of design: just three are 55 Broadway, the headquarters of London Transport; Piccadilly Circus station booking hall; and East Finchley station, with its famous figure of the archer.

Pick was a great exponent of design management. The new works extensions of the 1930s and 1940s represented a complete package of design and product innovation. Not only were the new customers using the Piccadilly, Northern and Central Line extensions given a new service, but this was combined with high quality design in graphics, stations and trains. This complete package was used to create a new identity for the then London Passenger Transport Board. Furthermore, design was used to focus the public's attention on the new service. Frank Pick had the advantage of an Underground which had a firm place in the market and growing demand. Essentially, he was able to use the management of design as a means of glorifying public transport service.

* Formerly Managing Director, London Underground.

Today's London Underground

Once more, the Underground finds itself placing great emphasis on design management. Design is now an essential part of a new competitive strategy. Changes in the business environment in general and that of the Underground in particular have precipitated this need. Customers want more than merely to travel from A to B. They demand quality of service and will respond positively to good quality, as successful retailers like Marks and Spencer have demonstrated.

In the early 1980s, a low point was reached in the level and quality of service provided to our customers. After concentrated efforts, a vastly more reliable service and an improved passenger environment have generated an atmosphere of change. Also, with the creation of London Underground Limited in 1985 as a wholly owned subsidiary of London Regional Transport (LRT), a new and separate corporate identity was needed.

Reflecting these new circumstances, the Underground has undergone a change in management style, with the introduction of disciplined project management and, associated with each project, an internal 'client' to whom the project manager is responsible. This not only makes the business more efficient, but allows the integration of design into the development of the product to be more easily controlled. First, the client must:

1 Determine the task and ensure the user's requirements, including design aspects, are met.
2 Obtain authority.
3 Ensure the investment is cost effective and well suited to the business as a whole.
4 Ensure proper project management is set up to complete the project to agreed design and timescale.

Then the project manager fulfils the defined requirements of the client, on programme and within authorized cost, in respect of planning, directing and controlling the project.

Product Innovation

The market for public transport is no longer a captive one. People have a real choice in both off-peak and peak periods between using their cars or the various forms of public transport. Furthermore, the passenger's perception of what should be available has widened. These changes in market environment have made the need for the effective management of design, product innovation and marketing far more pressing than in the days of Frank Pick. As a consequence, design has been identified as a way of making the

Underground's product match the marketplace, helping to attract passengers and so to benefit the business.

London Underground's product innovation policy has been to concentrate on two interrelated issues. The first has been the development of a complete service package for customers. The second has been a move in marketing strategy towards an increase in off-peak traffic. The development of the service package has concentrated on three particular features:

1   High quality customer-care attitudes at stations as a means of upgrading the product. Here the Underground has the opportunity to use the most successful features of others' efforts and to remodel them to fit its particular requirements.
2   Dynamic and constantly updated information. This makes wide use of modern electronics – including dot matrix indicators and centralized public address equipment.
3   Greater and more effective use of information posters and leaflets.

In developing these, the management of a coordinated design policy is leading to a product that is bright and appealing and which carries a strong London Underground identity in every aspect of its operation.

Similarly, efforts to increase off-peak travel have involved extensive reliance on good design, in particular the promotion of child travel (which, in turn, generates adult travel) and the development of innovative travel-cards. Also, there has been a series of poster campaigns where good communication through high quality typography and illustration has culminated in a recent return to sponsored art on the Underground.

The Development of Design

The need to ensure that design quality standards are maintained and costs controlled has led to a change in the emphasis given to design within the product planning framework. Quality standards are maintained by the 'client' (who controls project expenditure), but only after having been determined jointly with the LRT Design Working Group. This group is chaired by the LRT Design Director and includes the Marketing and Development Directors of LRT subsidiary companies, of which the principal is London Underground.

The LRT Design Director's role is to coordinate the development of design within LRT and its subsidiaries. The existence of such a post – involving a top-level designer steering the company's design policy – is concrete evidence of LRT's recognition of the significance of design. The Design Director advises clients and briefs project designers, but has no authority to issue direct instructions. In the unlikely event of a dispute between the Design Director and a client, the matter can be referred to the Board of LRT.

While the Design Director's terms of reference encourage the Underground to develop its own identity within the corporate framework, the Underground's design policy is the responsibility of its Marketing and Development Director (working through a Design Manager) whose main tasks in the design area are:

1 To apply LRT's overall design policy to the Underground.
2 To ensure that the London Underground branding is properly used.
3 To ensure that all aspects of design maintain a high quality.
4 To develop design expertise within the Underground, in conjunction with the LRT Design Director.

This relationship means that the London Underground Director who is charged with the implementation of design is also intimately concerned with its development.

The Design Tools

The Underground is currently reorienting its product. Traditionally, the train service has been seen as the ultimate product. Now it is recognized that the product must be viewed as the entire experience that the passenger receives. Therefore, product quality must be maintained not just in running the trains but in all the other facets of operation that customers see and experience. This includes pre-trip timetables and fares information; ticket hall, tickets, and frequency and reliability of the platform information; the train services; and a positive impression of cleanliness in general, staffing in general, the extent of overcrowding, the ability to sit down, and the appearance of the system. To develop and maintain a market-led approach to business, these crucial elements of the service package must be closely monitored and constantly improved.

To achieve this, one part of the Underground's strategy is to brand all elements, from stationery to trains. We are fortunate in possessing two very powerful tools – the Underground roundel and the New Johnston typeface. A stringent corporate identity audit carried out by LRT has confirmed the strength of these tools and has helped us consolidate our brand.

We have had to take a very disciplined approach to the use of our roundel, and its application as a brand. The corporate identity study has demonstrated this; and an example of the care that has been taken to continue the review is our detailed 'signing study', one aspect of which was to determine where misuse is occurring and how we might make improvements. Indeed, the study has recommended a purer use of our roundel and a revised Johnston typeface.

The Management Tools

Until the early 1980s the control of design, and its integration with product innovation, lay with the architect or the design engineer. Whilst these professionals are able to apply their particular skills to their jobs, personal design bias may intrude and the result may not necessarily fit in with the Underground's corporate ambitions. As part of an effort to gain greater control of projects, in all respects, the Underground has introduced the system of project management referred to earlier. One result has been the establishment of a 'client' for each job. In this role, the client maintains the overall interests of the business, whilst the project manager is responsible for completing a project to time and budget. The client's role has allowed more control over design, which has in turn given scope for more interesting design briefs. This has allowed the Boards of London Underground Limited and LRT to maintain an overall review role.

It is possible to compare the change we have managed to achieve, in terms of the management of design, with the process of buying a new house. By introducing project management, and determining a fixed role for the client, we have vastly improved our buying position. No longer do we purchase a mass produced dwelling off-the-peg, where we are only able to choose the colour of the bathroom suite. Now we are able to brief the architect and designer fully and then to review their work through its various development stages. This gives us a purpose built home, which has not had its construction standards compromised but matches our requirements fully.

The leadership and accountability of the client have enabled design to contribute significantly to the improvement of our product. This can be illustrated by a number of examples. Our station modernization scheme is a visible sign of the upgrading of our product. Three levels of identity have been incorporated into each scheme:

1   Station identity, by providing a 'sense of place' through, for example, decorated graphic panels.
2   Line identity, by dedicated coloured plates that directly relate to the Underground map.
3   Company identity, by using the corporate branding mark.

This scheme allows good design and innovative thinking at station level, but also takes account of two important corporate design considerations: the provision of information through line identity, poster sites and the use of dot matrix train indicators; and the establishment of a firm corporate identity by means of a universal brand (namely the roundel and the New Johnston typeface).

Another example concerns the programme of rolling stock renewals scheduled for 1990 onwards. This start date provides an opportunity to consider, during a six-year gap in rolling stock procurement, the design of one of the principal elements of our product. When the 1935 prototype stock was built its chief engineer said that everything was different on the new train, apart from the wheels and windows. The radical examination of engineering options on the 1986 prototype stock changed even these. Additionally, there has been a strong element of design experimentation. This was conceived, for the first time, by a firm of design consultants and was incorporated into three quite different prototypes. The Underground intended, through this process, to assist the promotion of the corporate identity, to develop new materials for colour and graphics, and to incorporate new features to improve the passenger environment.

We had the opportunity not only to test these new features before we ordered production trains, but also to survey the reaction of our customers. As a consequence, we hope this policy will acknowledge not only advances in technology, but a change in the design of our product and its relationship to our customers' requirements.

This policy of client control has been most important in all aspects of the management of design from our new Underground ticketing system to our advertising material. It has meant that the corporate aims of the business have been well represented in every aspect of our activities. We have been able not only to press for the highest quality of design, but also to retain a strong brand, thus enhancing the corporate identity throughout the company.

A Dynamic Process

While it is possible to be so positive about the design policy of London Underground, this does not mean that it is immutable. Far from this, we consider that design and product innovation should be carefully monitored to ensure that they accurately match the customer's emotional needs.

Through market research we have been able to establish, for example, that passengers prefer to have features recalling the street environment on deep tube platforms. Moreover, in a recent survey, market research has been used to determine what conditions customers perceive as being unsafe and what, in their view, would give an improvement. This information has been very important in creating a set of improvements that not only address the problem of crime and vandalism, but ensure that customers perceive that it is being dealt with.

It seems that passengers' expectation of the quality of the service package we should offer will continue to rise as they are presented with more and more sophisticated consumer goods elsewhere. The use of market research means that the Underground is able to match this growth in expectation with a corresponding development of its product and the continuing use of good design. With passenger expectation evolving, the management of our design effort is vital.

Conclusions

Design and product innovation are most valuable tools in attracting customers, but they must be very carefully managed. The Underground devolves its responsibility for the management of design to the 'client' to ensure that the final product matches both operational and customer requirements. Whilst the creative element is under the control of the design manager, thus allowing the company's corporate interests to be easily reviewed and maintained. Meanwhile, London Underground acknowledges that constant reference must be made to the customer in order to measure the success of existing ideas, and to determine how they should be developed for the future.

Many elements have played an important part in the commercial success the Underground is presently enjoying. Part of that success must lie with the good use of design. This has manifested itself not only in the bright new face of modernized stations, and good publicity material, but in the coherence given by the disciplined use of our corporate identity. The further application of design to the innovation of our product – the 'service package' – will help to bring even more success to London Underground.

4 Design Management at Olivetti

PAULO VITI and PIER PARIDE VIDARI
Olivetti, Italy

Contrary to popular belief, design management has little direct influence on the actual form of products or product groups, or on such graphic features as the logo, the trade mark and the printwork applied to them, the instructions and indications they carry, the messages they transmit, or the packaging. Nor does it directly affect the appearance of the vehicles that transport the products, or the surroundings in which they are displayed. What the design management function does do, however, is to provide an organizational framework within which all these things can flourish – along with many others.

There is really no effective difference between the design taking place within a company and the cultural tradition of which the company is a part. Whether they recognize it or not, as well as depending on a flow of research-driven new ideas, all high tech companies have some link with the history of civilization and with art trends. Olivetti's cultural commitments are deeply rooted in time. As an Italian company operating the world over, all this is more than just important; it coincides with the company's cultural mission and indeed with the aesthetic strategies initiated as long ago as 1908.

Design Independence

At Olivetti, design is a continuous process which enjoys a certain independence. Indeed, the managers who deal with design report directly to head office. It is very much a team-based activity, involving all the product engineers, the production engineers and the designers. It is significant that, in Italy, designers mainly graduate from faculties of architecture at the same universities and polytechnics that train engineers – which facilitates the dialogue between the two groups. Italian designers and architects also have a background of history and the humanities, and it is perhaps this kind of training that provides the basis for their extraordinary success worldwide.

Of course, in a company with traditions such as these, dialogue and joint effort are continuous. Designers are informed in many specialisms – such as ergonomics – as well as the central areas of design and technology. The decisions they take depend on the suggestions and proposals coming in from marketing, and on the requirements of production. Through experience built up over many years, they can select and encourage (should that be necessary) the more profitable channels of information.

Design is a matter of creating a substantial unity of image and identity (not just that of the organization) and of directing it towards common goals. Thus a state of harmony must be established between this objective – which we might call 'historical' – and the future objectives of the company, which are evolving all the time, and whose main aim is always a function of an overall strategy.

Realistically, one has to be aware that the image cannot replace the object; and that design represents at once the fundamental, formal and most indicative part of the product – and the most easily recognizable and often most powerful. Certain non-Olivetti people might conclude that design is not the object, but is merely its representation. Olivetti, on the other hand, has always avoided giving design an imprecise, debilitating role as a cover or as the bodywork of the product/object. When Olivetti designs, it combines and blends all roles – those of the engineers, of the designers, of the artwork staff – and there is absolutely no functional hierarchy, not even in chronological terms.

The outcome – in the purely practical terms within which any corporate objectives must always be seen – includes waste reduction, product optimization (in manufacturing methods too) and, above all, ongoing product improvement. It is for these reasons that a product embodying a large design content, when compared with an 'ordinary' product offering the same functions and price, will always be of superior quality, and its manufacturer will have a superior market image.

Design Management's Coordinating Role

Coordination is one of the fundamental roles of design management. The world of electronics demands the existence of a combination of fundamental, formal structures that are nearly always profoundly different one from another; compare keyboards with display monitors, for instance. They represent systems of forms, often independent, that can be entrusted to different designers. These systems are more important and longer lasting than the individual components of a particular monitor or keyboard, which are always changing and developing under the powerful drive of the major and minor inventions of electronics.

Design management has to back up fundamental interdisciplinary studies, stimulating the designer's insight and intuition. As a pure cultural operator, it is the designer who collects and synthesizes ideas coming from different levels: from poetry, from society, from taste, style and research. Design in a company like Olivetti means design for the future (and we are not concerned here with science fiction). So, design management also means a non-stop, responsible assessment of all possible tomorrows.

And there are other factors to be combined and harmonized through design management. A major one is the marketing sector. An equally important one is engineering design, both in the product and in production, all within an overall image/identity strategy. This same strategy has to be related to the requirements of a company that is by no means small, and whose market is international, both in scope and in product technology.

Design management must perform a moderating role. There can be no experimentation that does not have its roots in reality. Similarly, work that is based on an excessively large number of formal or technical hypotheses may need to be restricted. And the imaginative possibilities of those sectors of design that involve simpler technology may need to be encouraged in preference to those demanding more complex approaches.

Consultant Designers

Olivetti's routine policy is to use top designers, professional people from outside the company. These consultants are, however, in a rather unusual situation, for they are closely and continuously linked with the company and with its functions. Indeed, they are part and parcel of a structure that is essentially corporate – a structure that they use and within which they enjoy the status of managers. But they are consultants nonetheless, free to work for other companies too, which only improves the efforts of all, and brings into the company suggestions and experience that are often extremely valuable.

What is more, these designers are intellectual operators, driven by curiosity and always in close contact with the final users of their products. Of course, this is a somewhat unscientific form of marketing, but it often gets the designers' insight and intuition working much better than boring written reports. These same professionals very often make their debut in the world of design through Olivetti.

Highly indicative, for example, is the fact that two different sectors of the company work on keyboards: systems and furniture, and office products. In each, the designers' different styles and art forms can flourish to suit the needs of different end users. But the same ergonomic rules are followed, so it is essential that they collaborate to meet the requirements of the company and its marketing.

A further example that is worth a moment's reflection is the work station. Here, there is a real fusion of the company's culture and strategy with overall cultural processes, and with a general philosophy which can instantly be defined as fundamental for our common civilization: informatics. In Olivetti's case, work stations with an informatic or telematic content are, first and foremost, work stations for the office. Designing the work systems of tomorrow also means influencing the nature of the architectural structures that will house them and, in the end, influencing the very nature of our cities.

We may therefore conclude with confidence that the design of a keyboard or of a work station – rather than the 'spoon' cited by Ernest N. Rogers a few years ago – is the basic feature of the city of the future. This gives us fair warning about the kind of responsibility that rests with design and design management in today's great companies.

5 Design Management at Bang and Olufsen

JØRGEN PALSHØJ
Bang and Olufsen, Denmark

Introduction

Product design is concerned with realizing ideas and is the best means for creating differentiation between companies and/or products in the marketplace. To me, the idea or the concept of a product is very important. Design without excellent product ideas may have aesthetic value in its own right; however, as design also creates expectations about content, high aesthetic value alone may not be enough. Design used merely as differentiation to claim a higher price may succeed in the short term, but the success will be limited. Design as camouflage for a dull or superfluous product simply does not work.

From the early years of Bang and Olufsen, products have been carefully designed, but design management was not used as such until the beginning of the 1960s. The first result of conscious design management in the company was the Beomaster 900, the first transistorized stereo radio. The transistors were smaller than the valves used previously, thus allowing a new, modern, flat design. The Beomaster 900 was soon followed by the Beomaster 1000, which further explored the design advantages of the smaller transistors with their low heat output.

I am deliberately talking about design management and not just product design, because the launching of these products was harmonized with a renewed marketing philosophy and an advertising campaign that today we would identify as a lifestyle campaign. And this is to me the crux of the matter: product design which is not coordinated with marketing and communication is, in most cases, a waste of time – or at least an ineffective use of a very potent tool. Design management is first of all coordination of communication.

Design as a Continuous Process

At Bang and Olufsen design has neither beginning nor end. Although we only work with freelance designers, they participate continuously in the internal idea groups. Each product area has its own idea group consisting of specialists from R&D and product management. The group conducts general screening of the area and also considers reaction from the market as a basis for new inspiration.

The designers take part in all this – and when an idea for a new product begins to gel, they produce the first models of the product concept. They usually work in cardboard or similar simple materials instead of drawings, because visualizing size and proportion is the first and most important step in assessing the company's products. In the course of the following weeks, the designers will build new models to explore the ideas developed by the group during the process. We might end up with ten models or more; the later ones supersede the earlier, but are still based on the content and the possibilities of the original idea.

The fact that the designers are involved right from the conception of a new product ensures their influence on idea as well as on construction. Like other creative people, designers are born problem solvers. They have the gift of seeing difficulties as challenges and are often capable of perceiving problems from new angles – different from the engineers' scientific and systematic approach. Of course, designers cannot replace the technicians and specialists, who know the possibilities and limitations of materials, production processes, markets and distribution, but they can apply their knowledge of people. And designers can – better than in-house specialists – represent the consumer.

Involving designers from the very beginning ensures a broader and deeper understanding of the company, its ideas and its needs – and it gives designers the opportunity to reflect all this in the design. Another advantage of this continuous process is that we have checkpoints at different stages. Is the idea acceptable as such? Do we have R&D know-how and capacity? Is there a need in the market? Do we have the production facilities needed? At the earlier stages we do not spend a lot of time analysing each question, but just use common sense. At later stages we evaluate and analyse the same questions more thoroughly.

Decisions are thus taken in a kind of converging spiral. Asking questions about concept, design, economy, capacity and so on at every stage in the process means that we avoid ending up with a product only based on one good idea. We do not elaborate on a good marketing idea for months, only to discover that we lack the necessary R&D capacity – or vice versa. Of course, we have checkpoints where management decides if a project is acceptable or not. But by asking questions all the way through, we avoid wasting time on design projects which, when completed, are found to be all in vain.

Freedom and Control

All great ideas are created by one individual – but a modern, complicated product is not made by a single person. It is made in a fruitful association between a number of people who like to work together and who respect each other. Basically, designers and engineers differ in nature. They tend to dress in a different way; they speak differently; they even evaluate things in life differently. How do you unite people dominated by either the left or the right side of the brain? The very first phase in the life of a new product is always difficult. Different people have to find each other and start collaborating. Their job is to develop a new and excellent idea. It does not develop by itself. On the other hand it is not developed without a challenge.

Our experience is that, first of all, people need to get to know each other well. They need time to get acquainted. And it always pays to take plenty of time at the start. It is better to have all the relevant ideas on the table before you decide on one of them: the least expensive period is always the beginning of the process. Analysing the problem is often the most effective way to start. Knowing and understanding the problem completely might even create the embryo of the solution. The practical problem of managing design then is to find the right balance between freedom and control.

Total Design

Design management is often perceived as the coordination of the design of products with architecture, graphics, signage and so on. I argue also for coordinating product design with marketing and communication. Why not involve communication people directly in the design process, or let communication develop parallel to the product? After all, design is a communication process in itself. Parallel development ensures that the original ideas and intentions are described and retained – and expressed in all documentation. Furthermore, the communication specialists can better than anybody else, even the marketing or sales people, evaluate a product's ability to communicate. It is far easier to describe a product that is based on a good and characteristic idea. It communicates itself. It gives advertisements and brochures far greater impact and thus contributes to the company image.

Advertising people, especially copywriters and art directors, possess not only knowledge of marketing, which in this connection is extremely relevant, but also practical creativity which can often bring further qualities to a new product. An obvious idea is to build demonstration possibilities into a product. Or to consider if a product is photogenic. How does it appear on a TV screen? How does it work at an exhibition? Such considerations are routine for advertising people – but how often would an engineer have them in mind? Finally, it is my experience that if the advertising people find it difficult to grasp the idea of a new product for communication – you might have a problem yourself. You might have a me-too concept or a very limited idea.

I realize, however, that coordinating communication to this extent will be problematic in most companies, because design is usually fixed in the upstream organization and advertising in the downstream. It is often a fight to open up an organization to get a holistic view of new products.

How to Select Designers

At Bang and Olufsen we have no design studio – and no permanently employed designers. We only work with individual freelance designers. We have two very close teams, with whom we have collaborated for years, plus a new young team recently attached to the company. The decision not to have our own studio is partly a consequence of Bang and Olufsen's rural position (on the Danish west coast, far from Copenhagen and remote from any designer environment) and partly because we believe that designers need the inspiration that comes from working with different product categories and different problems.

Our ideal is for Bang and Olufsen's share of designers' time to be great enough to make us important, but also small enough to leave them free to tell us what they feel is the truth. In a company where product innovation is a must, as well as consistency in quality and attitudes to design, the mixture of designers associated on a long term basis and new designers on a short term basis seems to be fruitful. It ensures the right balance between continuity and renewal.

Selecting new designers may be the most critical task in design management. I think the only way to do this is to keep your eyes open constantly – and to take notice when you see a designer making something in a completely different product area which you feel would fit when you mentally transfer it into your own area. All designers have their own fingerprint – and this is the only indication you may have until you actually recruit them to work on your problems.

In my opinion, there is little point in briefing a designer by describing a target group as you do in advertising. Good designers are artists – and, like artists, are only able to create what they like themselves. So you have to look for the fingerprints. We do so by viewing exhibitions, scanning design magazines and accepting invitations to visit designers.

Is There a Need for a Design Policy?

No, and yes. I do not think designers need one. They need a deep knowledge of the company – far deeper than a formal policy or strategy can provide. But the rest of the company may need a design policy or strategy to understand the aims and methods. Everybody involved in the design process – or the total communication process – needs to know why and how the decisions on design are made. Therefore, design must be an integral part of the company culture – known, understood and accepted throughout the company. This is not only a prerequisite for creating the best conditions for design work in the company. It has another purpose: design of the company's products might even be the best way to create a strong company feeling amongst employees.

At Bang and Olufsen we formulated the *seven CICs* at the beginning of the 1970s (CIC means corporate identity component). They are being re-evaluated and may be brought up to date but, so far, they have been used as a kind of common internal strategy:

Authenticity   Faithful reproduction. The best sound or picture reproduction is that which comes closest to reality – as experienced by the human being and not by a measuring instrument.

Autovisuality (Harmony)   A mutual balance between functions, mode of operation and materials used. Design is a language, a communication between those who create the products and those who use them.

Credibility   In products, dealing and action. Product specifications are minimum data.

Domesticity (Relevance)   Technology is for the benefit of people – not the reverse.

Essentiality (Topicality)   Electronic products have no form given by nature. Design must be based upon a respect for man–machine relations. Simplicity in operation. Design is an expression of the time in which we live, not a passing fashion.

Selectivity   Bang and Olufsen is the alternative to mass production.

Inventiveness   As a small company we cannot carry through basic research in the electronic area. But we can implement the newest technology with creativity and inventiveness.

Avoiding Myopia

In any company you run the risk of becoming myopic. The world is outside; you, your company and your business are at the centre. It is wise to be aware of your weaknesses and strengths. But how can you avoid myopia? How can you view your situation from other angles? There may be many ways to do this. We have tried to create a design reference group to assist the company in evaluating our design development from time to time. It is a small group, consisting of independent members with high personal integrity. They represent art and culture in general, not especially design, trade or any specific profession.

Their comments and evaluation are only regarded as advice – and the group has no formal responsibility. It has been most helpful to Bang and Olufsen so far. The only limitation I see is that the group might gradually become too involved and contented with the company – and thus of less value. The solution to this will be to involve new people from time to time.

Conclusions

The biggest mistake in design is to ask a designer to solve your problems – and then relax, waiting for a superb result. You – and your employees in the company – have to get involved. Creativity is not bestowed only on designers.

You have to establish an environment of inventiveness, boldness and willingness to take risks. Even the best and most inventive designers need cooperation with your own creative people as sparring partners and specialists. Design in modern industry is teamwork, so even though you may hire the most inventive designers, you will need creative individuals in your own company as well. Creative people can be difficult in any organization. However, as Bernard Shaw wrote in *Man and Superman*, 'Maxims for revolutionists':

The reasonable man adapts himself to the world; the unreasonable one persists in trying to adapt the world to himself. Therefore all progress depends on the unreasonable man.

6 Design Maturity: the Ladder and the Wall

DAVID WALKER

The Open University, UK

Introduction

People at different levels within a company use designers – internal designers, external design consultants – they give designers a brief for a specific task or series of tasks, but the idea that design has any kind of overall coherence in a company's life simply doesn't enter anybody's consciousness. (Olins 1984)

Sometimes in writing about management it is easy to fall into the trap of offering something which is well ordered, coherent and static – a management utopia, tidy but quite unbelievable. The reality is, as we know, dynamic, subtle, untidy and *ad hoc*. For example, the way designers are used by organizations does not conform to one particular pattern.

A small company may be seeking the services of a designer for the first time; a medium organization already experienced in the use of graphic design may be taking a new step into interior design, or commissioning a new building; while a very large organization may be seeking a way of unifying disparate products into a complete and persuasive corporate identity.

So we have not only a large variety of specialist designers, but also various levels of involvement. This involvement of business with design can best be thought of as a ladder of design maturity, as in table 6.1. The ladder embodies the notion that the more mature an organization is, the more involved it is likely to be with the whole range of specialist designers. Conversely at the immature stage the relatively small company, inexperienced in design, will only have a foot on the first rung of the ladder. There will be relatively little contact with designers; for instance, the company might use a graphic designer to design a promotional leaflet and letterhead.

In other respects an organization may be mature – in age, in specialized experience, in staff training or whatever. The expression 'maturity' here is applied only to its experience of dealing with the design professions. It does not mean that the organization is young or inexperienced in other ways.

We could argue about the exact nature of each rung on the ladder and their relation one to another, but two things would remain true. First, as the organization matures it uses more and more different design specialists (consultants and in-house). Thus the bottom rungs of the ladder are not

Table 6.1   Ladder of design maturity

Design specialism	Nature of work
	Top
Corporate specialists	Corporate identity programme
Architects	Major buildings Landscape
Architectural and interior specialists	Factory/workshop/office environment
Interior design	Retail environment Shop fitting Furniture and accessories
Graphic and corporate specialists	Design manual/guidelines Visual identity Coordinated signs, stationery and print items (2D) Product brochures
Production engineering	Process improvements
Manufacturing specialists	Specialist advice, e.g. materials, manufacturing
Engineering and industrial designers	Coordinated range of products New product development Product improvements Product appearance/styling
Graphic designers	Exhibitions Instructional literature Retail display Packaging and sales literature Letterheading and logo
	Bottom

sawn away as the top rungs are reached; the use of design specialists is both a symptom of, and a contributor to, organizational complexity.

Secondly, the initial encounters with designers tend to be over the least costly items (such as graphics, print, display). Typically as an organization matures it becomes more ambitious in its use of design, and its investment, in both designers and designed products, enlarges into product ranges, work and office environments, corporate identity, etc. Hence, the ladder of design maturity can also be read as a ladder of increasing design costs.

A recent UK survey shows that external designers are used much more for the less costly fast moving consumer goods (devising graphics and packaging) than for capital goods and heavier engineering (table 6.2).

Table 6.2 How UK companies use consultants (percentages): the more costly the product, the less outsiders are brought in

Use of consultants	Fast moving consumer goods	Consumer durables	Repeat industrial	Capital industries
Design	53.7	30.1	18.8	7.2
Market research	60.3	37.0	26.8	25.4
Marketing	22.5	10.6	12.3	7.7
Training	24.4	15.7	22.5	25.4
Sales promotion	42.0	16.7	9.4	4.8

Source: Institute of Marketing (1984)

The Wall and the Context of Design

The territory of management is broader than the territory of design. Therefore, management provides the context for design both prior to briefing and after products have been completed. More than that, reactions to designed products form the basis of revisions to managerial objectives and in turn lead to new design briefs. Management closes the circle. Hence, the design context means:

1 Seeing the design process in its managerial setting.
2 Taking a broader view than is normal with designers – in fact, taking the broadest possible view.
3 Framing the elements of a design brief with respect to the organizational context.
4 Collating the responses to products (both new and existing).
5 Understanding the competitive field and design trends.
6 Closing the vital feedback loop between products and briefing.

The context of the design process can be split into two main elements: the internal organization (local demands and needs) and the external circumstances (general demands and needs).

It is easier to pay attention to the former than the latter. The internal nature of the organization is known (more or less), is accessible, is specific, can be influenced and is controllable (more or less). For example, think of the demands from departments parallel to design in an organization – from technical research, manufacturing or finance. Unfortunately these characteristics are rarely present in the external circumstances, which are imponderable, sometimes inaccessible, diffuse, fluctuating and out of control (more or less). Think of the actions of competitors, think of customer reactions, think of changes in government policy.

There is a clear conceptual division between the internal, controllable parts of the organization and the external, uncontrollable world beyond. To use a metaphor, there is a wall.

The managerial danger is to concentrate on the accessible, known criteria internal to the organization and to disregard the difficult imponderables of the external context. This would be a natural reaction – to solve the problems which can be clearly stated and ignore those which seem uncertain and shifting. Yet what happens if the internal problems are revealed as irrelevant in the wider context? What happens if internal solutions build up a momentum of their own which is well-nigh unstoppable? What happens if the external context shifts in a dramatic and unpredictable way? What happens if the wall is in the wrong place or falls away?

This sounds like the spectre from a manager's nightmare – but is exactly the position of many organizations in the dynamic of the real world. It is perfectly possible to have a well-managed, design-mature organization, with a clear strategy, a well-defined brief, imaginative and competent designers and thoughtful design criteria, but still to come to grief – because the strategy and criteria do not match the more obscure criteria of the market.

Conclusions

This may suggest that all the advice (in this volume and elsewhere) about internal design procedures, and about understanding design activities, is not worth very much if somehow the context is misinterpreted. Well, that is exactly right! The advice of this book is well intentioned – and should be followed. But under some circumstances the whole apparatus of management and design can sink without trace for reasons beyond the control of good conscientious managers and brilliant designers. The best that we can do is to foster a general awareness of the power of the context.

Commentators (Pilditch 1987) have stressed the importance of breaking down the walls *within* the organization, but there is also the danger of laboriously assembling a wall *around* the organization and ending up with a fortress against the frightening instability of the world. On the contrary, the perimeter between the inside and the outside of the organization should be flexible and even less precise. The shape of the organization should be reactive, adjusting to new external patterns, not a fortress but an amoeba.

References

Olins, W.: *The Wolff Olins Guide to Design Management.* London: Design Council, 1984.

Pilditch, J.: *Winning Ways.* London: Harper and Row, 1987.

Walker, D. J. et al.: *Overview: Issues,* P 791 Managing Design, Open University Press, 1989.

PART II

THE BUSINESS CONTEXT

7 Product Design and Company Performance

ROBIN ROY

The Open University, UK

Why is Product Design Important?

A key question for managers in industry is to judge which of various investment options is likely to give the best returns. In theory, investment in product design should be of particular benefit to a manufacturing company, affecting sales, market share, profits and growth, at least in the longer term. This is because design decisions have an impact on virtually every aspect of how attractive the company's products are in the market. In economists' jargon, product design influences both price and non-price factors in competitiveness; see table 7.1.

Table 7.1 The role of product design in competitiveness

Factor in competitiveness	Influence of design
Price	
Sales price	Is product designed for economic manufacture?
Life cycle costs	Is product designed taking into account costs of use and maintenance?
Non-price: product-related	
Product specification and quality	Design affects product performance, uniqueness, appearance, materials, finish, reliability, durability, safety, ease of use, etc.
Non-price: company-related	
Company image and sales promotion	Product presentation, packaging and display affects image and promotion
Delivery to time	Is product designed for ease of development and to meet delivery schedules?
After-sales service	Is product designed for ease of service and repair?

Design plays an even more dynamic role in competitiveness than table 7.1 might suggest. From the viewpoint of customers and users, as table 7.2 summarizes, different aspects of a product's design dominate at different stages of purchase and use.

Table 7.2 How design affects customers' views of a product at different stages of purchase and use

Phase	Product design factors
Before purchase	*Brochure characteristics* Manufacturer's specification, advertised performance and appearance, test results, image of company's products, list price
Purchase	*Showroom characteristics* Overall design and quality, special features, materials, colour, finish, first impressions of performance, purchase price
Initial use	*Performance characteristics* Actual performance, ease of use, safety, etc.
Long term use	*Value characteristics* Reliability, ease of maintenance, running cost, durability, etc.

Product design ought, therefore, to play a key role in the competitiveness of most manufacturing companies and form a central plank of their strategies along with production, marketing and sales. But is design actually such a crucial factor in company performance in practice? And if it is so important, why do many firms give design such relatively low status and priority, while others make the achievement of good design one of their corporate goals?

These are among the questions investigated in an international study of design management in several industries ranging from plastics to electronics conducted by the Design Innovation Group (DIG) at The Open University and the University of Manchester Institute of Science and Technology. Some of the findings of this study are summarized in this chapter.

Over 100 companies have been visited, ranging in size from under 20 to over 10,000 employees. In-depth interviews have been conducted with management, design, engineering and marketing staff to discover how they manage the design and development activity and integrate it with marketing and manufacturing.

The sectors surveyed included:

1 Engineering-based industries (e.g. domestic heating, bicycles, motor vehicles).
2 Design-based industries (e.g. plastics products, office furniture, decorative lighting).
3 Technology-based industries (e.g. electronic office equipment, computers, consumer electronics).

The sample included British firms and some of their leading competitors from Japan, Canada, Sweden, Denmark, Holland and West Germany. The foreign firms were chosen for their international reputation for producing well-designed products and for their success in export markets. Also included were several British 'design-conscious' firms which had won one or more awards, citations or prizes for design, or were considered as 'design leaders' by other firms in their industry. To assess the commercial impact of design, the business performance of the firms was measured over a seven-year period using several indicators including turnover, asset and profit growth, profit margin, and return on capital.

Management Attitudes towards Product Design

Design and company strategy

There are some firms for whom producing 'good design' is a major objective. For such firms design is a key part of corporate image and business strategy.

For example, the philosophy of the American-owned office furniture manufacturer Herman Miller is that 'there is a market for good design', meaning that innovative, modern furniture that has been created by eminent architects and designers will sell and make a profit. Likewise, the management of the design-conscious British manufacturer Hille International said that 'our goal is to make well-designed furniture', citing their range of moulded polypropylene chairs, created by design consultant Robin Day, which has won several design awards and sold over 10 million units since it was introduced in the early 1960s.

In contrast, there is a substantial minority of firms – roughly a third in our surveys – whose managers believe that product design is not worthy of much time, effort or money. In such firms, design is typically undertaken as a minor part of another job in the company such as marketing or production. Some managers are hardly aware that design decisions are made or who is responsible for them. Products may end up being designed literally on the back of a cigarette packet or by default, for example by a toolmaker making a mould for a plastics product.

Most managers fall somewhere between these extremes, believing that employing specialist design expertise is one of several possible strategies for achieving commercial objectives. So for example most furniture and heating firms in our sample defined their objectives in terms of profitability and growth, and adopted various design and other strategies to meet these objectives, including:

1 Attempting to design new or updated products to match changing market requirements, often through contacts with purchasers, specifiers and end users.
2 Moving up-market through investment in design into higher value or more sophisticated products.
3 Diversifying into new product areas or moving into new markets, such as exports.
4 Attempting to reduce costs, for example by designing for economic manufacture, computerization and automation.

In the plastics products sector there was a striking difference between the strategies of the 'design-conscious' firms and typical firms representative of the industry as a whole. All the design-conscious firms said that making up-market products of good design and high quality was part of their strategy for expansion, compared with only 20 per cent of the typical firms. The remaining typical plastics firms either had no expansion strategy or planned to grow by diversifying or introducing new manufacturing technology.

In all the sectors studied, the commercially successful firms were not necessarily those that had won awards and citations for good design. However, statistically the design-conscious firms did perform significantly better than the others on certain business indicators such as profit margin and return on capital (Roy 1987).

Fast growing, profitable firms tended to be those where managers recognized the importance of design to the business derived from a genuine enthusiasm for what they were making. Thus managers of successful plastics firms would be enthusiastic about plastic as a material, rather than regarding it as an inferior substitute for wood or metal, and would themselves have expertise in designing in plastics or recruit designers or consultants who had such expertise.

The successful firms were also very committed to satisfying their customers. They did not get so carried away with enthusiasm for the product as to make achieving technical perfection or gaining design awards an end in itself unrelated to costs, customer demands, or changing markets and technology. Thus, a bicycle firm which had been run by cycle enthusiasts was more successful after being taken over by managers who were also cyclists but were willing to design products that balanced technical performance with price and style.

The most financially successful firms in our survey were therefore those which had clear business objectives and a strategy for achieving them that balanced their own particular strengths – in design, marketing or manufacturing – with the requirements of their customers.

What managers mean by design

Clearly what managers mean by design will have an important influence on how they use it. In general we found that in commercially successful firms the senior staff had a broad understanding of design, while in less successful firms the managers had a much narrower view.

For example, in the plastics products sector there was a contrast between the design-conscious firms, whose managers viewed design in terms of several factors – fitness for purpose, making products that sell or make a profit, ergonomics, production efficiency, etc. – and the typical firms, one-third of which defined design as 'shape' or 'visual appearance' alone. Likewise in other sectors, there was a contrast between firms whose managers had a broad understanding of design, requiring that several departments or individuals collaborate in order to meet multiple requirements, and firms that saw design as being concerned just with the initial concept, technical performance or styling, and so could be a specialist function (see table 7.3).

Table 7.3 How managers in British firms defined 'design'

Broad meaning	Narrow meaning
Office furniture 'Design is finding a visually satisfying form that fulfils the requirements of function and manufacturing, satisfies me, and can be sold at a profit'	The concept on the drawing board
'Design encompasses styling, production engineering, etc. The design department is where everything has to be considered. Its operation depends on the company pulling together'	It's all about fashion!
Domestic heating equipment 'Literally everything from idea to production, including brochures and installers' manuals'	The creation of new ideas Combustion engineering
Electronic business equipment 'Design encompasses product specification through detailed hardware and software design ... Design has a global role in the company. There's no dividing line between marketing and design'	Technical design Initial creation or idea

Having a broad understanding of the meaning of design may not seem very important, but the firms that did have such an understanding also realized how design decisions did not just determine concept, form and performance, but also influenced all the other factors that contributed to a product's competitiveness in the market.

Design and competitiveness

What factors did the managers in the firms we surveyed believe gave their products a competitive edge?

Figure 7.1 shows how British furniture, heating and electronics firms and their leading foreign competitors responded. Despite the widely reported view that British industry attempts to compete mainly on price, only 16 per cent of the British managers felt that price alone was the key factor. The corresponding figure for the foreign firms was 11 per cent. But it was notable that the foreign firms tended to attribute product competitiveness to more factors than their UK rivals. In particular 'technical performance' was crucial to all the foreign firms, with a high proportion also including product quality, delivery and after-sales service.

In the plastics products industry, the design-conscious firms all said that quality, design and performance were much more important than low prices in securing markets. But the commercially successful ones also recognized that price was important in some markets. For example Netlon, manufacturers of extruded plastics mesh (figure 7.2), found that price was more important in selling their product for packaging than for other uses such as garden netting.

It was relevant therefore that more foreign than British furniture, heating and electronics firms mentioned 'value for money' as giving their products the competitive edge, and that 78 per cent of the foreign firms priced their

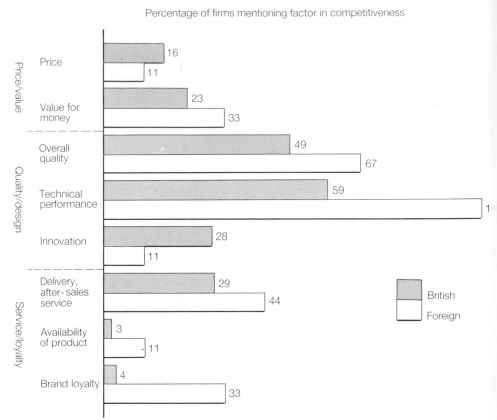

Figure 7.1 What British and foreign firms said gave their products a competitive edge. The British sample included 42 firms in office furniture, domestic heating, electronic business equipment and computing; the foreign firms comprised 9 world leaders in the same industry sectors

products flexibly depending on what customers in different markets were willing to pay. The Danish firm Lego, for instance, sold its hugely successful plastic construction toys worldwide on their high quality and design, but priced its products flexibly for different countries depending, amongst other things, on what rival products were available.

Thus, the commercially successful firms did not view competitiveness in terms of a single dimension. Their managers and designers understood the relative value of different price and non-price factors in different market sectors and how these were changing. For example in the bicycle industry, price is a key factor at the lower 'utility' end of the market; fashionable styling and image are crucial for children's and teenage bikes, while technical specification and choice of components and materials are more important to buyers of up-market sports bikes.

Successful firms therefore identify more precisely which sectors their products are aimed at and employ in-house or external expertise – industrial designers, development engineers, packaging designers, etc. – to meet the requirements of customers in those sectors. Flexible pricing and designing products to achieve economic manufacture are also approaches used to

Figure 7.2 Netlon extruded plastic mesh. Netlon plastic mesh is one of the few innovative products to succeed in both commercial and design terms. It was made possible by an innovation in extrusion technology that used the property that plastics weld together when soft. Netlon mesh is now made from a variety of polymers for a very wide range of applications, from fruit packaging and garden netting to earth reinforcement for civil engineering and filters for kidney machines. The firm is constantly seeking new applications for the technology and has won several awards, including a Design Council Award and Queen's Awards for Technological Innovation and for Export Achievement *Photos*: by kind permission of Netlon Ltd

provide products which customers in a particular market consider good value.

Effective Design Management Practices

There is not space here to detail all the design management practices that we found were associated with successful companies. So below are three which were of particular importance in distinguishing fast growing, profitable firms from the rest.

Sources of design ideas Where did the firms we surveyed get information for planning and ideas for designing new and updated products?

Most firms attempted to get information about the markets they were designing for, but those with high growth rates generally gathered marketing and customer information from as many different sources as they could afford.

For example, a successful small British hi-fi manufacturer with limited budgets defined the design parameters for a new product (figure 7.3) by talking to its dealers and to technical journalists, consulting published marketing reports, reading the specialist magazines, examining the competition and monitoring consumer comments on its other products. A successful British lift truck manufacturer used similar sources and in addition monitored developments in related industries and invited buyers and users to its test track to try out new designs. The two Japanese electronic business equipment manufacturers we visited went furthest; they had a 'special monitoring system' – a panel of customers and users who were regularly consulted on what new products they wanted.

Secondly, new and updated products in these successful firms were usually the result of group decision making rather than being the brainchild of the chief executive or another individual. In particular, new product ideas arose from a group significantly more often in the foreign firms than in their British competitors. Thus the Japanese special monitoring system was part of an elaborate system of product planning in which marketing, design and engineering staff were all involved.

There are of course exceptions. Occasionally a chief executive or other individual has a talent for identifying market opportunities and conceiving products that meet those requirements. A notable example is Alan Sugar of Amstrad Consumer Electronics, with his highly successful concept of low priced complete packaged systems in audio, TV/video and personal computers. This 'go it alone' method is good for creating innovative design concepts. It is however riskier than a team approach and can lead to spectacular market failures as well as successes, a good example being some of the products produced by Sir Clive Sinclair's companies.

Figure 7.3 Arcam Alpha amplifier. In the audio equipment market the most important factors in the budget hi-fi sector, apart from sound quality, are visual appearance and the quality of finish and construction. To enter this sector, A&R Cambridge Ltd together with a design consultancy, Cambridge Industrial Design, developed the Arcam Alpha stereo amplifier to have audio performance rivalling that of A&R's more expensive designs, but styled to appear to design-conscious consumers and capable of being assembled in half the time to achieve a target retail price of £130. Thus design can be used strategically to move down-market as well as up. Managerially, the key to the successful collaboration between A&R and Cambridge Industrial Design was a very detailed design brief and frequent meetings during product development. For more details see Roy et al. (1987) and Open University (1988)

Photos: A&R Cambridge Ltd

Thirdly, the commercially successful firms generally adopted an evolutionary approach to product development using competitors' products and technology as a starting point for their own products and design ideas. As is shown in figure 7.4, the foreign furniture, heating and electronics firms modified, improved or even imitated competitors' designs more often than their British rivals, some of whom felt they always had to start from scratch in design. The foreign firms also used suppliers more often to help with product development.

This suggests that for the majority of firms which cannot aspire to be design/technology leaders, adaptation, improvement or even imitation of competitors' products or design ideas and joint development with suppliers can be successful methods of design.

Of course, this means not that innovative design and technology is not required, but that it is usually difficult to establish a profitable market for a major innovation. We came across several examples of novel products, such as an ergonomically improved computer keyboard, that were technically excellent and well designed, but which had failed to find a market. We also encountered examples of innovative designs, such as Redring Electric's plastic jug kettle, which managed to establish a new market but were then almost squeezed out by improved designs from rival firms. On the other hand we found some successful innovations, such as a technique of making extruded plastic mesh which had led to an extremely successful range of products (figure 7.2).

The adoption of an evolutionary design approach does not mean that creativity is unimportant. 'Creativity lies in integrating available components in a new way', one designer said. Even when working on a variation of an existing product, designers in the successful firms produced several concepts in sketch and model/mockup form for management and/or customers to evaluate. A variety of techniques were used to stimulate creative thinking, including brainstorming, meetings between managers, marketers and designers 'to thrash out ideas', and product workshops to which dealers, buyers and end users would be invited.

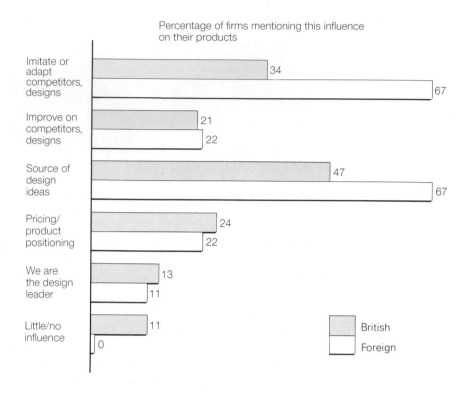

Percentage of firms mentioning this influence on their products

Figure 7.4 How British and foreign firms' products were influenced by competitors' designs. Samples as for figure 7.1

Thus for most firms, being highly innovative is not necessarily the path to success. But to be a successful fast follower or improver also needs designers with creative ability to make the detail design improvements that give products a competitive advantage.

The design brief   Of the firms we surveyed, 84 per cent drew up briefs for product development and in 64 per cent this was a formal document. But 38 per cent of UK heating and 13 per cent of UK furniture firms said they did not provide even an informal brief: 'There is no brief; ideas are based on the technical director's understanding of the market' was a typical comment. This was because in these sectors 'new' products were often modifications of existing designs, so only a verbal brief was considered necessary. Or the firms were small and very informal briefing was used. Thus the head of a photographic firm briefed his chief designer: 'I was using our forceps last night. They kept slipping into the tray. I think we can do better than that, don't you?'

Although highly informal methods can work in some situations, what differentiated the most successful firms from the remainder – more than any other factor – was the care that management took in drawing up a comprehensive brief at the start of any major product development project. While less successful companies tended to provide a brief that merely gave the required function and price of the product, the rapidly expanding, profitable firms were significantly more likely to include at least the following additional information:

Evidence of market demand
Details of target market and customer requirements
Relevant national or international standards

and, where relevant,

Guidelines on appearance, style and image
Guidelines on ergonomics and safety.

Probably the most striking differences were between the British and the world-leading foreign firms, both in the responsibility for the brief and in its form and content. Although not all foreign firms used formal briefs, developing them was almost always a group responsibility. In many British firms, the brief was drawn up by the chief executive with advice from senior colleagues. One consequence was that the foreign firms generally produced specifications that contained more comprehensive sets of product requirements. Figure 7.5 shows that 'selling price' was the only element of the brief included by more British firms than their overseas competitors. Foreign firms were also significantly more likely to provide briefs that included sketches, or other visual representations of design ideas, combined with written and verbal instructions. (Remember, though, we are comparing typical British companies with world-leading foreign competitors.)

The vagueness of the briefs provided by some UK firms, while apparently giving much freedom to designers, often inhibited creativity and wasted time. This was because design effort had to be spent defining the problem, sometimes by producing unsuitable designs which had to be modified, rather than solving it.

Employment of designers

Several reports, such as the Corfield Report on product design (NEDO 1979), have advocated that there should be a qualified design director on the company board. In our survey 74 per cent of the furniture, heating and electronics firms had a board member qualified in design or technology, although only 22 per cent had a director responsible for design alone, while in 26 per cent the chief executive was responsible for design/development as one of many tasks.

The Corfield Report also urged firms to devote more and better qualified human resources to design/development. We found significant intersector differences in the mean proportion of a firm's total staff employed in research, design and development, ranging in the UK firms from 2–3 per cent in furniture and plastics, about half of whom had a degree or equivalent qualification, to 20 per cent in electronics, of whom 90 per cent were professionally qualified. In the large Japanese companies the percentage was even higher; one corporation employed 5000 scientists, engineers and

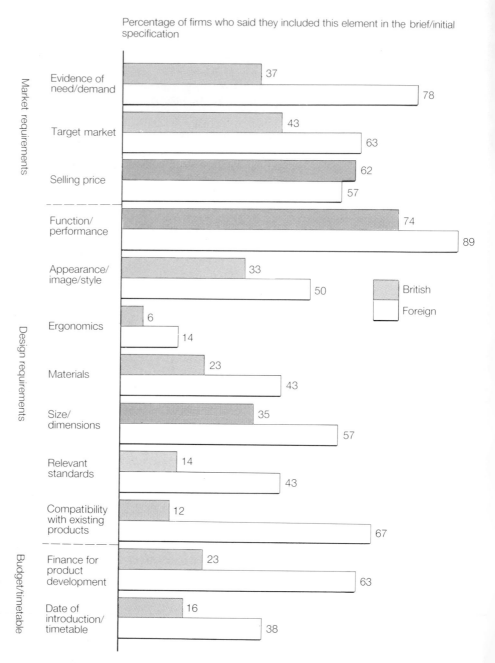

Percentage of firms who said they included this element in the brief/initial specification

Market requirements

Evidence of need/demand — British 37, Foreign 78

Target market — British 43, Foreign 63

Selling price — British 62, Foreign 57

Design requirements

Function/performance — British 74, Foreign 89

Appearance/image/style — British 33, Foreign 50

Ergonomics — British 6, Foreign 14

Materials — British 23, Foreign 43

Size/dimensions — British 35, Foreign 57

Relevant standards — British 14, Foreign 43

Compatibility with existing products — British 12, Foreign 67

Budget/timetable

Finance for product development — British 23, Foreign 63

Date of introduction/timetable — British 16, Foreign 38

British
Foreign

Figure 7.5 What British and foreign firms included in the brief or initial specification for product development. Samples as for figure 7.1 *Source*: Open University, Introduction to course P791, © 1988 The Open University

designers (25 per cent of its total staff), almost all with a degree or higher qualification.

The difference in human resources for undertaking design/development in competing firms could therefore be very great. In one heating firm the technical director would carry out most design/development work, while in others similar products would be developed by a large team of combustion engineers, testing staff and industrial designers. With some exceptions, the greater investment of design/engineering expertise was reflected in significantly better business performance. The successful firms generally employed a higher proportion of their staff in research, design and development than the average for their industry. Firms which employed in-house industrial/product designers also performed significantly better financially than those which did not.

But there were also successful firms, especially in the furniture industry, which did not employ any in-house designers, relying instead on outside consultants. Indeed, among the Scandinavian furniture manufacturers, use of consultants for industrial design was universal; usually they were recruited through design competitions held by the client firms. The Danish toy manufacturer Lego uses the same method to find creative people for its in-house design team. Overall, while 86 per cent of firms had in-house designers, 82 per cent also used consultants for specialist tasks, such as designing in plastics or software engineering. Some smaller firms had overcome their reluctance to using professional designers with the availability in the UK since 1982 of government subsidies to enable them to employ design consultants.

The key factor was not whether designers were in-house, consultants or both, but that design was viewed by management as an investment justifying the best professionals the firm could afford. Olivetti, for example, retains internationally known consultants for industrial/product/graphic design, while Philips has its in-house Concern Industrial Design Centre. And both employ many in-house staff for research and engineering development.

Given a strategic commitment to design, management has the task of recruiting designers or consultants with the right skills, ensuring that they are properly briefed, and meeting frequently with designers during product development; all of which determines whether designers' expertise and creativity is used successfully.

Conclusions

Our investigations support the conclusion of others (e.g. Lorenz 1986; Pilditch 1987) that product design and development is an important, and often unrecognized, element in the commercial success of manufacturing companies, which has to be integrated – through team effort – with other parts of the business, especially marketing and manufacturing.

Companies that invest in research, design and development are often also good at other aspects of business; indeed there seems to be a positive feedback effect between business success and good design. But it is by consistently doing everything well, while achieving excellence in at least one aspect, say product design, that certain companies have become leaders in their industry. Other companies have failed commercially because they placed too much emphasis on some aspects of their business, say R&D or manufacturing, while virtually ignoring other key elements, such as design.

Acknowledgements

The author would like to thank the following members of the Design Innovation Group who contributed to the work on which this chapter is based: Dr Margaret Bruce; Georgy Leslie; Dr Jenny Lewis; Dr Stephen Potter; Dr John Towriss; Dr Vivien Walsh. The research was supported by grants from The Open University and the Joint ESRC/SERC Committee.

References and Further Reading

Bruce, M.: The design process and the 'crisis' in the U.K. information technology industry. *Design Studies.* January 1985.

Design Innovation Group: *Design-based Innovation in Manufacturing Industry: Principles and Practices for Successful Design and Production.* Report DIG-02. Milton Keynes: Design Innovation Group, The Open University, January 1986.

Lorenz, C.: *The Design Dimension.* Oxford and New York: Basil Blackwell, 1986.

NEDO: *Product Design: Report by K. G. Corfield.* London: National Economic Development Office, 1979.

Open University: *Case Studies: A&R, Hewlett-Packard.* Unit 2 of Open Business School course P791 Managing Design. Milton Keynes: The Open University Press, 1988.

Pilditch, J.: *Winning Ways.* London: Harper and Row, 1987.

Roy, R. and Wield, D. (eds): *Product Design and Technological Innovation.* Milton Keynes and Philadelphia: The Open University Press, 1986.

Roy, R.: Design for business success. *Engineering.* January 1987.

Roy, R., Walker, D. and Cross, N. C.: *Design for the Market* (a multimedia learning pack). Watford: Engineering Industry Training Board Publications, 1987.

Walsh, V. M. and Roy, R.: *Plastics Products: Successful Firms, Innovation and Good Design.* Report DIG-01. Milton Keynes: Design Innovation Group, The Open University, August 1983. Summarized in Walsh, V. M.: Plastics products: successful firms and good design. In Langdon, R.: *Design Policy. Vol. 2: Design and Industry.* London: Design Council, 1984.

Walsh, V. M. and Roy, R.: The designer as 'gatekeeper' in manufacturing industry. *Design Studies.* July 1985.

8 New Responses in Design Management

DAVID L. HAWK

New Jersey Institute of Technology, USA, and the Stockholm School of Economics, Sweden

Introduction

Contemporary businesses are undergoing rapid transformation. This is not from any internal desire to change, but in response to radical shifts in the context of day-to-day operations. Traditional theories of management fail to either predict or explain the shifts. Change is not unique to our era but the danger in mistaken responses is now greater. The consequences of misapplications of advanced technologies have grown exponentially. Social systems and population groups that can be adversely impacted by business mistakes are large. In addition, fewer physical sanctuaries and frontiers are available for escape when problems seem to overwhelm. A small but growing number of executives seem concerned about the situation.

Business encountered profound changes in the eighteenth century. Industrialization allowed production of large quantities and varieties of goods and services. Product objectification, engineering rationalization and mass production opened up new opportunities for business activity. Much of the potential in these three areas has since been realized, although the latent problems that accompanied industrialization continue to emerge: the diseconomies, externalities and instabilities that travel in the shadows of focused goal-attaining, the downside of human activities. A transformation is needed and there are signs that it has already begun although its direction and structure are unclear. One argument is that it is the result of purposeful human design for a post-industrial society. A counter-argument is that it is simply the downside of the previous era and the next era still awaits new ideas for its design. The first option is obviously the more desirable. The contents of this chapter reflect the possibility of both options.

The central focus here is design and its management or, more poetically, the fit between human activities and their contexts. Using the fitness concept, there are three important aspects of design:

1 The quality of design as it fits within or relates to a given context.
2 The capacity continuously to redesign the process to upgrade the ability to find fit through learning.
3 The capability of management to encourage quality and redesign.

Design is a neglected area of business activity. While often discussed in business circles and magazines, this is in the restricted sense of resultant products or packages, not as a fundamental business activity. The approach here sees design as a comprehensive and integrative activity, reflecting emerging business practice and a small body of business literature (Churchman 1978; Ackoff 1974; Emery and Trist 1965). Since the approach is uncommon, it is important to define three aspects of comprehensive design:

1   Objective seeking behaviour that identifies and selects ends to be attained while retaining the possibility to change ends: for example, changing from producing the largest possible automobiles to producing automobiles of the highest possible quality.
2   Specification of activities required to achieve the ends as identified and defined: for example, identifying the attributes of automobile quality, gaining access to resources, establishing a production process and managing all three.
3   Communication to others of the objectives sought and activities required: for example, providing clear representations of the product, production process and management system.

A Design Context in Motion: 1945–85

1945–75   European and US business operations expanded into large international organizations after World War II. The advantages of vertical integration of resources were combined with the potentials of horizontal market reach. The resulting enterprises defined the cutting edge of business. Success with products and services depended upon informing customers (through expanding advertising) and providing timely deliveries of dependable products (via the growing distribution systems from the expanding production facilities).

Marketing, sales and distribution systems were created and enlarged. Employment in these provided the surest path to corporate advancement. Production operations were important but only as a necessary basis to the more interesting and dynamic activities of financial management of the bottom line. Those who ran production spoke a different language. Product design was even further down the hierarchy and carried out by people who spoke yet another language. Qualitative aspects (of the organization, product or service) were either unimportant, lost in gaps between departments or assumed to exist simply in whatever the organization was and did. Straight-line projection of growth was the guiding ideal for the overall process. This was achieved through proliferation or acquisitions of competencies. One image of this era is the manager, holding a hand over the telephone receiver, calling out 'I have a customer on the line, complaining about our product. What is it we make?' The seventies were a turbulent time for business. There were incredible overnight successes, followed by dismal failures – all achieved while pursuing a consistent set of unchanged goals. Resources for industrial production became highly uncertain. Industrialization ideals began to encounter dilemmas.

1975–80   Additional troubles were initiated for business-as-usual. Foreign competition further destabilized the situation. Some foreign companies had learned how to improve on the shortcomings of traditional business practice. They created new firms with flattened hierarchies, efficient distribution systems, new principles of production, new attitudes about management, new ideas about design, and ideas about integrating all of these together; foreign firms were setting a new pace. The result was an imperative to change, but the direction was unclear. Should traditional firms try to be more like their new international competitors, or should they do something altogether different? Both strategies were tried. The evidence suggested that change is essential, expensive, capable of making a bad situation worse – and generally confusing. Hard working organizations and employees could no longer count on success. Management could no longer focus on bottom-line numbers to increase quantitative throughputs. Employee loyalty was eroded through bankruptcies, mergers, buyouts and reorganizations.

1980–85   Competition was growing, and growing more difficult. Markets had become fragile and in question by the eighties. Distribution systems were found to be archaic, hierarchical and unreliable. There were far too many intermediaries in distribution and middle managers in marketing and finance. Buyouts and cleanouts took advantage of this situation. When the context of business changed, so too did the rules of corporate employment, advancement and organizational success. Something unusual was at work. Standard methods and theories no longer worked. The context that once was the playing field for business had become a major actor in the game itself. The stage set had become the major actor, a possibility first raised by Emery and Trist (1965).

1985–present   All prior rules are now under question. Advertising is distrusted by the markets buried under it. Products fail to perform as intended and services often offer none. Lawyers and courts are expected to fill the gaps but generally fill their coffers instead. Uncertainties with human, material, capital and energy resources are large. On top of this, there are uncertainties with management of information flow. A chief executive could previously count on a hierarchy of management communication, and if this broke down could assume authoritarian control until the situation stabilized. Increased application of industrial democracy ideas, quality of working life experiments and computerized communications networks have made the hierarchy as shaky and fragmented as the context in which it operates. Taking authoritarian control in a company with distributed information systems may only ensure greater instability (Pava 1983).

Managers are trying to find a new theory to explain what is already happening in practice. The current work and thinking of several international management consulting groups illustrates this. One recent example was a Tom Peters Group symposium held outside London to discuss issues similar to those outlined here. It included top-level managers from numerous European companies including ICI, Volvo, Hilti, London Life and Apple of Europe. The topic was a critique of the Peters excellence theme in the light of rapid changes occurring in the context of business operations (written by Hawk and Hedund, 1986). Should excellence be revised, upgraded or dropped? It was concluded that it is important but insufficient for success. Too many companies that were considered excellent had gone out of business between 1982 and 1986. Peters has since published an additional best selling management book (1988) which should be consulted for an elaboration of how best to respond to the chaos that surrounds excellence.

The potential for organizational turbulence, first discussed by Emery and Trist (1965), seems on schedule. How best can this contextual instability be responded to? Achieving this will involve many issues and activities. Foremost will be appreciation of and talent in design. One example of how this might translate into practice is offered in the following section.

Using Design to Redefine and Manage Context

In the 1940s, Texaco Inc. found that significant and growing resources were being invested in achieving a consistently high standard of refined product. This was a source of high business cost and resource inefficiency. There was the possibility of even more significant problems at some future time when it would be necessary to use lower qualities of crude oil. Up to 20 per cent of the potential value of a barrel of crude could be lost during its refining. An additional amount of each barrel was of such low quality that it was unusable. By the close of World War II, it was clear that there was a problem with the traditional internal combustion engine. The optimum strategy was felt to be the design of a more efficient engine that could run on lower quality fuels. Achieving this end would be advantageous to all concerned.

A project to study the feasibility of reaching this objective was established by Texaco. It began with the operating characteristics of the internal combustion engine. A proposal was made to redesign the engine so that it would meet the project objective; the result was a stratified charge engine design. The guiding principle came from the fact that fuel pumped into the combustion chamber was unevenly distributed. If there was only one spark per explosion, it had to be set for an average fuel mix. The new design proposed a way to stratify multiple charges which could take advantage of the variety within the cylinder. The new engine design could burn more efficiently, with less pollution and with a capability of operating on lower quality fuels. The design had the potential to revolutionize engine design and performance and refining operations. We can now only speculate on how this might have helped avoid many future problems associated with pollution, resource consumption and product liability.

It was taken to the market. First stop was the US Department of Defense. They had first articulated the problem to which the stratified charge engine was a response, so they would obviously be interested in the redesigned engine. While interested, they felt it was up to their major suppliers, the truck and tank manufacturers, to produce engines. Their role was consumption, not production. The next stop was at two of three major car producing companies. Reportedly, they were visibly unimpressed and uninterested. They argued that they employed hundreds of the best trained engineers, and if the stratified charge engine was such a good idea it would have been invented by their people. Besides, even if it was a good idea, it could disrupt a system that was already performing well. Consumers hadn't asked for this design change, so why worry them? The traditional engine design had become a contented cash-cow. With a couple of research papers summarizing the design, it was placed on the shelf.

Twenty years later a Stanford engineering professor took one of the papers off the shelf and began new research building on the stratified charge idea. At about this time, the Honda Motor Company was getting ready to enter the world of major car production. They began their journey by rethinking the ideas behind car design – with a particular emphasis on the engine. This was due to the company founder and chief executive, who had a passion for engines. A management system was set up that encouraged brainstorming sessions about 'best' design. During these arguments, someone located the information on the potentials of the stratified engine design.

Honda bought the rights to the stratified engine design as it had then been developed and improved it for mass production. The result was the remarkable Honda CVCC engine of the late seventies and early eighties. It had three important qualities:

1   Durable and unbothered by the lower quality fuel that might be put through it.
2   The most energy efficient, mass produced, high performance engine of the day.
3   The cleanest running production car and the last engine in the US to have to install pollution control devices (catalytic converters).

What can be learned from this and related examples of design and its purposeful management? It shows that management can indeed respond to contextual changes prior to them overwhelming the operations and control aspects of the firm. It illustrates that design activity is the key. The growing significance of design in responding to contextual change is also seen in recent literature analysing the continuing success of certain national approaches to doing business. The following are comments from a summary of a study (Gomory and Schmitt 1988) about the relations between science and products. Design was the proposed linkage. The authors describe research that juxtaposed Japanese and US approaches to design.

In the United States, the design phase of the cycle of development has traditionally concentrated on the features and performance of the product rather than on the process by which it will be manufactured. We design the product first and then tackle the job of how it is to be made. Yet the eventual cost and quality of the product is inseparable from the way in which it is made. If a product can be made easily, its cost will be low and, most probably, its quality high. A complex product, with many features and elements for product performance, but designed without regard to the intricacies of making it, becomes a product of high cost, questionable quality and uncertain reliability. Yet there has been a strong propensity in US engineering to follow this course, a course that separates design and manufacturability ... By contrast, the Japanese are oriented to simplicity in their designs ... Their designs embody state-of-the-art technology and easy-to-use functionality in attractive designs, which tend to be manufacturable ... Although low manufacturing costs is a natural corollary of simple designs, the Japanese nevertheless also make it an explicit objective of design ... The US designer, starting with a certain tradeoff between cost and performance, proceeds through subsequent design changes to add features and performance, and ends up with higher cost. (The new GM Baretta, in going from an $8000 car to one costing $11,000, is a clear example of this tendency.) The Japanese designer holds to the designated costs; no matter how many design cycles there may be, the initial cost barrier remains inviolate. Japanese designs are not only strongly constrained by the cost of the original product but are also optimized to permit further cost reductions in successive generations of the product.

The authors go on to point out that an additional aspect of Japanese success is the close tie, or the lack of distinction at all, between the engineers engaged in manufacturing and those engaged in development. This is one of many factors that contributes to rapid development and manufacturing cycles.

The Japanese approach is not a panacea for design, but points to several ways in which they are learning to deal with: new management approaches, change processes, design ideas, resource efficiencies, manufacturability and integration. Design and its management are the key.

A Basis for Improved Guidelines

Companies that intend to survive in a dynamic context must have design competence in three areas: management, product and process. Management design is traditionally known as organizational design (or simply OD). Product design comes from the tradition of industrial design (ID), as well as many other design professions in engineering, architecture, the arts, etc. Process design is usually associated with the discipline of industrial engineering (IE). Any approach that presumes to offer comprehensive guidelines for the management of design must account for the activities of all three and not stop at the borders. The argument is that if problems are linked then the methods for solving them must also be linked.

Problems in the design of management

The modern manager must learn to handle an increasing number and array of issues. The usual way is to add expertise by hiring additional staff or consultants with distinct competencies in what are thought to be the major problem areas. Another approach is to reorganize and retrain current employees to upgrade and expand their areas of competence. This second strategy results in fewer parts to integrate and encourages the discarding of irrelevant past issues to make room for new contingencies. It is insufficient to set up an organizational structure, based on either a formal information hierarchy or a functional matrix. Formal distinctions between those that design, produce, sell, finance and direct are generally unhelpful in conditions of change. Integrated resources are needed to respond to opportunities and emergencies that arise without warning.

Problems in the design of products

The desired end of any design activity is creating a whole which manifests potentials beyond those explainable by the parts in abstraction. Systemic properties are at the heart of what is both desirable and difficult about design. The success and failure of goods, services or organizations can only be evaluated against some context. How A relates to B and D, but not to C only has meaning in terms of a context. Potentials and problems of a product need to be seen against a larger context.

Problems in the design of processes

Production processes can serve virtually any end. They have achieved astounding results during two centuries but, unfortunately, the consequences of having achieved some of these results has become costly. Part of this is due to design that dissects continuous processes into distinct parts, where what comes before and after is left out of the thought process. This emphasizes analytic forms of information and discredits synthetic forms of knowledge. Long known to be a problem in product design, this dilemma is now an even more serious problem in process design. It is supported by management systems that use quarterly profit-sheet results to direct and coordinate. Improvement requires better models of design and management; a number of guidelines are offered as the foundation for a new model.

Guidelines for the Design of Management

Evaluation of performance

Doing a good job in managing good design is often distinct from being promoted for seeming to do well – like the 'Teflon manager' who never lets the results of a returning bad decision stick. In contrast to traditional management logic, those who do a good job over the long term may be those who learn to concentrate responsibility (often on themselves) for that which goes wrong and distribute responsibility (to others) for things that work out well.

Learning how to learn   Great learning can take place through making mistakes. Management needs to learn how to learn and how to help others to learn. This is especially important when design and redesign activities are involved. Learning through mistakes is fundamental to the education and practice of design. Again, this notion is in contrast to traditional logic but is critical to achieving quality and reducing defects.

Attitudes towards others   Management needs to specify the model of human behaviour on which its model of design management is based. Are humans and what they do seen as mechanical components that can be predicted, prodded and placed – or do they have undefined and non-rational capabilities that await enhancement and encouragement (Davis 1972; Rhinelander 1973)? Employees and the results of what they do are substantially different under the two models.

Guidelines for the Design of Products

Meaningful limits   Limits are the most important aspect of quality design and they must be limits that clearly make a difference. It is said that without limits nothing is possible, While this is true, the limits need to be well selected; they must be the crucial limits, articulated at the outset. Nature is a good source of the limits that matter; the second law of thermodynamics offers several limits for design. As well as reminding the designer of the lack of free lunches, the entropy law suggests that all design acts are irreversible and thus matter very much. One result of taking this approach is that a clear definition of efficiency becomes part of the design process. This definition encourages the wise use of resources in product production, operation and life cycle. More details of this thesis can be found in work using the second law of thermodynamics to guide production (Hawk 1986; Georgescu-Roegen 1971).

Measures of success   Efficiency is more helpful than productivity for comprehensive design. Productivity has long caused serious difficulties in evaluating success. Where intensively used, it gets reduced to a single variable that can be measured and maximized to the discredit of many supporting variables. Total factor productivity is a better alternative, similar to comprehensive measures of efficiency; and efficiency has the advantage of being a more general term. (Productivity could, and often does, point out that, for example, a toxic waste by-product costs $0.50 per gallon of product to eliminate from a process. But comprehensive efficiency points out that it costs $5 per gallon to dispose of what has cost $0.50 per gallon to avoid during production.)

Testing assumptions   Most of the creative potential in a situation is hidden in its assumptions. The deeper you go into underlying assumptions, the more alternatives you will find. There are good reasons for not going too deeply, such as completing the project, but there are costs associated with remaining at the surface.

Guidelines for the Design of Processes

Beyond mass-production

As products become more diversified and varied, the processes for making them can converge. This contradicts the major truth of the industrial revolution. Evidence can be seen where new production processes are becoming similar between industries, yet all moving towards the ideal of single copies. For example, a US Navy spare parts production plant in South Carolina turns out 100,000 different parts, yet with an average run of only four copies of each part.

A holistic framework

Organizational design can usually be carried out in a two-dimensional (horizontal–vertical) framework; product design requires three dimensions (space); and process design, to be successful, needs four dimensions (a space–time continuum). The dilemma is that the complexity of four dimensions encourages managers to resort to a simpler arrangement, such as a one-dimensional assembly line. A more robust and interesting process design can be seen in a four-dimensional sphere for autonomous work groups that decide how, where and when they will work.

A need for a common language

To integrate process design into the same framework as design of management and product, it is imperative to use a language that integrates instead of disintegrates. One way to do this is to conceive of design as organization. This encourages conceptualizations that can relate people, products and production processes into a whole. Another way is to experiment with alternative modes of representation. One interesting example of this is the home building factory where drawings are no longer allowed because they were found to emphasize unduly certain parts of the organization and production.

Conclusions

There must always be a general rule that governs the development of guidelines and encourages their upgrading or replacement. Most such rules are simply statements of common sense, but since sense is not always so common, we must continually restate it. One general rule is seen in T. S. Eliot's observation in *Little Gidding*:

We shall not cease from exploration
And the end of all our exploring
Will be to arrive where we started
And know the place for the first time.

References and Further Reading

Ackoff, R.: *Scientific Method: Optimizing Applied Research Decisions*. New York: Wiley, 1962.

Ackoff, R.: *Redesigning the Future*. New York: Wiley, 1974.

Churchman, W.: *The Systems Approach and its Enemies*. New York: Basic, 1978.

Davis, L. and Taylor, J. (eds): *Design of Jobs*. Harmondsworth, Middlesex: Penguin, 1972.

Emery, F. (ed.): *Systems Thinking*. Harmondsworth, Middlesex: Penguin, 1969.

Emery, F. and Trist, E.: The causal texture of organizational environments. *Human Relations*. Vol. 18, pp. 21–32, 1965.

Fabricant, S.: Problems of productivity measurement. *Measuring Productivity*. New York: Unipub, 1984.

Georgescu-Roegen, N.: *The Entropy Law and the Economic Process*. Cambridge, Mass.: Harvard University Press, 1971.

Gomory, R. and Schmitt, R.: Science and product. *Science*. Vol. 240, pp. 1131–2, 1203–4, 27 May 1988.

Hawk, D. L.: Setting a paradigm for building economics. *Habitat International*. Vol. 10, no. 4, pp. 5–21, 1986.

Lawrence, P. and Dyer, D.: *Renewing American Industry*. New York: Free Press, 1983.

Pava, C.: *Managing New Office Technology*. New York: Free Press, 1983.

Peters, T. J. and Waterman, R. H.: *In Search of Excellence*. New York: Harper and Row, 1982.

Peters, T. J.: *Thriving on Chaos: Handbook for a Management Revolution*. New York: Knopf, 1988.

Rhinelander, P.: *Is Man Incomprehensible to Man?* San Francisco: Freeman, 1973.

9 Design as a Strategic Management Tool

BRIGITTE BORJA DE MOZOTA
Université René Descartes, France

Introduction

New trends in management emphasize the importance of the image of a company and consider the image as a corporate resource. This interest in managing the image of a company explains the development of design management and of design strategies. Design, here, is understood in its conceptual meaning as a technique that can be used either in communications strategy (graphic and environment design) or in product development strategy (packaging and product design). Whatever the final output, design as an innovation technique is always a problem solving activity.

Like advertising, design is a tool managers can choose to use or not. All companies are designed but only some are designed by designers. How design works on a company's profits, what is meant by a design strategy, how design relates to marketing, why design should be introduced during the opportunity identification phase as an idea generation tool: these are the questions addressed in this chapter. The search for significant differences is the international challenge. Designers have cards to play with because they are often the only ones able to see and draw what managers can only describe with words and figures.

Design and Company Profits

Innovations produced by designers are *signs*. A sign is the resultant of a three-dimensional system and of the complex set of relationships between the attributes of each dimension: structure, function and symbol. All signs or innovations have a three-level communication system:

1 The *innovation-itself* system means our communication with the structure of the sign, its technological dimension and its syntax.
2 The *innovation-user* system means our communication with the function of the sign, its pragmatic dimension and its utility.
3 The *innovation-environment* system means our communication with the

symbolism of the sign, its semantic dimension and the connotative and denotative factors it conveys about our relationship with others.

Hence, if all innovations are signs in their visual, physical aspects, any other environment artefact is also a sign. And the environment is a system of signs with bilateral influence on any new sign or innovation produced.

Design, in order to be considered as an efficient management of innovation technique, has to create value. The signs (graphics or products)

Figure 9.1 The sign value concept. The variables are as follows:

AV active variables of design, i.e. the three dimensions of structure, function and symbol

UV uncontrollable variables of the environment

EV essential variables, or criteria for the success of the innovation, as chosen by management originating the project

conceived by designers have to be profitable. Therefore, we can define the 'sign value' concept as the equation of design technique efficiency, as in figure 9.1. This concept explains how design efficiency can be measured. It also explains why design is part of marketing. Design and marketing are both concerned with the exchange relationships between a company and its environment. Design reinforces the consensus that exchange forms the central phenomenon for marketing study by initiating a visual exchange system and by considering the aesthetics of a company or of an innovation as a set of controllable variables.

In order to identify whether design is a means of increasing the probability of commercial success, eleven case studies were chosen for an experimental survey. These design projects were all recent repositioning innovations realized by experienced multidisciplinary design firms. They cover all aspects of design and of the economy. Findings were based on evidence successively obtained from the designer responsible, then the managers or marketing directors of the firms commissioning the projects. The projects were:

Visual identification system for Baccarat (design: Lonsdale)
New logo for P. A. Consultants (P. A. Design)
Packaging body lotion for Nivea (design: Carré Noir)
Motor oil packaging for Shell (design: J. L. Barrault)
Locomotive symbol for French railways (design: MBD)
Document shredding machine for Secap (ENFI Design)
Heat pump for Airwell (P. A. Design)
Electronic central heating control (design: MBD)
Sportswear collection for Lee Europe (fashion design: PROMOSTYL)
Bank agencies for Caisse Epargne Ecureuil (design: Lonsdale).

The structure of the survey was based on a number of assertions. The failure rate of innovations is high. Design will therefore be considered efficient if it increases profit by minimizing the risks of failure. These risks include poor management of innovation; non-participative management, with no control procedures; poor management of ideas, such as an inaccurate innovation match with the image of a company, or errors in the innovation marketing strategy; and failure to meet consumers' needs, perhaps by incorrect pos-

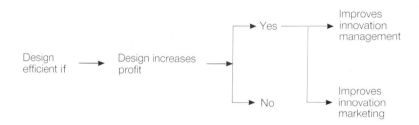

Figure 9.2 Survey summary

itioning or because newness was not perceived by the market. Figure 9.2 summarizes the survey.

The following results were obtained (see figure 9.3).

Design increases profits

Eight out of eleven projects responded. In five projects out of eight, the ratio of design cost to profit fell below one in less than 18 months. Interestingly, profit increase is achieved not only by expanding sales or by a drop in manufacturing costs, but also by economies made on advertising costs. Design costs are often considered to be negligible compared with the total research budget for a new product.

Design promotes participative management of innovation

Design encourages the setting up of multidisciplinary work committees where all the company's main functions are represented. This horizontal team is one of the strengths of design. Typically, it includes the manager, the R&D department, the manufacturing and the marketing directors.

Design does not improve management control of innovation

Design has no effect on the risk of innovation failure. Control innovation procedures are still often restricted to a final prototype test. Only three (out of eleven) projects showed a continuing follow-up of the creative process, taken rather as a way of consensus than a way of controlling designers.

Design improves the management of ideas

Studying the environment visually, as a system of signs, design does stimulate new ideas and improves the company's information about its environment. Ideas arise by associative thinking and transfer operations (Lorenz 1986). Examples include new shapes deriving from the need for coherence with the environment, transfer of technology from different fields having similar structural problems, and transfer of visual culture trends from artistic fields.

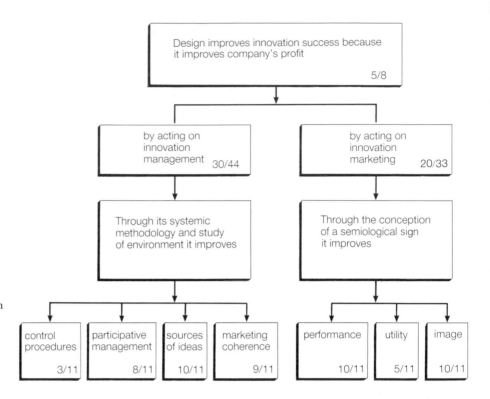

Figure 9.3 Design's impact on innovation. Results of eleven design projects over seven final categories (77 possible responses); only eight firms responded quantitatively to the first (profit) question

Design improves the image of the organization

Design studies the organization as a visual system. It encourages common themes between each marketing policy, tending to create either a global communication process or a more coherent marketing mix.

Design improves the innovation's performance

Acting on the innovation structure, designers change its physical characteristics. The difference is perceived as an improvement in modernity and performance. This outcome is obtained by designers through deliberate investigation of components (value analysis – even in graphic design) in order to change, reduce or simplify. This investigation is sometimes followed by the introduction of new manufacturing processes.

Design's action on the utility of the innovation is not proved

By changing the innovation function system, designers act on the man–machine relationship. Differences were perceived as an improvement of the utility of the innovation only in five projects out of eleven. Where the difference is significant for the consumer (or user), the result is obtained by designers using ergonomics (change in commands) and/or changing the site and size of typography and symbols to improve the reading distance.

Design improves the innovation image

Acting on the innovation's symbolic dimension, design changes its image characteristics. The difference improves the preference score of the innovation. This outcome is obtained by designers amending the colour scheme and/or the detailing of shape; new connotations are associated with a new shape, a new picture in the environment.

The conclusion of the survey was left to the managers who commissioned the design projects. Asked to rank by importance design's contribution to innovation success, they reported that:

1. Design improves the coherence of the organization image and its differentiation from competition (score 23/25).
2. Design creates innovations closer to consumer needs (score 20/25).
3. Design improves the organization's profit (score 15/25).
4. Design improves the management of ideas (score 12/25).
5. Design promotes participative management of innovation (score 9/25).

This conclusion suggests that managers have a poor understanding and perception of how design works on a company's results. The action of design on marketing quality is widely perceived but its action on the management of innovation is underestimated. This survey illustrates why the design function has to be managed and why a design strategy has to be chosen to achieve an appropriate design mix.

Design and Strategic Positioning

Managers have a design problem whenever their decisions end in a visual output. Whether or not they decide to use the expertise of designers to optimize this decision is their choice, but they cannot avoid the necessity to translate the strategy into various visual artefacts. Managers introducing design in their innovation strategy develop, as a consequence of this technique, a new sequential decision process and opportunity identification process for new product (or service) development.

New product and service development is usually a five-step decision process (figure 9.4): opportunity identification, concept, testing, introduction and profit management. Opportunity identification is the definition of the best market to enter and the generation of ideas that could be the basis for entry (Urban and Hauser 1980). Sources of ideas are innovations evolving in the market and innovations issuing from R&D and management. Because, in their creative process, designers use both consumer needs and technological changes as sources of ideas for a change in product shape, their technique is a useful coordinating tool between marketing and R&D.

Designers in the diverging phase of idea generation study visually, perhaps using photographs, the various products existing in the company's competitive environment. This visual audit of the product universe is a useful

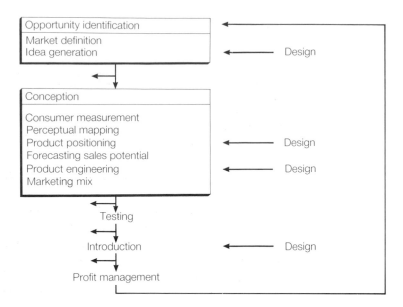

Figure 9.4 New product development process with design inputs
Source: adapted from Urban and Hauser (1980)

tool for product differentiation strategy, especially with respect to the product function and the semantic message. A thorough study of colour different-iation or materials and product controls can provide numerous new ideas for product reformulation. A visual chart, with pictures of the products and its competitors, can help the efficiency of traditional brainstorming sessions. A systematic search of product differentiation can be done following the three dimensions of all shapes: structure, function and symbol. Design, like marketing, is the art of differential strategy.

Design also changes the concept phase in the new product development process. Besides converting ideas into a physical entity, it introduces, through roughs and sketches, new product ideas in product positioning and consumer measurement evaluation. Then, when the final direction has been selected, design works with production on product engineering and the production process. In this phase also, designers provide a link between market studies, marketing mix and production constraints. Figure 9.4 shows how and where design is involved in the new product development process.

To understand the strategic nature of the design technique we can refer to the three basic strategies developed by Porter (1980), shown in table 9.1.

These three basic strategies relate to three design strategies:

Design-to-cost   Design impact on productivity is required. Basic strategy: cost domination.
Design-image   Differentiation involves creating market power. Basic strategy: differentiation.
Design-user   The company concentrates on becoming the specialist in one market segment. Basic strategy: concentration.

Table 9.1 Basic business strategies

	Competitive advantage	
Strategic target	Uniqueness of product perception by buyers	Low costs
Whole sector	Differentiation	Cost domination
Particular segment	Concentration	

Source: Porter (1980)

Managing the Visual Information System

Considering design as a management tool means using the fact that a company and its environment are visual systems. This has the immediate consequence of modifying the idea a company has of its environment. Traditionally the environment of any organization is described as cultural, social, political, economical and competitive. Design broadens its contents to artistic fields, giving a better understanding by identifying avant-garde trends rather than looking backwards. The analogical thinking used by designers requires the systematic illustration of the company's competitive environment, and enlarges the scope of the company's competitors to include companies manufacturing products of similar structural shapes. Thus cross-pollination must become part of the company's information system. For example, a TV set is not visually very different from an electric oven; both are 'boxes' and technical solutions can be transferred from one field to the other. But design also involves the visual coherence of the product in its final environment; therefore the company information system should visualize this integrative function. For example, a food company could find differentiating ideas by a study of kitchen decoration trends.

A new visual information system develops qualitative and futuristic marketing approaches which companies will need more and more. Fashion trends, especially colour trends in the apparel industry, are of interest to companies in the vehicle or electrical appliance industries. If creating a new shape implies cross-pollination, why not manage this systematically and encourage the transfer of shape solutions from other fields as a new method for producing ideas for product differentiation? This explains much of the success of Italian design on world markets.

Managing design across divisions: transversality

Research into strategic management emphasizes the value of developing links between R&D, marketing and communications. These functions tend to be considered as equal forces in the company's final strategic development. Partition between marketing and R&D, in particular, has proved inefficient and dangerous. Design is the driving force behind reconciliation, since both technology and market trends in consumer needs are sources of innovation in the conception of a new visual output.

This explains what is sometimes referred to as the transversality of design and the creativity of managers in stimulating synergy through committees. Design, marketing and R&D are equal partners in the success of strategic positioning. And the design methodology, if managed, can form a bridge between engineers and marketing.

A new consumer behaviour paradigm

Introducing design as a management tool reinforces the idea of marketing as a communication rather than a control process. Design communicates with the autonomous contemporary consumer directly through purchasing behaviour; freshness and a sense of pleasure in original or beautiful objects are among the qualities desired. In consumer purchasing behaviour, aesthetics is a useful guide to segmentation (where aesthetics is considered as a global sign and not as an additive attribute). When evaluative judgements depend primarily on highly subjective phenomena, preference scales of objective criteria – quality, durability, performance – are insufficient. Such evaluations do not take into account the potential interactions between attributes in affecting consumer response to the final visual output. This is consistent with the view that aesthetics is a holistic and integrative system and, therefore, that objects should be judged in representational forms and on the neglected grounds of sensory pleasure and emotional response. Aesthetics, with its visual expression through sketches, drawings or models, becomes a valuable tool with which to explore the multidimensional aspects of any innovation and the perception mechanisms of users.

Traditional marketing philosophies, considering products as bundles of attributes, are inapplicable; new methods are required because aesthetic stimuli exist only as wholes. In this holistic approach to consumer behaviour, drawing becomes a useful means of market study and market positioning.

Design Mix and Design Strategy

A *design strategy* is the allocation of resources granted to designers by management in order to make the company's positioning visible. *Design mix* means implementing and dividing the design effort and budget between the visual components of a company. These components are: environments (buildings, showrooms, offices, factories, etc), communications materials (stationery, instruction manuals, literature, advertising, etc.) and the products (or services). Dividing the design effort between these visual components amounts to choosing a design mix between four basic design policies:

Graphic design policy
Packaging design policy
Product design policy
Environmental (architectural) design policy.

Because design policies are applied to environments, products and communications materials, they are linked with basic marketing policies. There-

Table 9.2 Coordination between design mix and marketing mix

| Design mix | Price | Marketing mix | | Sales |
		Product	Advertising PR	
Graphic		X	X	
Product	X	X		X
Packaging		X	X	
Architecture			X	

fore coordination between marketing mix and design mix is necessary, as shown in table 9.2.

Issuing a design strategy means managing the design mix coherence with the visual differentiation and positioning chosen by the company management in its competitive markets. The visual differentiation, whether structural, functional and/or symbolic, determines all design policies. Selecting a company's design strategy or design standard means managing and controlling the contradiction between design policies and marketing policies. It means clearly specifying the driving force granted to design creativity. Choosing a design strategy also means deciding the following:

Visual dimension to rely on for differentiation
Extent of design responsibility for innovation success
Selection of a design mix
Design budget
Selection of a design organization
Design role in the company.

Three Basic Design Strategies

Design-to-cost strategy This strategy (figure 9.5a) concerns mainly companies relying on cost domination (table 9.1), giving preference in their strategic development to technological advance and manufacturing process competence. This production-driven development justifies a visual positioning, giving preference to the structural dimension of the product. The design strategy relies on a design mix using product design policy and architectural policy. The objective given to designers is summarized in one phrase: 'make it cheaper', either in the manufacturing process or in the product components system.

Design-image strategy This strategy (figure 9.5b) concerns companies relying on market power for their competitive advantage. This differentiation strategy (table 9.1) agrees with a design strategy based on the semantic dimension of products and documents. Value is added by differences in product perception and message. Design is close to communications strategy and advertising. Design responsibility can be varied, such as new packaging to echo a new TV campaign or

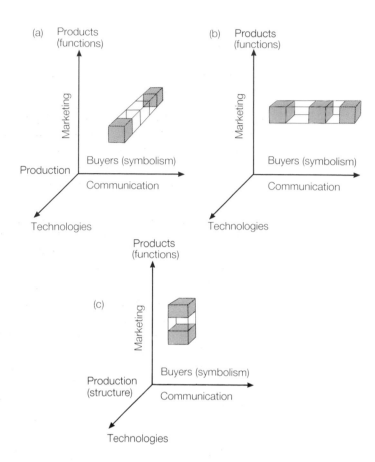

Figure 9.5 The three basic design strategies: (a) design-to-cost strategy (b) design-image strategy (c) design-user strategy

new products to complete a range and emphasize a brand image. This approach is common in the food industry and the 'appeal' industries of cosmetics, cars and domestic appliances. The design mix is mainly packaging design, and restyling in product design. The role of design in the organization is to reinforce the marketing strategy and to finalize a product that is of better quality because of extra aesthetic value, because of visually emotive factors stemming from the resulting fit of all product performance factors.

Design-user strategy   This last basic design strategy (figure 9.5c) is suitable for companies developing a concentration strategy (table 9.1). Companies concentrate their competitive advantage by being the specialist supplier to one market, to one type of user. The strategy relies on knowing better the needs of this market. Design therefore should rely on differences arising from improvements in the functional performance of the product or service. The design objective is to become better by producing a unique or more appropriate product, perhaps one that works better or is easier to handle. The design mix is primarily product design. Designers' methods are close to those of the ergonomists. The budget allocated to design is often included in R&D.

Conclusions

To conclude, a checklist is offered to assist the analysis of design management activities.

1 When did you last initiate a design audit in your company? Do you control annually all design investments in your company? Do you have control procedures to check the return on funds invested in graphic, packaging, product and architectural design? If your answer is no, or if you do not use designers, follow the checklist below.

2 Ask yourself or ask a well-established design consulting firm (with inter-disciplinary competence in graphic, product, architectural and other design) to conduct a design audit. Provide a brief emphasizing the basic philosophy of your firm. What business are you in? Who are your main competitors? What is your product (or service) strategy? Do you insist on high quality standards – or do you aim at being the first, the leader with a low price policy? Do you have a communications strategy? Can you show documents issued to personnel, press, suppliers and customers? In the last three years? What are the contents of your messages? What image do you want to convey?

3 Analyse the results of your product strategy and your communications strategy. Is there a visible coherence between the two? Do you have, in general, a common design strategy, similar messages and visual codes in communications design strategy and product design strategy (including the graphics on products)? If the answer is no, or there is poor synergy, a more detailed analysis will have to be conducted.

4 First, a design audit team should be set up comprising various specialists:
Buyers of components, stationery, building refurbishment
Market researchers
Design managers or design consultants
Finance or managing directors
Publicity and communications executives.

5 This team will plan the various operations, dividing the work between them. Action required will include:
Meetings with all commercial and production executives
Reviews with sales force at different points of sale
Visits to all buildings and production outlets
Centralization of all written documents used by the firm.

6 Information to gather for the final diagnosis:
Product strategy   Place of design. Ratio of investments between design activities, especially ratio of investments during the opportunity phase. Is design involved in the strategic information system? Ratio of design costs to R&D costs. Checking of costs of bought-in items with the aim of cutting these costs. Ratio of design costs and expenses involved in buying packaging, labelling, stationery, etc. Checking of any committee responsible for coordination and control of new product launches.

Information system   Is there a permanent visual audit made by designers of the competition at the distribution site? What documents are given to the sales force?

Communications strategy   Ratio of investments between design expenditure, especially on corporate design, and communications costs. Percentage of design devoted to advertising, public relations, exhibitions, architecture at new or renovated sites, etc. Signage costs. Is there a person responsible for the coherence of all communications design?

7   Diagnosis and solutions. The design audit team should draw up a summary chart along the lines of table 9.3. Finally, this design audit should create conditions of permanent control of the design strategy:

(a)   Appointment of a design manager.
(b)   Description of this position: responsibility for achieving design strategy objectives.
(c)   Setting up with top management of an annual design strategy with the appropriate design mix, i.e. allocation of budget between the four different types of design in table 9.3.
(d)   Selection of designers and control of their work/objectives.
(e)   Coordination of all elements of the firm's strategy: marketing, R&D, personnel, training, refurbishment of buildings, etc.
(f)   Responsibility for visual coherence between product and communications strategy.

Table 9.3   Design audit summary (percentages)

	Graphic	Package	Product	Architecture
Strategy/design mix				
Ratios of design investments				
Investments involved related to design: cost components				
If cost domination strategy: evaluation of cost reduction by design				
If marketing differentiation strategy: evaluation of quality and image upgrading possible				

References

Borja de Mozota, B: *Thèse doctorat sciences de gestion*. Paris, 1985.

Lorenz, C.: *The Design Dimension*. Oxford and New York: Basil Blackwell, 1986.

Porter, M.: *Competitive Strategy*. New York: Free Press, 1980.

Urban, G. L. and Hauser, J. R.: *Design and Marketing of New Products*. Englewood Cliffs, NJ: Prentice-Hall, 1980.

10 Twelve Principles of Design Management

JENS BERNSEN
Danish Design Centre, Denmark

Use Design as a Management Tool

Ever since IBM's Tom J. Watson coined the phrase 'good design is good business', business executives have in principle accepted the idea that design is one of the keys to success in industry. Yet it is a big step to move from accepting this idea in principle to practising it. Making good design a reality in an industrial enterprise is a question not only of goodwill but also of skills.

Good design is not just a commodity that can be bought directly and brought to the company from outside. It is something that must come from within the company too, by activating and managing the creative resources and skills of the company itself. How do you generate good design? How do you buy it? And how do you make design a success in practice?

The name of the game is *design management*. The idea is to use design as a management tool. Good design in a company is a question of good management. Yet good product design, graphic design and environmental design are not the only results of good design management. A commitment to good design may have implications which go far beyond design.

Design is concerned with the relationship between people and things. Therefore design is an excellent way to define the quality of products and of company communication. In fact, in products and communication it is impossible to define quality (except in very general terms) without a design concept.

This makes design a tool for defining and communicating the goals of a company in operational form, a means of establishing, visualizing and communicating corporate aims. By creating an environment geared to goal-oriented action, design is a tool which makes the creative resources of the company come to life, with all the prospects this holds for the growth and development of the company itself.

Get Your Definitions Right

A basic point in design management is to get the definitions right. Design is a name for both the creative process and its outcome. Design as a problem solving process pursues two goals:

1   To achieve the simplest possible solution to a problem – without violating the complexity of the task.
2   To adapt the design to the user and not the other way around.

Good product design means unifying

Purpose
Manufacturing
Engineering
Function
Beauty
Relationship to the environment.

Beauty is a quality often associated with design, yet beauty is not a goal which can be pursued independently of the other requirements. Usually it is the other way around: the beauty of a product is the natural result of seeking the simplest possible solution to a complex problem, a solution which respects all rational demands and bears a human imprint.

Design management in a company also means graphic design or, as some prefer to call it, communication design. Good graphic design is first of all a question of recognizing what communication is all about, namely knowing

Who you are
Who you are talking to
What you want to say
The nature of the media you are using.

A lot of poor communication design occurs because people have ignored these points and have confused the basic definition of communication. The art consists of integrating the word and the image in the graphic communication to match the way the human eye and mind perceive what they see.

The single most important point to make is that product design and graphic communication must form a unity. In any business, design is a question of building a clear product identity. We may take this a step further and say that product identity is part of corporate identity, which is the sum of

Products
Communication
Environment design.

To a design-based company all of these should be an expression of the same idea.

Make Commitment to Good Design a Boardroom Issue

Most companies have a blurred image of design. The products of a company often seem to come from ten different companies, with brochures designed by three different agencies all wanting to make their own creative contribution. On top of this, stationery may be produced by several different printers and ordered by as many company staff members. The result is an image of total confusion and often sheer lack of quality.

This does not happen because somebody wants it to happen. It happens because no one has overall responsibility for design. Although design relates to everything the company is and does, nobody is really responsible for it. Make the commitment to good design an issue in the boardroom.

This implies that there should be at least one member of the board who takes a personal and qualified interest in design. It also implies that design should be on the agenda at board meetings and that the plans of the company should include a policy for design. This will give company design initiatives support from the top – crucial to the implementation of any design project.

One person should be appointed to be in charge of the implementation of the design policy of the company both at the project level and in day-to-day quality control. This design manager should be a highly qualified generalist, be high ranking in the firm and have the authority to work anywhere within the company. And finally, top management should use design as a means of visualizing and communicating the goals of the company. These goals should be explicit and they should be communicated in writing. There should be no uncertainty about the design policy of the company, and it should be known by all the staff.

Introduce Design Step by Step

An industrial company which has not previously worked consciously and professionally with design has dozens of stimulating design tasks waiting to be tackled at the same time. The best advice is: don't. Initiating several design projects simultaneously in a company which has no experience with design is likely to lead to disaster; high costs, unconvincing results and in-house resistance to the very notion of design.

Start with a single project – on a modest scale. Start with a project which can produce visible and useful results within a limited time and at a modest investment. Make your first design project a success. Then, even if it is on a small scale, it will contribute to 'selling' the idea of working with design in-house and thus pave the way for larger projects in the future.

By starting with a small scale design project, the company can develop its design and design management skills in a gradual way through the natural process of learning by doing. The choice of the first design project is critical. A good candidate has the following properties:

1 It can produce improvements in economy or manufacturing or technical performance.
2 It illustrates the man–machine or the man–communication link.
3 It calls for the best of the company's own skills and probably involves cooperation with an inspiring freelance designer.

Finally, the first project should lead to bigger projects in a natural fashion. It should provide the direction for future development and yet avoid provoking premature decisions which can block the development of future design projects.

The selection of the first design project within a company should thus be made while considering the major design tasks which lie ahead. Planning should include the sequence in which these tasks should be undertaken.

Apply this sequential structure to each project too. Divide projects into steps, each ending with a milestone. Use these milestones as decision points where intermediate results are presented and evaluated and where decisions can be made as to whether or not to proceed with the next step. This provides a simple and safe way of monitoring the progress of projects.

Use Design to Create Unity of Purpose

The introduction of a new product often follows a 'random walk' pattern. First, a prototype is designed without any clear idea of who the users of the product will be. This prototype is then modified for manufacturing in broad outline. A designer may be called in at the last minute to add a few cosmetic touches which must respect all the limitations the product has already had imposed upon it. After that, brochures are produced, then a user's guide or a service manual. Packaging is added almost as an afterthought. And last of all, with almost no connection to the rest of the process, the marketing people concoct a plan for the introduction and marketing of the new product. The end result is a product which looks more like a working prototype than a professional tool. It is a product with an idea which never emerged or was lost in the development process.

Good design management sees the product and its accompanying communication as the expression of one and the same idea. The best advice is to develop the single components of the project in a *parallel* rather than a *sequential* fashion. Start by developing an idea which applies to all parts of the project. Prepare a rough draft for all components of the solution – product, packaging, user's guide, brochures and advertisements – at a very early stage of the project. And use the fact that this requires many complementary skills to make people meet.

This creates more than just a unity of purpose for all parts of the solution. The contact between professionals who rarely meet around a common problem may in itself generate an abundance of ideas and impulses for the development of the project and thus bring the in-house creativity of the company to life. In this way, working with design produces a sense of unity not only for the product but for the company as well.

The Swiss watch Swatch is a case in point. Swatch beat the competition at its own game by good design and good design management. The success of the Swatch can be attributed not only to its name or designs, to its brochures or advertising photography or packaging, or to the fact that it uses a combination of design and high technology to manufacture an endless variety of different yet coherent designs. Swatch's success derives from the *unity* of all of these expressions of one and the same idea.

Look for Challenges to Stimulate Innovation

Manufacturing a product which the market wants is, of course, a key to success in any industry. Yet identifying what the market wants is not just a case of doing a market survey. The market's perception of what it wants is greatly influenced by what is already available. If the results of market research are taken too literally, a company may well end up introducing yesterday's product – two years from now.

The aim is to introduce a product which makes the user say: 'This is what I always wanted, but I didn't know I wanted it until I saw it.' Such products rarely, if ever, spring from a market survey. They come from perceiving a real but unidentified need and translating it into a new product.

Product innovation is first and foremost a question not of developing new technology but of using the technology we already have in new and innovative products. New technology is, in many respects, a set of solutions looking for problems. Uncovering a problem or a need can be a highly creative act; some of the most innovative designs are as much problem statements as they are solutions.

Maybe somebody from Sony saw the youngsters on rollerskates in New York, carrying their 'ghettoblasters' on their shoulders, preventing their freedom of movement. Or maybe this person simply dreamt about using a private, high quality music machine on an aeroplane. At any rate, the Sony Walkman has become one of the world's most striking successes in consumer electronics. This is not just because of its technology but because it reveals and interprets a real need.

Another example concerns Leitech, a small Danish mechanical engineering company. A few years ago the company reviewed how much time it took to check thread quality and measure hole depth. Realizing that others must experience the same problems, the company decided to design a tool that does both things in one operation. Most engineers could have designed the tool from this simple problem statement, yet only Leitech saw the problem. Today the Leitech thread control gauge is the company's biggest money maker.

State Your Goals in a Design Brief

Initiate any design project by stating the goals of the project in a design brief. The design brief serves a dual purpose:

1   It identifies and communicates the goals of the project.
2   It serves as a frame of reference for the evaluation of the solutions.

Writing a design brief is an art in its own right. A good design brief includes the essential demands specified in a fashion which really describes them and not – as it is too often the case – the solution.

Some years ago, Christian Bjorn designed a high pressure cleaning unit for Gerni A/S. The initial brief stated that the unit should include a drawer for a user's manual. Christian Bjorn commented that this was not the real issue and did away with both the drawer and the user's manual. He simplified the operation of the machine so that it was nearly self-explanatory and had instructions silk screen printed on top of the machine.

A good design brief should include only the essential demands, not all kinds of demands. You can flood people with so much information that they ignore it all. So concentrate on what is really important, on the demands which are most likely to trigger the creativity of the designer and the development group.

In 1984 Ole Sondergaard designed a new logo for DAMPA A/S, a Danish manufacturer of acoustic ceilings made from perforated aluminium sheet metal. The brief, prepared by the Danish Design Council, stated that the new logo should visualize the acoustic associations of the name with the DAMPA product. Ole Sondergaard chose a bold Helvetica for the DAMPA typeface and formed the letters with dots (raster screen), in much the same way as holes are punched into the DAMPA panels themselves. In addition, he reduced the type weight from the beginning to the end of the logo, thus simulating the way the word DAMPA is pronounced (with the stress on the first syllable) and the acoustic operation of the product.

Design is a problem solving process and, like all creative undertakings, also a goal seeking process. Therefore the goals set out in the initial design brief should not be regarded as final. The design process itself is often instrumental in revealing what the real goals of the process should be. Therefore revise the design brief from time to time as new and stimulating goals are revealed.

Identify the Big Idea

All striking designs have this in common: they are all, in one way or another, an expression of a 'big idea'. So when undertaking a design project, identify a big idea. Then make this idea a main feature in the design, something to remember it by. This big idea could originate from several sources:

1 It could be a reflection of the most prominent function of the design.
2 It could be an intelligent interpretation of a limitation of the design task.
3 It could be a feature which stresses the personality of the design via any – or several – of the senses.

When Bang and Olufsen launched their classical turntable with a tangential arm in 1972, they introduced a revolutionary idea: a gramophone which plays the record the same way it was recorded. They stressed this big idea in the design by adding an extra arm parallel to the pick-up arm to carry the light sensor. But by introducing the extra arm rather than building the light sensor into the pick-up arm they visualized and dramatized the fact that the pick-up performed a parallel movement, which would not be so obvious on a gramophone with only the single arm.

Or take the Olympus XA camera. The idea is that it fits into a shirt pocket and still takes 24 × 36 mm frames. This requires the lens to be hidden by a lid when the camera is in the pocket. The camera emphasizes this feature with its vaulted lid. The lid shows the function and adds a characteristic feature to the design which can be experienced both visually and by touch.

Any big idea stands and falls by its implementation. Lack of attention to detail or sloppy manufacturing can ruin any design. Impeccable attention to detail separates the work of the professional from the work of the happy amateur. It is also here that a designer can make an important contribution to the quality of the finished product.

Sometimes the detail *is* the big idea. When the company now known as 3M started out in business, their product was sandpaper. 3M soon discovered that the crucial factor in achieving a high quality product was neither the sand nor the paper but the glue which was used to connect the two. 3M never succeeded as a manufacturer of sandpaper but made an extraordinary success out of manufacturing glue, and subsequently a range of adhesive products. 3M made a detail the big idea of their business.

Accept the Limitations of the Task

The limitations of a design project may spring from the task itself, or they may even be self-imposed. Consider for example one of the biggest design projects a company can undertake – the implementation of a programme for its visual identity.

An identity programme for a company is a set of standards for the design of the logo of the company and its main brands, its trade mark (if any), its use of colour and its typography. It is also a set of rules for how these elements should be used wherever the company manifests itself visually.

The success of an identity programme is, to a great extent, a question of sticking to standards, of going along with self-imposed restraints. This does set limits to creative freedom, but is not necessarily a drawback. Limitations can be a major source of creativity. Too much creative freedom is, in fact, the reason why so many companies make a mess of their visual identity.

The real skills of a designer show in a capacity to accept the limitations of a task and to make them into a virtue. Some years ago, the Danish Design Council commissioned the graphic designer Henry Anton Knudsen to design the DD symbol. The designer put two Ds of his own design side by side, but it didn't look quite right. The Ds had a tendency to merge, so you had to keep pulling them apart. Then at one point in the discussion, this comment was made: 'Yes, but why don't you go along with this tendency and tie them together with some sort of string?' This was, of course, the solution. So Henry Anton Knudsen went home and did it.

What goes for graphic design goes for product design too. Some of the most striking designs stem from identifying or choosing a set of limitations and then making the design spring from them. Designing an aeroplane, for example, is a fight with the forces of nature which takes engineering design to the limits of what the materials will allow – often resulting in striking beauty, not as a goal but as a by-product.

Or take one of the most successful designs ever, the VW Beetle. Even the smallest part of the body was, in a sense, a reflection of the entire car. Again this was an exercise in choosing a set of self-imposed limitations and working within them.

Sometimes rules have to be broken, either out of sheer necessity or out of impulse. It is the mark of the quality of the design programme and the designer that such necessary deviations from the rules enrich the result instead of being a drawback.

Make Design a Dialogue between Complementary Skills

No idea is born in isolation. It is in the nature of ideas that any new idea is – in one way or another – a combination of two existing ideas. New ideas are often created by people who can pursue two lines of thought, or two ways of working with the same thought, at the same time.

Designers often possess this capacity in their skills in making drafts. A draft is a visualization of an idea and at the same time a vehicle for developing ideas. Via their drafts, designers not only put on paper what is in their minds; they also establish a dialogue between the sketch and the thought. They set up a mechanism for developing ideas. This type of dialogue can also take place between people of different skills and it can produce the same results – the development of the ideas they are working with.

Set up dialogues between complementary skills throughout the development of the project. Such dialogues are a major source of creativity. The development of a new product has a sense of built-in possibilities for such exchanges. Dialogue is like a ping-pong game between viewpoints which illuminate each other and stimulate the development of the product concept.

Work with the design of the product and the design of the user's manual or the brochure for the product simultaneously. This may give useful impulses in both directions. If, for example, it proves difficult or impossible to write a good user's manual for the product, or to make a sales point in a brochure, then there must be something wrong with the product. The dialogue spots such problems at an early stage while it is still easy to make adjustments. In addition, dialogue helps the entire project achieve coherence.

Engineering and use are another pair of viewpoints which should engage in a dialogue during product development. The discussion should be between the company's own staff or freelance designers on one side, and a qualified and unbiased user on the other. Physical models or mock-ups may be used in the same fashion to establish a dialogue between the line of thought represented by the same design in two and three dimensions.

Seek Identification between User and Tool

The element of identification is crucial to any design, be it a physical product or tool, a piece of graphic design, or even a non-physical tool such as a computer program. Will users identify with the tool? Will they see it as a natural extension of their own skills and purposes? It is exactly this identification between the user and the tool that is at the core of the world's most successful designs. This is true whether the combination is the carpenter and a hammer, the photographer and a Leica M, or the Copenhagen taxi driver and a Mercedes 300D.

The key to understanding the success of these products is not their technology as such. The Leica M is not the world's most advanced camera, yet it tends to bring home a disproportionately large number of the world's best photographs. Or ask why a Copenhagen taxi driver chooses a Mercedes rather than one of its cheaper competitors, and you may get a lecture on man–machine identification which is better than can be given here.

The key to the success of these product designs is the perfect fit between the tool and the mind and senses of the user. The product does the essential things so well that the user can forget the tool and become absorbed in the task. Many products fail because they do not match the mind and the skills of their users. Products which are technically possible are often introduced today without serving any real purpose. They may include more functions than can be put to any sensible use, and still not do the basic things right.

Seek identification between the user and the tool all the way through the design process. Regard the product you develop not just as a commodity for some impersonal market, but as a tool for creative individuals rather than consumers. And use your insight into the man–tool relationship to decide which technology you use, and for what. Contact with users of your new product will stimulate a decisive leap forward in the development of the product. So will contact with a designer from outside the company.

In some way there should always be a skilled user looking over the designer's shoulder, adding a contribution to the design process. Companies sometimes lose this capacity because they know their product too much from inside the company. They have lost the capacity to experience the product the way the user will.

Create Positive Feedback between Image and Identity

The image of a company is at least three different things:

The image the company thinks it has
The image the company actually has
The image the company wishes to have.

Each of these three images may, in turn, be many images. Building – or contributing to – the image of the company is a vital part of design management. Both the products and the graphic communication of a company have dual functions. One is direct: their prime functions. The other is indirect: how to convey the image of the company. This image definition should be part of the brief for both product development and communication design projects.

Make management's definition of the image the company wants to achieve part of the design brief of any project. This definition, based necessarily upon what the company is, should at the same time set the direction for the future. A product design – or a programme for the visual identity of a company – which does not reflect the essence of the company and its products is a cosmetic fraud and cannot be maintained in the long run.

On the other hand, a product design and an identity programme which convey an image somewhere between what the company is and what it aspires to be may contribute to the development of the company itself.

Bang and Olufsen's Chief Engineer, E. Rorbaek Madsen, a master of both engineering and industrial design, said in 1985 when looking back on his work: 'When I started at Bang and Olufsen almost 40 years ago, we didn't know about project groups. Each problem was one you solved by yourself. But we were still a team, working with great enthusiasm. I think what kept the enterprise together was that from the beginning the two founders called B&O "The Danish seal of quality". This influenced our attitude to things. We wanted to be in the forefront.'

The pursuit of excellence via design is not a democratic process, and yet it is a process which involves the entire enterprise. The introduction of good design, in everything a company does, is a gamble between image and identity. It is a game which requires the active participation of the entire company.

The art of design management is to create positive feedback between image and identity and to pursue the ideal of quality via design as a means of bringing you closer to what you want to be. And when you reach this goal, aim even higher. The sky is the limit.

11 Design as a Business Strategy

COLIN CLIPSON
University of Michigan, USA

Introduction

The design activity carries responsibility for the success of the whole enterprise. The information interface, management structure and the way that designing is organized and used within the whole enterprise must give adequate acknowledgement to the functional requirements of the total system. Design of the product itself is the key to successful manufacturing. It must be basic, rugged, perhaps elegant, but certainly simple and not fussy. The design must solve the specific problems addressed without adding other problems. This can only be satisfactorily accomplished when the whole activity from the design pad through to the customer accepting delivery is viewed realistically and managed as an integrated system. (Gardiner 1986)

Industry leaders in manufacturing use engineering, product, graphic, interior and building design as resources essential to achieving their business goals to increase profits and market share. They employ these various forms of design in a timely fashion to meet the needs of a particular product or market. Herbert Simon, the Nobel laureate, described designing as the activity that helps change existing conditions into preferred ones. Designing is the means by which we translate ideas, technologies, manufacturing potential, market needs and corporate financial resources into tangible, useful outcomes or products.

Designing in its many forms should be viewed then as a component of strategic planning, as part of the information systems that guide organizations, as part of any productive environment for work, manufacture, sales and service and, of course, as part of the product itself. Designing is central and cannot be effective as an afterthought.

Design is More Than the Product

Ask most designers what they mean by design and they will start by describing the visual and physical attributes of a product such as a chair, a building, a poster or the arrangement of physical objects in a space. To understand a design process and the activity of designing fully we must operationalize it and understand that it is much more than the physical form or appearance of the product. For example:

1 To make the best possible use of business resources the design of the product should be seen from the outset as an issue of economic producibility, where every element of the product is designed for ease of manufacture, assembly, handling, shipping and use. Philippe Villiers has stated that the well-designed product is a predictable product.

2 When we buy a do-it-yourself product, the parts assembly process, the assembly instructions, packaging, warranty, maintenance manual and spare parts are all important to the success of the actual product to be used, and must be designed for and marketed to a consumer whose capabilities and requirements have been carefully studied.

3 The graphics on an Exxon service station and a Sony exhibit at an international product fair are each designed as part of a system for meeting corporate goals.

4 More and more manufacturers stress the after-market benefits of their products as part of the item being sold. This is particularly true of products which are technically complex and where the user is dependent upon experts for maintenance (e.g. cars, computers, video equipment).

5 Corporate annual reports and internal communication publications present a coordinated vision of the company, its achievements and its markets.

Hence designing, in a more comprehensive sense, is taking care of and accounting for all those aspects of translation of need, innovation, cost, manufacture and ultimate safe easy use of products. Starting from scratch, designing takes place in all aspects of the particular product development, from identifying the need for improvement or a new product, to seeing to it that the product works properly in the hands of the user. It is also true when products are 'conspicuously consumed' that more than the product is sold; the consumer takes on the image and emotive qualities projected by the product and its manufacturer. This is as true of a workaday object like a Craftsman toolbox as it is of 'high design' objects from Bang and Olufsen. Design performance may be approached in a number of ways, but a commonly agreed set of indicators remains elusive in terms of practical methodology. The following factors may be considered.

Appropriateness of purpose, arrangement and materials All designed products should be fit for their purpose or stated need, and the parts or components arranged so that there is a 'goodness of fit' between the design and its producibility, the user and the function of the product, the product and its operational environment. This goodness of fit and sense of purpose distinguishes effective design from poor design. When there is a poor relationship or mismatch between the components of a product or between the product and the user, there is usually inefficiency of use, significant user adaptation to make the thing work, modifications to the product and, in some cases, product failure or accidents in use. Mismatches between stated purposes and product function ultimately lead to significant modification, poor sales performance and eventual obsolescence – to say nothing of litigation.

Integration of design

Complex functional relationships (and often barriers) exist in many firms between R&D, manufacturing, marketing and the product development managers. The relative influences of these groups have a marked effect on the ultimate performance of the product. Unless their efforts are integrated to design a cost effective, manufacturable product, those efforts will be dissipated and often conflicting. While design for manufacturing (DFM) and design for assembly (DFA) are today's catch-phrases, both are essential strategies for competitive business. Both have to be integrated into the enterprise by appropriate organizational design of multifunctional teams.

Quality, value and design

The key to success in business is to provide a product or service of superior value to the end user at a competitive price, and in such a fashion that the end user will perceive and appreciate the value of the product or service in everyday use. While this seems to be a deceptively simple guideline to an age-old need, it is nevertheless elusive. Many US manufacturers were conditioned by the throwaway mentality of the post-war years, and the notions of value and quality had very little appeal to their marketing minds until recently. Take for example the automobile and the houseware industries. Advertising had created a retail climate in which the replacement of products often became an annual event and consumers began to feel inadequate without the latest model. In a domestic market once dominated by American cars, the US auto industry now faces severe competition from abroad and a much less faithful consumer clientele; there is a foreign automobile in most California garages. Kitchens once filled with the proverbial American gadgets are now infiltrated by Cuisinart, Braun and Sanyo.

The lesson of the 1980s is that products must have quality and added value that can be identified by the customer. A product has attributes that can be translated into value or lack of value. For example, if a product turns out to be efficient and useful, saves time and effort, and is economical to operate, it has enhanced value to the user. If the product or service is conveniently available to the consumer when he or she desires it, this is obviously of value. If the product carries some additional qualities such as refinement in detail, exclusiveness and rarity, it will also receive additional attention and value from potential users. These attributes, combined with safety, well-being and ease of use, are perceived by consumers as having high value and quality. Failure to provide one or more of these attributes in a product or service results in mismatch between the user and the product. The consumer starts to look elsewhere. Similarly, if too much attention and corporate resources are used to talk about these attributes through advertising rather than to build them into the product, the result will ultimately be the same – a dissatisfied customer.

Consumers seek value in their purchases by matching quality and price along with the symbolic, emotive and status qualities of the product. This might imply a particular style or fashion as well as the degree of workmanship and quality. Quality in itself will not guarantee profitability unless specific market factors are taken into account, and if the price for quality and design is in excess of consumers' expectations. The quality and perceived value of the product in question have to be at a competitive price and have to reach the market at the right moment. In some sectors of US business, there is a real quality perception gap between the manufacturer and the consumer. Manufacturers' claims for increased quality are not borne out by consumer perceptions identified by market research. During the last decade we have witnessed significant shifts in what is of value to consumers.

The Whirlpool Corporation report on consumers in the 1980s shows that the most important indicators of quality are as follows (percentages):

Workmanship	44
Materials	16
Safety	14
Made in USA	14
Guarantee	7
Styling	2
High price	1
Status brand	1
Advertised brand	1
Packaging	0

The first two indicators were largely ignored in houseware products until the 1980s.

Efficiency and durability

When the principles of functional arrangement, fit and material choice and technology are in accord, there is a greater likelihood of high performance and efficiency. One of the major tests of product performance, given a feasible cost relationship to R&D, manufacture and distribution, is how efficient, durable and economic the product is during use. In most consumer products there is an expectation of trouble-free use and reasonable life, even when there is a low price tag attached to the product. There are exceptions to this. For example, some products of the fashion trade are treated as ephemeral even when they are relatively expensive; however, they are expected to provide immediate access to style, a new image or symbolic qualities whose value is immediate rather than long term. Other examples are popular music products and fad products.

Problems occur when products do not meet expected levels of efficiency, durability and economy and when follow-up service or guarantees have not been honoured by the producer. A product with good after-market service can survive with only modest performance. When a product is hyped or its capabilities are exaggerated, there is potential poorness of fit between the product (or service) and the end users.

Symbolic or stylistic value

All outcomes of designing are imbued with some symbolic or emotive content; this applies equally to a product, a building, a communication or a service. When we talk about designed objects or communications, we also talk about form and we frequently talk about style.

After functional, economic and engineering considerations have been defined, we will still be left with emotive issues of 'design quality' and the proverbial debates on 'good forms' and 'high and low design'. Visual style and emotive factors engendered by the resulting fit of all other product performance factors are often a powerful stimulus to consumers. In consumer personal products the emotive content is high, while in industrial goods it may be much lower; however, it is always present.

Of all the performance issues this is probably the most perplexing and difficult to assess, and measuring the emotive content of a product remains elusive. The evaluation of 'good' and 'bad' design (or good and bad 'looking' design) is an endless designer debate, and one which is most difficult to relate to other areas of business performance when it is divorced from the lifestyle preferences of consumers; it becomes a question of the good or bad taste of one consumer group or another.

Organizational Issues

Design is an integral part of the organizational structure and is composed of design processes, methods, production and designed products and outcomes. The scope is expanded to include planning, looking ahead and solving problems that are procedural as well as end product oriented in their implications. Designing is seen, then, as an organizational activity, as well as the forming of end products like a chair, an automobile or an office building.

The pursuit of such operational definitions implicitly examines the absence of designing as well as its presence in organizations of various types and management structures. Thus, poor designing refers not just to a product that does not work but also, more comprehensively, to an organization that is badly set up to reach its product development goals, and in some sense or other is unable to meet the needs of chosen markets with available skills and resources. Such organizations can be said to be poorly designed. When Max DePree of Herman Miller elaborated on his notion of 'theory fastball', in which he stressed the need for an integrated baseball team not just a fast pitcher, he was using a simple analogue for a well-designed organization in which necessary structures, processes and people are aligned to translate innovations and new ideas into marketable products (DePree 1981). Good ideas do not travel very far unless they are carried through by a well-designed development process.

Measuring Design Performance

In studying various types of organizations and the ways design is used to improve their competitive edge and effectiveness, it is necessary to identify those levels of corporate activity to which designing in its various forms makes a contribution. These are identified below, and the absence of well-integrated design in any one of these levels should signal an audit of a company's approach to product development:

1   As a part of strategic planning in the organization.
2   As a part of the information systems that inform, monitor and control the organization.
3   In the planning of productive environments for work, manufacture, sales and services.
4   In the products, services and information that are introduced into the marketplace by the organization.

When design is not an integrated partner at each of these levels and the development system falters, we see the kinds of mistakes occurring that have become only too frequent across industries. Effective designing can be shown to make money for a corporation, and it can be shown to save money. It is equally clear that ineffective design increases cost for the firm, can be wasteful and, in some cases, dangerous to health and safety.

In a market-sensitive economy, products and services are continually being evaluated and modified or replaced with more products and services that meet the changing needs and demands of the consumer market. Designing plays a significant role in realizing the goals through improved design of manufacturing and assembly. Effective designing produces carefully developed products and services that reach the market on time, are of value and are dependable to the consumer. Effective design, by definition, solves the problem in such a way that the consumer finds more value in this particular product than another. Ineffective design creates problems for the company and the consumer and wastes money, time and resources. Poor design causes dissatisfaction, delays, accidents and legal complications for the company. Automobile recalls are a good example of this.

Designing in the best sense is as economical as possible with materials, energy and form. Effective designing looks for ways to make design development, production and use of the product simple. In progressive organizations, all employees, not just managers, are involved in saving money by designing more effectively. Efforts to control and reduce the life cycle costs of hospitals, offices, cars and aeroplanes, for example, have come into focus over the last ten years. The economics of designing have become an increasingly important issue since the initial energy crisis of 1973 and the import crises of the 1980s. Design can help increase economic performance in the marketplace and can help keep cost down in the company. Both can increase the competitive edge.

Design and Economic Performance

Some examples of the impact of design on business performance demonstrate its bottom-line importance.

There is a growing business interest and body of knowledge on the impact of design on productivity in business environments, both blue and white collar (Olson 1983). Recent studies (American Productivity Center 1982; Steelcase Inc. 1983; Rand Corporation 1983) draw attention to the role of well-designed physical environments in improving productivity in a wide variety of work settings. Optimum productivity depends upon the integration of three types of resources:

Human
Tools and systems
Built environments and workspaces.

These studies support the thesis that the design of the workspace has a significant effect on the quality of work and the satisfaction of workers. One of the studies from the American Productivity Center claims that productivity improvement programmes put in place by 140 US organizations, and including those three types of improvement, reported productivity improvements of 9.5 per cent. The annual savings by pervasive use of such improvement programmes would be enormous – approximately $95 billion per year.

Hewlett-Packard has started a company-wide review of the facility needs of all types of workspaces, with the aim of improving the performance and delivery of all HP work and manufacturing environments (Clipson 1984–8). Hewlett-Packard has also introduced a company-wide international programme of design systems coordination that affects the dimensions, configuration, ergonomics, colour and finishes of computer products and peripherals. The programme, known as Rosebud, has had far-reaching effects throughout HP divisions and an impact at all levels of the company, from strategic planning to detailed industrial design work, from manufacturing to materials management and distribution. Packaging savings alone were estimated to be as much as $50 million in one year (Clipson 1984–8).

A design audit carried out by the Michigan Department of Management and Budget and the University of Michigan helped reduce business paperwork by at least 18 per cent in state agencies. In part of the study with the Michigan Department of Natural Resources, 171 business forms were reviewed; as a result, 47 of them were eliminated or redesigned in combination. In the same programme, the Department of Commerce reduced paperwork 32 per cent by eliminating 242 business forms and redesigning the paper form system. Officials have estimated that the reduction will save Michigan businesses more than 80,000 hours of paperwork per year. Paperwork redesign and reduction is a major component of the government office productivity efforts. The project makes the state the first to compile a computer inventory of all forms required by all state agencies dealing with business (IBROM 1983–4).

Leading US businesses are in the vanguard of innovative plant development. AT&T has developed a printed circuit board plant at AT&T Technologies, Richmond, Virginia that receives, via computer, design and manufacturing instructions from up to 12 Bell Laboratories throughout the US. These instructions are used to turn out circuit boards on demand to meet their market needs. Changes in design can be achieved within 24 hours and without interrupting production.

Both INTEL and Apple Computers are developing new and innovative facilities at a dramatic rate. With design, development and procurement compressed to a few months in some cases, their plants are designed to absorb the rapid changes in design and manufacturing that are characteristic of the industry. The Macintosh plant at Fremont, California is able to assemble a Macintosh with its 450 parts in less than a minute. Labour costs account for 1 per cent of the cost of making the computer. Parts are moved to and from assembly and testing by automated programmed handling systems (Clipson 1984).

United Technologies, with an annual construction and equipment budget of $480 million per year, has inserted a design process audit into its design and construction business. The company uses an independent group of reviewers, internal staff and the commissioned designers to review design requirements, according to the building programme, at the 10 per cent and 30 per cent stages of the design development. If the project is assessed as being over budget or not meeting programme requirements, then adjustments are made by the team. Savings from this simple, two-stage audit of the design process on three projects were $30 million in the first year of implementation (Clipson 1984).

Innovative organizations, whether they are just starting up or are mature businesses, share certain organizational and behavioural characteristics. They tend to be non-bureaucratic, to place a great deal of emphasis on various forms of employee participation or part ownership, and to promote fluent communication between divisions and departments. To do this they have had to create divisional and disciplinary overlaps and connections so that ideas flow across, not just up and down, a division. Good examples of this are Donnelly Mirrors, 3M and Herman Miller. Innovative organizations regard designing in its various forms as an integrated corporate strategy and evaluate design in terms of economic and functional performance. At the same time, these corporations are quick to point out the 'tangibility' that various forms of design give to the corporation through visual forms – corporate identity, graphic communication, buildings, packages – all making up a visual expression of corporate culture.

Conclusions

Innovation in the corporate setting comes from many starting points and processes. Some, like Braun, have found their sources almost exclusively from within; others, like Herman Miller, have developed a tradition of design consultants from outside. Both work because there is an understanding of and commitment to design at the top, and this is known at all levels of the organization. A key element of these firms is their ability to challenge traditional assumptions about business practice. Because its primary function is translation, designing is the only means by which an organization can continually convert technical, production and market needs into successful products. Making design a central factor enables a company to develop products that are manufacturable, provides economies of standardization while retaining product differentiation, and promotes high value and uniqueness in the marketplace. It is the only process through which the technological, ergonomic and stylistic specifications can be configured into viable products, services or environments. Thus it should be at the centre of all business activities.

Note

Issues are drawn from the findings of 'The Competitive Edge', a research initiative funded by the US federal government and by some private sector organizations. The principal focus has been to research and document the role of various forms of designing in business and to elaborate upon the strategies by which design can become a central force for improving performance. Over 400 businesses in the United States and abroad have been surveyed and a research team has carried out detailed case studies of 16 firms, some very large like Hewlett-Packard and others with as few as ten employees.

References and Further Reading

American Productivity Center: *White Collar Productivity. The National Challenge.* 1982.

Clipson, C.: *Competitive Edge Business Survey.* Ann Arbor: University of Michigan, 1984.

Clipson, C.: *Competitive Edge Project Case Study Manual.* Ann Arbor: University of Michigan, 1984–8.

DePree, M.: *Theory Fastball.* Palo Alto: Stanford Conference on Design, 1981.

Gardiner, K.: *IBM Advanced Manufacturing Systems Publication.* New York: IBM Technical Institutes, 1986.

IBROM: *Business Paperwork Reduction Project.* Michigan Department of Management and Budget and University of Michigan Independent Business Research Office, 1983–4.

Olson, V.: *White Collar Waste.* Englewood Cliffs, NJ: Prentice-Hall, 1983.

Rand Corporation: *Advanced Office Systems: an Empirical Look at Utilization and Satisfaction.* 1983.

Steelcase Inc.: *The Impact of Open Office Furniture Systems on Employee Productivity.* 1983.

Whitney, D. E.: Manufacturing by design. *Harvard Business Review.* July–August 1988.

12 Policies, Objectives and Standards

MARK OAKLEY

Aston Business School, UK

Introduction

Too many companies make the big mistake of rushing into design projects without first considering the implications of what they are doing. It is even true to say that managers sometimes get so carried away by their enthusiasm for creating a new or improved product that it may be only after many months of hard and expensive work that they realize they are heading in quite the wrong direction.

One such case comes to mind. The managing director of a firm manufacturing catering equipment announced that work should start on the designing of a new version of one of the products. A small team was hastily assembled and for the following 12 months went through the motions of creating the new product. Eventually, it became apparent to all concerned that the exercise was getting nowhere. Concepts and prototypes failed to impress the sales and marketing staff, the manufacturing department did not want to get involved at all – and the managing director just could not understand why there was no new product.

Only at this point did it become clear that the basic problem was that no one had thought out what the new product should be; there was no specification to guide the designers. They had assumed that they were required to provide a new product with similar performance to the original one – similar appearance, speed of operation, capacity and so on, but more 'up to date'. In fact, as was discovered once the various personalities in the firm finally came together to discuss the project, fundamental changes in the catering industry were giving rise to a whole new set of customer needs. The managing director had indeed recognized this when initiating the design project, but had failed to articulate the essential requirements or to seek the views of others in the firm about what they considered to be important. Furthermore, no thought had been given to how the firm should relate to this changing market in terms of the types of products it should offer, their prices, design features and so on.

Such lack of awareness and preparation is not unusual, especially in older, less responsive companies. In this case, the company was able to withstand a year's wasted work and subsequently redirected its design efforts towards a successful conclusion. The lesson is that design projects should be set up only after proper consideration of a range of matters. For want of better terminology, this preparatory work may be described as design policy making – understanding and defining the context and constraints within which design work will be carried out.

A recurring notion in this book is that managers tend to treat design with less thoroughness than other activities for which they may have responsibility. Design policy making is a prime example. Few managers would, for example, invest in major new plant or make radical changes to the distribution system without a precise assessment and statement of the company's needs, expectations and goals. Yet, as in the example just described, design work may be embarked upon with virtually no preparation – and with little appreciation of the potentially massive costs and implications.

Key Design Policy Issues

There are all sorts of companies operating in many different environments, so it is dangerous to be dogmatic about the precise items which need to be addressed in every case. However, it is quite possible to summarize main areas of concern and to provide examples of the kinds of questions which managers should be asking themselves.

What is the role of design? First, and most important of all, managers must come to terms with the true importance of design to their enterprise. There are many businesses today where design is the most critical factor in achieving success. If products are not at least as good-looking and competent in operation as those of competitors, poor sales will be the inevitable result, probably at uncomfortably low prices and profit margins. The same may be true of other aspects; printed materials, business premises, vehicle liveries and much more may have to be designed to reach or exceed the standards that customers have become accustomed to expect.

However, it is possible that in some cases the correct conclusion may be that design actually plays little or no part in the life of a company. It is difficult to think of examples! Perhaps monopoly or near-monopoly suppliers of products and services may be the most likely candidates – together with the few professions and industries where effective restrictive practices still operate regardless of customer interests.

The truth is that, for nearly all companies, design is important. The task of managers is to decide the real extent of this importance in relation to all the other factors which influence customers. Thus, managers must consider design compared with, for example, after-sales service provision, speed of delivery, payment methods (including credit terms) and all the other issues likely to be involved in a purchase decision. Similarly, careful appraisal should be made of the evident importance attached to design by main competitors. Then, in the light of this information, it will be possible to make a judgement about what are likely to be appropriate levels of design effort and expenditure.

Which types of design are important?

The analysis should also highlight the types of design and design expertise which are necessary – and the relative 'quantities' of each that will be needed. For the majority of product-based companies, the priorities are likely to be the technology and styling of the product. For service-based companies, the design of operating systems, the environment in which the service is provided and supporting print materials may be the main areas of design attention.

Care should be taken to reach an objective conclusion about these matters. Perhaps the biggest danger is to fail to realize that customer needs may have changed. In earlier times, the design emphasis for a particular product may well have been on certain technical aspects, but now appearance or comfort might be much more important. A manager from one of the large car companies recently commented about the difficulty of convincing colleagues that a good stereo system might be more important than a further refinement of the suspension system, given that many drivers now spend more time waiting in city traffic jams than travelling at high speed on twisting roads.

The problem in this kind of case may be that the dominance of certain skills and interests may influence too greatly where the design effort will be put. This can raise tricky problems about how to dispose of excesses of some skills while recruiting more of others – but this must not be an excuse for failing to correctly identify real design needs.

Is our company seen as design leader or design follower?

As well as deciding where the balance of design efforts should lie, managers must also take account of customers' expectations of the design results which will be achieved. In other words, do customers perceive our products as representing the leading edge or the state of the art, to use two prevalent descriptions? Will they be disappointed if we fail to provide further radical design results if that is what they have come to expect from us?

A commitment to always set the trend, whether in appearance or technology, may have some very important consequences in terms of the high level of resources which may have to be allocated. It is also a policy which may become increasingly costly to pursue when novel results become more difficult to achieve on a regular basis or if other producers decide to compete head-on. For many companies, a me-too approach to product and corporate design is entirely sensible. Much as designers may espouse creativity and originality, the fact is that, apart from very rare cases, all design results derive from existing items and largely represent incremental developments.

Indeed, straying beyond the norms of current taste or fashion – or remaining outside because of a failure to recognize these prevailing norms – can have serious unwelcome consequences. If managers misjudge current trends, they may find themselves with products which are conspicuous by their lack of conformity. This distinctiveness may be an advantage but if unintentionally achieved is more likely to be a drawback. Some companies are able to hedge their bets by offering several different ranges – some incorporating current or 'classic' design features, others genuinely experimental and trend setting.

In many companies the choice will boil down to the risks involved. Being a design leader is likely to mean living with high levels of risk: greater expenditure, less certainty about customer response, the possibility of spectacular failure – but also the chance of a really outstanding success. Being a design follower means less risk and lower costs – and steady, determined attention to design can also lead to the attainment of a substantial and secure market share.

What can we expect design to achieve?

The claims made for the results that design can achieve are often exaggerated and overlook the importance of other activities within companies. Design will not produce market miracles if management's general competence is poor, quality levels are low, the workforce is badly trained and lacking motivation – or the company just does not know where it is going. What design *can* do is to reinforce other activities and courses of action.

For example, if managers have identified particular segments of the market where they intend to concentrate efforts, designs can be produced which match ideally the characteristics of the targeted customers. The use of design is also most important where some kind of image adjustment is required. As fashions and lifestyles change, companies may feel they need to present a different face to the world. This may include new types and presentations of products, perhaps together with a revised corporate identity.

The question of integrated or fragmented design approaches may arise. The former is usually applicable to large companies operating in fields where customers want to be reassured that they are dealing with a totally competent, secure supplier. Computer companies are good examples of this and they take great trouble to ensure that all parts of their operations meet well-defined design standards.

Other types of companies may favour a fragmented design approach. Huge conglomerates may not wish to be identified with the vast assortment of products and services which they provide. In such cases, a variety of different, often completely uncoordinated design approaches is encouraged. Customers may be more tempted to buy a loaf of bread apparently lovingly made by hand in a rustic bakery than one made in a giant automated factory. The careful design of such a 'factory' product and its packaging can help to convey the desired impression. Whether such use of design is always in customers' interests and can be justified is something which individual companies and their managers must decide for themselves.

Sometimes companies and even whole industries can make dramatic switches in design approaches. Some years ago, for example, large breweries were busy creating vast chains of public houses and applying much the same corporate styling to each. This integrated design approach has now given way to the promotion of uniqueness; pubs are having their history and individuality returned to them (even where they had none in the first place!).

What is the rate of design change in our market?

In addition to considerations of radical or incremental, integrated or fragmented design, it is also necessary to know the speed at which design changes are taking place. The general trend is for products and services to have shorter useful lives in the marketplace. This is most evident in the case of fashion items, where last year's or last season's offerings just will not be accepted by customers. Now, increasing affluence and the growth of competition are tending to generally speed up the rate of design change. Where previously a design revision might have been needed, say, only once every five or six years, now sales may fall if a new design is not introduced every two or three years.

At the same time, increasing technical complexity may require design projects of longer duration than in the past. So managers may be faced with the prospect both of having to design more frequently and of needing to start projects further in advance of launch dates. Failure to understand the prevailing rates of change can lead to what may be termed 'design gaps' – loss of sales and revenue caused by introducing new designs too late.

Because of the need to start design projects in good time, waiting for a signal (typically a fall in sales) that a new design is required is often too late and just does not allow sufficient time to produce the revised version. Managers must study their markets, and the products or services within them, so that they can predict both the correct timing of design activities and the resources that will be needed to achieve time and output targets.

Should design be done within the company or outside?

The issue of resources raises the question of where designing should be done. The basic decision is whether the work can be done in-house or whether it would be better to contract out, perhaps to a specialist design consultancy. Sometimes both approaches may be used together.

If the decision is to do some or all of the work in-house, it is most important to ensure that the right skills and experience are available, and a design audit should be carried out prior to the start of each major project to confirm this. If there are weaknesses it is better to know about them before work starts rather than find out at an advanced stage in the project. If audits are carried out on a regular basis, managers will be able to maintain the correct balance of skills by devising appropriate training programmes for design staff and by appointing new people in the right areas.

Where it seems more sensible to have design work done outside the company, two main considerations arise. One is how best to organize and maintain proper control over the subcontracted work – but the first is how to set about choosing a consultancy to do the work. As part of their general design policy making, top managers should draw up guidelines relating to the identification and selection of outside design specialists suited to the needs of their firm.

Setting Design Objectives

It is surprising how often design projects – even whole design programmes – are set up without a clear statement of the objectives that need to be achieved. Too often, design objectives are confused with design standards. Designers and others may feel that if they have achieved design results of a certain quality (however that might be measured), they have fulfilled their obligations. Such attitudes may have their roots in design education and professional institutions where commercial realities tend to have little influence; preoccupations are with absolute indicators rather than relative ones.

In the real world of industry and commerce, the main measure of success of a new design is whether it appeals to enough customers to generate acceptable levels of turnover and profit. Whether the design is considered good or bad in comparison with some reference standard is of secondary importance and may even be irrelevant or misleading.

Unfortunately, because senior managers frequently fail to define design objectives, designers may feel they have no alternative but to work towards standards which they *believe* to be relevant. When this happens, products may end up over-designed or over-engineered – out of balance or too expensive for the markets in which they are to be sold. They may lack mass appeal, even though they may receive praise from design critics and possibly win design prizes.

The attainment of high design standards is not in itself a bad thing providing the objectives are satisfied at the same time. So what sort of objectives should be set and in what form should they be described?

Usually, design objectives should be expressed in terms of the end results which are being sought in the market rather than in terms of specific design characteristics. For example, if the need is to 'increase market share by attracting customers in the next youngest age group' then the objective should be (tightly) phrased in that form rather than as a requirement to 'use brighter colours' or any similar instruction.

Design objectives may be expressed in terms of:

1 Attainments sought (e.g. increase market share).
2 Existing recognition to be consolidated (e.g. reputation for safety or traditional values).
3 Change of position in market (e.g. move to higher price range).
4 Exploitation of new technology.
5 Projection of new image or identity.

It is the job of the designer or design team to decide how to meet the objectives; that is why they are employed. If managers set objectives in terms of design characteristics, they are pre-empting the design process and may be eliminating many potentially useful approaches.

Design objectives need not be exclusively concerned with marketing aspects; other company needs and aspirations may be addressed as well. Often, ease of manufacture or the minimizing of manufacturing costs are very important issues which can be greatly influenced during the design process. Similarly, the solving of distribution or storage problems might be included, or perhaps new legal requirements.

Setting Design Standards

If design objectives tell designers *what* needs to be achieved, then design standards provide guidance about *how* results should be obtained. As just indicated, great care must be taken to avoid specifying standards which are so tight that designers have little freedom of approach. Generally, it is better to issue standards in terms of boundaries or levels within which results are expected to lie.

So, for example, if the new product is a pocket radio to be used by teenagers, one of the design standards might be that the product must be capable of falling at least 0.5 metres onto a hard surface without damage. Framed in this way, the standard does not restrict the designer's freedom to explore the use of different materials, shapes and methods of construction.

Types of standard   Standards are likely to fall into three categories. In the first are national or international standards, which may be advisory or mandatory (often advisory standards are effectively compulsory because customers expect compliance). These are particularly concerned with safety and with products where customers need some reassurance that predictable performance or conformance with other equipment will be provided.

The second category consists of company-wide design standards, which are typically concerned with aspects of appearance or manufacturing quality. Certain colours, shapes, typefaces, logos, etc. may be elements of a carefully developed 'house style' which the company normally requires to be incorporated in any new product. Varying degrees of licence may be allowed, and this may be explained in some form of corporate design manual. Manufacturing standards may also be recorded in manual form and may include such aspects as preferred material types, tolerance limits, assembly methods and so on.

The third category of standards comprises those which relate to the specific product which is the subject of the current design exercise. Such standards may be concerned with many different aspects of the product – and the more complex the product, the more standards there may be. Thus, vehicles and aircraft manufacturers may well draw up details of standards running into many volumes – or, more likely these days, occupying large amounts of computer storage space.

Specifying standards   Where there is no previous experience of drawing up standards, managers may find the task rather daunting. Usually, the process is relatively straightforward when dealing with technical aspects. Desired results may be specified in terms of physical quantities which can be easily checked or measured. Performance standards for a new motor car engine, for example, can be issued in the form of power output, fuel consumption, working life, etc. Other possible technical standards might relate to maintenance (e.g. time required to replace parts), interchangeability (e.g. with other components) or noise generation – all of which can be readily described and measured.

The real difficulty for managers lies in dealing with design standards relating to softer issues such as appearance and comfort. Here, they are likely to be faced with conflicting opinions, not just about which levels or degrees to specify, but about whether one approach is better than others. These are the dangerous waters of taste, fashion and judgement – and the response of many managers may be to leave such issues unspecified in the hope that the designers will take care of them.

Sometimes, in these circumstances, designers will indeed produce solutions which are entirely satisfactory and meet with the approval of customers, critics and company personnel. More likely, though, the outcome will not be considered satisfactory; managers may feel they have been presented with something they 'did not quite have in mind', or, if the product does reach the market, it may be deemed to be out of date, uncompetitive, lacking distinction, etc. Further design work and more dialogue between managers and designers may eventually produce a result which is acceptable, but with much greater cost and use of time than if a good job of setting appropriate standards had been done in the first place.

Although difficult, it is not impossible for managers to draw up standards for non-technical aspects. The starting point should be any guidelines already available relating to the company's general identity. These may indicate preferred shapes, colours, visual images, etc. in addition to detailed standards for typefaces, logos, symbols and so on.

The next step is to consider whether the company is already associated with any distinctive elements of styling or if there are other aesthetic features of products upon which customers are known to place some value. If so, and it is felt that these should be preserved or developed, they must be carefully described and, if possible, quantified in some way. For example, products might have a reputation for 'looking solid', which may be a function of the dimensions, shapes and surface finishes normally used.

Thirdly, managers should look at other competing products to see what common elements of style there may be. It may be possible to identify clear subgroups of products aimed at different sectors of the market. Products may be exhibiting certain trends or may be evolving in ways which enable predictions to be made about the characteristics which will be prominent at some future time – perhaps just the time when the newly designed product will be launched.

Because of their training and experience, designers can give managers invaluable advice about such trends. Whilst most people may see only certain unremarkable features, designers may be able to detect and describe underlying changes which are taking place. Companies that do all their designing using their own staff may benefit by commissioning outside design consultants to provide this kind of advice.

Finally, and this can only be part of a continuing design programme, standards should be related to independent judgements of quality and success. Examples include design award schemes and competitions; a resolve to achieve recognition can help focus attention on the levels of design attainment necessary for success. Similarly, regular use of panels or clinics made up of customers, critics, designers and others can help establish markets and provide guidance for future work.

Conclusions

Adequate preparations for design work can have greater effect on eventual results than the quality of designers employed or the skill with which projects are managed. Responsibility rests firmly with senior managers to ensure that, first of all, they think through the internal and external issues relevant to designing in their companies. This will ensure that a proper *policy framework* will exist to guide design activities. Then it is necessary to set appropriate *objectives* towards which progress can be directed and *standards* against which results can be measured. In today's highly competitive business environments, failure to provide this level of managerial direction is not just foolish but should be seen as a major neglect of corporate responsibility.

PART III

CORPORATE APPROACHES TO DESIGN MANAGEMENT

13 Developing a Corporate Approach*

ALAN TOPALIAN

Alto Design Management, UK

Introduction

In this age of mass communication and distribution, productive capacity and manufacturing expertise are no longer the keys to economic power. Many developing nations have demonstrated that they can deliver quality products and undercut the manufacturing costs of industrially advanced nations.

As organizations redistribute their resources to take greater advantage of world economic and trading conditions, a new hierarchy of skills and control is emerging. Research and development, design and marketing are becoming accepted as the principal functions by which new knowledge and wealth are generated, and therefore as critical to the profitability and survival of these organizations. Indeed, expertise in these fields is an important characteristic that differentiates the industrially advanced from developing nations.

Since the mid 1970s, evidence has been mounting – particularly in the United Kingdom and North America – that the principal source of poor design standards in business is poor management. For many managers, design represents unfamiliar territory: they tend to adopt a superficial approach, often leading to indifferent results. If higher design standards are to be attained, it is important to increase awareness among managers of the contribution design makes to success in business, to convince them that an indifferent approach to managing design is incompatible with a professional approach to management, and also to help them develop the skills necessary to improve performance.

This chapter attempts to analyse some of the critical factors in developing a corporate approach to design management in business organizations.

Different Perceptions of Design in Industry

Consider first how organizations vary in the way design is perceived and handled. At one extreme, there are those in which design is not seen as representing a separately identifiable set of activities with a tangible contribution to profitability. If asked to indicate where the principal design effort is concentrated, managers may point to the quality control laboratory

* This paper was presented at the first Olivetti Design Management Symposium, Munich, in June 1984.

where materials and products are tested, the tooling shop of a factory, the drawing office in the production department, new product development in the marketing department, the public relations department, or the stationery office. In some instances, managers even believe that their companies' design work is done by print suppliers – 'to specification'.

Such a *laissez-faire* approach is characterized by departments dealing predominantly with their own design needs with little consultation between them and few constraints imposed. In the absence of any coordination and with the use of many different suppliers, the common ground between apparently unconnected design activities is obscured; design knowledge is dispersed around the organization. If design activities are sporadic, disparate and relatively unimportant, a case could be made for allowing departmental managers to continue to handle their particular design requirements without formal coordination. In addition, if a clear visual identity is not seen as important, it could be argued that a centralized, separately identifiable design function is not necessary.

But as the volume, complexity and importance of design activity increase, so does the need to integrate the work experience and operational interests of different departments to build a common approach. The piecemeal, compartmentalized execution of design activities becomes progressively inefficient. Organizations in this situation find it difficult to raise understanding between departments and to harness common interests. Typically, they engage in stop/start design strategies which tend to be inordinately influenced by interdepartmental conflict and the personal whims of executives.

There is a lack of continuity in activities, investment, standards and personnel involved, and hence rarely any consistency in decision making or results achieved. Little cumulative learning takes place, so it is not uncommon to find 'the wheel being reinvented' time and again with mistakes repeated. Such organizations are also more likely to get stuck in a 'design rut', partly because many of their energies are dissipated in overcoming these inefficiencies. The quality of design lags behind the requirements of the market, and competitors make dramatic inroads into market share. Naturally, corporate spirit suffers.

A more integrated approach to design may evolve in small steps. Consultation between departmental managers becomes more formal; increased contact on design matters leads to a greater awareness of design activities and a clearer perception of the scope and contribution of design. Senior managers take a greater interest and become more frequently involved. As experiences are shared through discussion or joint projects, a common language emerges together with an understanding of the attitudes and information needs of different departments and team members.

By analysing procedures – such as the briefing of specialists, the programming of work, and how progress is monitored – examples of better practice are distilled and recorded for future guidance. Budgets are established specifically for design work. The numbers and calibre of design staff rise; a wider range of expertise is recruited and used on a consultative basis. Departments begin to work in parallel at all stages of design projects, rather than in sequence with each department executing one stage and then handing over to another. Design specialists are brought into project teams working on a wider range of problems and their contributions are sought earlier in the problem solving process.

In other words, design becomes more visible and is moved progressively towards the centre of decision making as it is absorbed into the wider perspectives of staff in other departments. Agreement between departmental managers could also lead to a rationalization in the way design is administered: first, through cooperation on all work in one or two categories; secondly, by entrusting design projects to a handful of key individuals. So the authority of the person who deals with product design, say, is extended to encompass packaging; the person in charge of producing promotional literature might be asked to deal with exhibition design as well. Eventually all design activities may be handed over to one individual – perhaps initially as a subsidiary responsibility, then as a major responsibility, sometimes even a sole responsibility. Understandably, as the status of design rises, the status of such individuals should rise, too, as should the level to which they report and have direct access in the management hierarchy.

At the opposite extreme of the spectrum, there are organizations in which a professional approach to design pervades almost every activity and grade of staff. It would be inconceivable for a product or any other output to be produced without the introduction of an appropriately qualified designer early on in each project. To these organizations, design represents a form of investment like any other, where profitability has justified the resources committed to the establishment and maintenance of the necessary standards. Design expertise and performance are considered significant corporate assets and will feature prominently in company plans. Design activities will be coordinated with the extensive use not only of standard operating procedures but also of strict corporate design guidelines. Designers will frequently report to the chief executive or an executive member of the board. They will have ready access to senior management in several departments. Their terms of reference will often allow them to seek out problem areas and propose projects. Senior managers will keep in close touch with the design developments of their organizations, often acting as champions of this work.

Paradoxically it is not in these organizations that one encounters the most rigid design management regimes. If anything, there tends to be a feel of freedom in the corporate atmosphere – but this is deceptive because control of design activities and standards is tight despite appearances. The explanation is that companies which have cultivated an enlightened, professional approach to design over many years tend to move from operating a *laissez-faire* style to one of total centralization in order to impress the new regime on staff and introduce rigid controls into operations. As procedures and standards are proved, management and staff suitably motivated and trained, experience accumulated and confidence raised, a strategy of gradual controlled decentralization follows. In the process, managers and key staff acquire greater discretion within set limits which they understand fully and with which they identify absolutely.

The Essential Features of a Corporate Approach to Design

From this review, the essence of a *corporate* approach to the management of design can be summarized as follows

1   Recognition that design represents an identifiable set of activities with an important contribution to long term profitability, and thus needs to be managed rigorously.
2   An awareness of design activities within an organization and how these relate one to another. There must also be a clear understanding of how design decisions and activities link with those in, say, marketing and production, as well as the contribution design skills make to the operations of that organization. This knowledge should permeate several levels of employees – not simply the management grades.
3   A formally accepted view on what the management of design entails, and how design activities might best be managed.
4   A clear hierarchy of design responsibility, with executive responsibility for design formally assigned to a senior manager.
5   Some degree of specification on a corporate basis.
6   A strong sense of self-control at most levels of staff, backed by an appropriate degree of corporate supervision of design activities.
7   Design must become an integral part of the thinking and decision making within all departments, just as marketing, production and financial issues must become integral to the thinking of design specialists.

The Scope of Design Management

No attempt at developing a corporate approach to design can succeed unless the design function is brought into focus, for relatively few managers in industry have a clear idea of what design work goes on in their organizations. The word 'design' itself conjures up so many different meanings and emotions that a common attitude or level of understanding should never be assumed, even among managers of the same organization. It is important, therefore, to clarify the principal categories of design in business, and also to separate out design activities from the other more established functions such as marketing and production. For unless this is done the management of design will never attract the kind of attention necessary to initiate effective action.

For the majority of business organizations, the principal categories of design activity are product, graphic, exhibition, interior and exterior environmental design (including architecture). Account must be taken of all these categories because they are often closely interrelated in the total presentation of an organization, its products and services to target audiences. An organization that concentrates attention on one category while neglecting its responsibilities in the others cannot manage design effectively. For example, a company that spends substantial sums on the design of products yet invests nothing to improve the environment in which employees work will eventually limit the quality of product it can manufacture as well as damage its reputation as a caring employer. The company that designs attractive packaging and promotional literature for ineffective products does little to sustain its reputation, as customers are unlikely to be fooled for long by second-rate products. Thus the professional management of design activities involves the integration of, and uplifting standards in, all categories of design. Those who promote design management simply as a more rigorous approach to innovation or new product development reveal a fundamental misunderstanding of what managing design involves.

The second prerequisite for developing a corporate approach to design management is an understanding of what is involved in managing design effectively. Despite substantial common ground in the management of different functions in business, the detail of design management is not the same as that of marketing or production management. Without a grasp of such detail, the quest for design excellence will be in vain. So what is the scope of design management? This can be mapped out by examining the management of design at two interrelated levels: the corporate design management level and the design project management level.

At the lower project management level, the issues derive essentially from the shorter term, relatively confined problems encountered during the administration of design projects. In corporate design management, the issues centre on the longer term implications of the relationship between an organization and its environment, and the contribution that design skills and activities make to this relationship. The key issues are summarized in table 13.1.

Table 13.1 Key issues encompassed by design management

At the corporate level	At the project level
• The contribution of design skills to corporate profitability	• The nature of the design process and different types of design project
• Design policy and strategy formulation	• Design project proposals and the briefing process
• Design responsibility and leadership	• Selection of designers
• Positioning and integrating the design resource within organizations	• Bringing together and managing design project teams
• Devising and introducing corporate design management systems	• Planning and administering design projects
• Establishing and maintaining corporate design standards	• Costing design work and drawing up budgets
• Corporate design and design management audits	• Project documentation and control systems
• Sources of new design investment opportunities	• Design research
• The legal dimension of design	• Presentation of design proposals
• Evaluating major design investment decisions	• Implementation of design solutions
• Design management development programmes	• Evaluation of design projects
• Design and the manifestation of corporate identity	

Responsibility for Design

Who is responsible for the effective management of design in business organizations? There are actually four kinds of responsibility to be considered.

Responsibility for individual design projects

This is not fixed at any particular level in the management hierarchy but varies according to the nature of projects; levels also vary between companies in relation to priorities, experience, skills and so on. An essential feature in the efficient management of design activities is to assign projects to the appropriate levels.

Projects arising out of problems with characteristics which lie predominantly to the left of the columns in table 13.2 tend to be ones which

Table 13.2   Factors affecting the nature of design problems

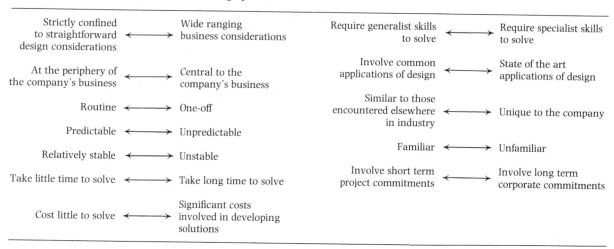

Strictly confined to straightforward design considerations ⟷	Wide ranging business considerations	Require generalist skills to solve ⟷	Require specialist skills to solve
At the periphery of the company's business ⟷	Central to the company's business	Involve common applications of design ⟷	State of the art applications of design
Routine ⟷	One-off	Similar to those encountered elsewhere in industry ⟷	Unique to the company
Predictable ⟷	Unpredictable	Familiar ⟷	Unfamiliar
Relatively stable ⟷	Unstable	Involve short term project commitments ⟷	Involve long term corporate commitments
Take little time to solve ⟷	Take long time to solve		
Cost little to solve ⟷	Significant costs involved in developing solutions		

can be executed relatively low down the hierarchy. In these cases, procedures and standards can be tightly specified. Work programmes and budgets can be drawn up relatively accurately and ought to be amenable to rigorous control. Project teams tend to be restricted to a handful of members, and interest confined to one or two departments.

Not least, the consequences of failure are relatively limited. There is little justification for senior executives becoming deeply involved in such cases, though they will almost certainly be kept informed of the principal decisions. Thus the addition of, say, an internal materials requisition form to the range of company stationery may be the responsibility of the stores supervisor or production foreman in the factory, or the chief clerk in the company office. Responsibility for the design of an exhibition stand for a minor trade show might be assigned to a junior marketing executive at head office. Routine updates of products and sales catalogues may be the responsibility of product and sales managers respectively, and so on.

Problems with characteristics which lie predominantly to the right of the columns in table 13.2 can involve a fair amount of exploration in uncharted territory; consequently uncertainty is high. Project teams tend to be large to encompass the necessary range of expertise and reflect the interests of several departments affected by the work. The consequences of failure could be severe. The development of a completely new product concept, the introduction of computer-based techniques into the manufacturing process, the move to corporate offices, and the revision of visual identification systems are examples of projects which encompass these kinds of problems. In all these cases, responsibility should rightly be vested at a high level – perhaps with the chief executive, marketing director, corporate facilities director or production director.

Such responsibility does not necessarily entail a personal involvement in the day-to-day administration of projects. The prime feature is the sanctioning of principal decisions – especially in relation to levels of expenditure and solutions adopted. Other features include ensuring that problems are properly defined, that the right project teams are brought together, that communication between the team and other interested parties is effective, and that approved procedures and specified standards are followed. In sum, it is essential to provide the leadership necessary for projects to be handled effectively.

Executive responsibility for design
This encompasses overall planning, administration, monitoring and control of the design function. In planning, the responsibilities cover the formulation of design objectives, strategies and development programmes which address the corporate objectives and strategies of an organization and form part of its strategic plans. In monitoring, responsibilities include the examination of how the organization will be affected by design trends and developments elsewhere, and the comparison of design performance with policies and set programmes as well as with the competition. In administration and control, the responsibilities relate to the translation of plans into reality in order to achieve corporate objectives. If a corporate approach to design is to succeed, it is essential that a senior manager is formally charged with executive responsibility for design. The level at which such responsibility is vested again varies between organizations. Ideally, executive responsibility for design should be a full-time commitment of a board director who has equal status to colleagues who head the marketing, production and finance functions.

Ultimate responsibility for design
This does not vary either by type of organization or between industries. Ultimate design responsibility lies with the individual who has the greatest discretion and authority to make or approve design and design-related investment decisions without formally being required to refer to others. In

effect, this individual is accountable for all the design work undertaken by and on behalf of the company. In business organizations this responsibility normally rests with the chairman of the board or chief executive, depending on the type of company and the working relationship between these individuals.

What else does this responsibility entail? Ideally, the person with ultimate design responsibility will demand, without concession, that the design function makes a full contribution to the achievement of corporate goals. This means insisting that design is fully integrated with the other functions of an organization, that design is always one of the factors taken into account in investment decisions, and that design specialists are consulted in all such decisions. The ultimate design responsible will mould the atmosphere within the organization so that design skills are respected and the appropriate design talent is employed and blossoms. Design policies will be translated into reality throughout the organization, and the board, managers and staff will be made fully aware of the breadth of design activity within the organization and understand the standards adopted. To sum up, the ultimate design responsible sets the hallmark of professionalism with which design is managed.

Collective board responsibility for design

Boards of directors of business organizations have a collective responsibility for design as the body which supervises and supports the executive. This responsibility entails a commitment to the professional management of design: making sure that executive design responsibility is assigned to a senior manager, preferably a member of the board; insisting on the establishment of a rigorous design management system; monitoring the productivity of resources invested in design activities; ensuring that the organization has adequate policies for key outside relationships (for example, a clear and effective visual identity with customers and investors, or close liaison with suppliers on the design of bought-in components); supervising design matters which are of particular corporate significance (for example, joint ventures with other companies to develop major new products, or the acquisition of major new design facilities such as CAD systems); and finally, ensuring that the organization has access to the top independent design authorities in fields which are of interest (recognized design experts, specialist design groups, R&D institutions and so on).

Corporate Control Through Specification

It should be stressed that a corporate approach to design management does not necessarily imply a strictly centralized approach. Total centralization – with all design work channelled through a single design department at head office, a single independent design group or other supplier – is rarely an efficient or convenient option, and very few larger organizations adopt it.

Indeed, a truly corporate approach cannot be achieved if the design function is totally centralized; self-discipline and control, and hence a degree of decentralization, are fundamental features of this approach.

Nevertheless, given the realities of professional management and design practice, some central control of design is essential. In its simplest forms, central supervision and control may be carried out retrospectively by a standing committee which reviews the results of design work at set intervals, or pro-actively by requiring that important project decisions are sanctioned by a central authority. It may also be effected through the appointment of an appropriate senior executive to oversee every project from start to finish.

More effective corporate control tends to require the establishment of a sensible design management infrastructure encompassing

1   Specific design objectives and strategies which address the achievement of the corporate objectives.
2   The appointment of appropriately qualified individuals to key positions.
3   Corporate design standards.
4   Guidelines on preferred procedures.
5   Detailed budgeting.
6   A comprehensive audit and evaluation procedure.

The link between design and business policies is critical for effective control. It is very difficult to evolve effective design strategies when corporate objectives and strategies are not clearly defined. Similarly, the achievement of corporate objectives will be that much more difficult if companies fail to think through the implications of their business strategies in design terms and plan ahead to ensure that the necessary design resources and solutions are available to them when required.

Consider this basic example: a company seeks to raise profits by improving margins on turnover from 20 to 30 per cent. What design strategies might contribute to the achievement of this objective? The first might be to alter the design of existing products so that simpler processes and less specialized materials are used while maintaining prices. A second option is to redesign the work environment to enable productivity to rise. A third is to improve the design of promotional material, thus achieving higher sales per dollar spent. Finally, a new range of products might be developed featuring higher added value which can be sold in markets where demand is more stable.

Sensible specification on a corporate basis is another aspect of the control mechanism; this often derives from a full understanding of an organization's corporate identity and the clear positioning of products and services in the market. A consistent presentation to target audiences can be sustained through the specification of quality and performance standards relating to, say, properties of materials used, quality of finish, and safety and service-ability of products; clarity of visual identification and information on promotional literature and packaging; the structure and content of comprehensive design project proposals; and the visual attractiveness and physical comfort of working environments. The creation or adoption of stringent standards which are internationally recognized can do much to enhance an organization's reputation and open a wider range of markets to its products.

A high design profile in corporate success is largely a matter of involving the right people in design activities. Corporate guidelines on the selection of key personnel can greatly raise the chances of ending up with appropriately qualified and motivated individuals. Given the critical (often detrimental) influence managers have on the outcome of design activities, these guidelines ought to relate, first and foremost, to managers who are given responsibility for design activities, then to design specialists. An approved list of carefully vetted design specialists and suppliers can be a valuable tool for ensuring conformity to corporate standards and results of a consistently high calibre.

With the right standards and the right people, the final area of corporate specification concerns the regulation of how these people go about their design work. Guidelines on preferred procedures should cover most aspects of design activity such as diagnosis of problems, briefing of design specialists, presentation of design proposals, programming of design work, monitoring of the progress of design projects, and the evaluation of results achieved.

However, it should always be kept in mind that the prescription of standards does not automatically confer on them the stamp of corporate acceptability or imply adoption by staff on a personal basis. The potential differences between personal, professional and corporate standards are crucial and are echoed in many aspects of business management. Put very simply, the differences may be represented by heart and mind: professionally an individual's mind might indicate one response to a situation, while at a more personal level the heart might suggest otherwise. In the design field, it would seem that personal standards among non-design trained executives often overrule professional standards, and that personal standards are too frequently assumed to be identical with corporate standards, leading to catastrophic results. Thus any programme to raise design standards is essentially a continuous series of battles for the hearts and minds of executives and staff.

Finally, effective control of design activities will never be achieved without the recognition that expenditure on design has to be carefully planned under clearly identified design budgets, with design development programmes formally set out in strategic plans. Lurching from crisis to crisis, carrying out work with funds extracted unwillingly out of a hodgepodge of other budgets, demeans the status of the design function and is no formula for achieving design excellence.

Conclusion

Design management is a living, developing discipline. It is not instituted on a once-and-for-all basis, but has to be carefully nurtured and rigorously supervised to cope with changing circumstances. The clearest indication that an organization has adopted a professional corporate approach to the management of design is when design enters the bloodstream of an organization and is integrated fully into the fabric of that organization. This implies that design data form part of the central core of information on which the organization operates, and that design issues always receive serious consideration before decisions are taken. It also implies that design specialists and staff from other disciplines work in parallel through all stages of a wide range of activities, not just design projects, and that this approach embraces several grades of employees.

However, this interdisciplinary approach to problem solving goes beyond multidisciplinary teams to what might be described as empathetic multi-strand thinking. Team members do not simply bring their particular specialisms to the solution of problems, but actually think through problems in an integrated manner: that is, there is a design strand in the thinking of non-design staff, just as there are marketing/production/finance strands in the thinking of design specialists.

Finally, a corporate approach to design cannot be instituted unless top management has the necessary will and stamina to demand improved performance and enforce higher design standards – particularly from middle managers, who tend to be most involved in the day-to-day administration of design projects. Token interest among senior managers is no guarantee that subordinates will be galvanized into effective action, or that standards will be maintained. Elegant rhetoric that is not backed by practical commitment rarely fools staff.

Further Reading

National Economic Development Office: *International Price Competitiveness, Non-Price Factors and Export Performance*. London: NEDO, 1977.

Roy, R., Potter, S. et al.: *Design and the Economy*. London: Design Council, 1990.

Topalian, A.: *The Management of Design Projects*. London: Associated Business Press, 1980.

Topalian, A.: The role of company boards in design leadership. *Engineering Management International*, pp. 75–86, March 1984.

Topalian, A.: Organizational features that nurture design success in business enterprises. *Proceedings of the Second International Conference of Engineering Management*, pp. 50–7, Toronto: September 1989.

Topalian, A.: *Design Leadership in British Business: the Role of Non-Executive Directors and Corporate Design Consultants*. London: Alto, 1989.

Topalian, A.: *Proposed Syllabus for Design Management Courses*. London: Alto, 1990.

14 Managing the Product Innovation Process

BRIAN WILSON
Traqson, UK

Introduction

This chapter is about managing innovation – the process of taking an original concept through market assessment, design and development stages to the point of a product being sold profitably in the marketplace. It is a highly complex process, involving the integration of a wide variety of functional activities within an organization into a cohesive and effective whole. It is critically dependent upon good management.

International comparisons tend to show that some countries – notably West Germany, Japan and the USA – consistently achieve high levels of industrial productivity and competitiveness. The relatively poor performance of other countries is associated with low levels of capital investment in manufacturing processes but, even more importantly, with failure to develop sufficient high added value products. Amongst the factors which impede successful product development, a recurring one is the number of managers from companies, large and small, who seek to exploit inventions with little if any real checking as to whether there is likely to be a market or not.

The dawning of a real sense of market awareness can sometimes have spectacular results, as in the case of one very big research and development department. After many months of reappraising its role, it decided that it should not be attempting just to invent new products; it should be developing a market research capability and using its wide range of technology and expertise to meet market needs – once they had been defined. From a position of low credibility and morale caused by its inability to come up with new products, it had, within a couple of years, achieved high regard, high levels of morale and a whole range of successful products.

The idea of 'push versus pull' applies very much in this area. If one tries to force products onto people which they see as not pertinent to their needs (push), they will inevitably reject them. If, on the other hand, their needs have been identified and they are then offered products which satisfy those needs, they will be keen to buy (pull). But it would be unrealistic to imagine that firms will always start out with the market in mind. Inventiveness cannot be constrained, nor would it be sensible to attempt to do so. However, there is no doubt that, before a decision is made to develop an invention into a product, some fairly extensive market research is required. The risks in developing a product and in hoping for the best are just too great; as a result of such behaviour, many is the small company that has ended up in liquidation and many is the larger company which has greatly weakened its position.

Another essential factor in product development work is the extent to which other functions in the company participate in, and contribute to, the design process. For example, a prime consideration for designers may be the degree of automation which should be designed into a manufacturing process. Should it be a dedicated one – or a flexible one, able to accommodate a range of future products or short, high variety runs? Advice from senior management will be needed but designers must also consult a range of departments: frequently, production or marketing people are faced with problems which could so easily have been avoided if they had had a chance to put their requirements to the designers.

The financial people have to be considered as well. The company's accountants will be management's watchdogs if the project is being financed in-house – or equally critical external financiers may be involved if outside funding is being sought. Either can easily bring a promising proposal to a quick stop. Somehow the project manager has to convince those who have to take a wider business view that, technical elegance or excitement apart, the investment of a substantial sum of money is going to give a good return. The financiers will wish to see a soundly constructed investment proposal, to examine the assumptions which have been made and to assess the areas of uncertainty and risk. It is critical that the proposal be presented in a professional way. If not, it will have a greatly reduced chance of success.

Managing Innovation

Hence the most critical aspect of all, as far as the product innovation process is concerned, is its management. Unless the various interactions between departments are taking place effectively and people have a broad view of what is going on, and of their part in it, it is likely that functionalism will rear its ugly head. It is also probable that the designers, who have the lead role, will concentrate only on their particular contribution, without any thought to or concern for the problems which they may create for other people – or, in the end, for the marketability and serviceability of the product.

The main problem in managing the highly uncertain innovation process is how to bring together the various inputs in a cohesive and structured way. There has been a gradual movement this century from a view of one right structure (pyramidical, hierarchical and functional) whatever an organization's activities, to a recognition that the optimum structure is influenced by the nature of a company's operations and underlying technology. The American organizational theorists Lawrence and Lorsch (1967) have taken their thinking further and have brought two insights to bear.

One is that you cannot consider an organization as a total entity; it is made up of a number of constituent parts, each of which needs to be structured in a way appropriate to its own particular range of activities. The second is that the determinant of structure is not primarily technology, though this bears on it; rather, it is the degree of uncertainty which needs to be faced by the particular unit in its day-to-day operations. For example, a research department lives with a great deal of uncertainty and, to be successful, requires a very open, interpersonal style and a flattish rather than pyramidical structure. There are experts at all levels in such an organization, each contributing to decision taking. At the other end of the spectrum lies production which, working to clear schedules with well-known technology, has to cope with much less uncertainty and may be best organized in a more formal way. Other departments fall between, with design and marketing towards the more uncertain end.

Though such differentiated structures and associated mini-cultures are essential to any organization's success, they are not the whole story. Each differentiated part of a larger organization inevitably develops its own style, own jargon and own ways of doing things – and such differences are a formula for conflict. Lawrence and Lorsch found, based on a great deal of well-validated research, that effective organizations are those which have not only differentiated their structures but also set up integrating and conflict managing mechanisms (see figure 14.1).

Bringing this theory to bear, it is clear that, for successful innovation and product development, there need to be effective means of drawing together people from a wide range of highly differentiated departments. There is no doubt that a good integrator – a proactive, experienced, well-informed and well-respected individual – can be a great asset to a company. However, to rely on one individual – to encourage, cajole, coerce or do whatever is necessary to get the required degree of cooperation – has the inherent weakness that the integrator is the only one who bears the responsibility and carries the can. Such an approach is limited in the possibilities which it generates for getting a shared ownership of the need for a successful outcome.

A well-chaired committee goes a little beyond the good integrator. However, committees all too often become talking shops, with departmental representatives seeing themselves as little more than purveyors of the party line, with no freedom to make commitments on behalf of their departments. Any show of understanding of wider issues by a representative is all too often seen from a narrow, departmental point of view as lack of loyalty to the department.

Something much more is required; an effective matrix structure is probably most likely to succeed. Matrix organizations have been promoted as a way of better coping with turbulent environments and with all the uncertainties which they generate. Senior managers, operating within traditional

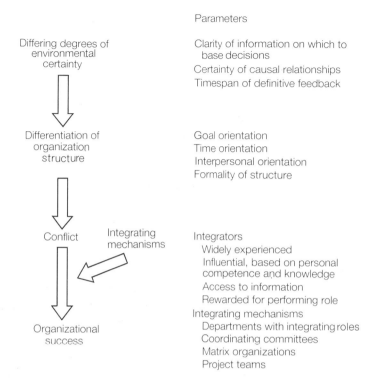

Parameters

Differing degrees of
environmental
certainty

Clarity of information on which to
base decisions
Certainty of causal relationships
Timespan of definitive feedback

Differentiation of
organization
structure

Goal orientation
Time orientation
Interpersonal orientation
Formality of structure

Conflict

Integrating
mechanisms

Integrators
  Widely experienced
  Influential, based on personal
  competence and knowledge
  Access to information
  Rewarded for performing role
Integrating mechanisms
  Departments with integrating roles
  Coordinating committees
  Matrix organizations
  Project teams

Organizational
success

Figure 14.1 Theory of
integration and differentiation
Source: Dresdner Bank
AG, Frankfurt

bureaucratic, hierarchical structures can find themselves overwhelmed by the volume of decisions which they are required to make. They need to get more collaboration between departments, with fewer differences coming up to be resolved at their level and more decisions taken at lower levels in the organization. A typical matrix organization retains the functional/ departmental structure but establishes cross-linking teams under the leadership of a team manager. This manager will have a representative of each major functional area in the team which, as a whole, may be charged with overall day-to-day control of its business or product area – as well as with developing new products, producing long term business/product development plans, and preparing soundly based investment proposals for senior management's decision.

Though this sounds fine in theory, it is not quite so easy to put into practice. The establishment of such a structure leads to a change in the power balance, with functional heads often feeling most acutely that their power and authority is being undermined by the upstart team managers. For success, the new structure must be established with great care. Teams need clear remits; policy guidelines for their operation must be established; and the dual roles, both vertical (within the function) and horizontal (within the business area), have to be recognized and properly rewarded. Above all, an *esprit de corps* must be developed amongst team members and recognition must be given to their powerful influencing role based on the breadth and depth of knowledge between them which, for their area of business, is unique within the organization.

Product Development Teams

Whilst such a structure is good for creating, amongst other things, a clear recognition of new product needs and opportunities, in itself it does not go far enough. It is not capable of taking new product ideas and translating them into successfully marketed products. What is required is a further development of the matrix structure, the creation of dedicated product development or project teams.

Major projects have long been a thorn in the side of many companies. The traditional, sequential way of handling projects just has not worked. Typically a technical department has developed the design brief and sub-contracted the work to the relevant engineering sections (mechanical, electrical, instruments, etc.), each of which has been expected to achieve its defined contribution, with coordination provided by the technical department project manager. Once the designs have been completed, they have been passed to procurement and to production (or construction, in the case of a capital project) who, in turn, have then been expected to perform.

As anyone knows who has been involved in such a project, things seldom proceed smoothly. Because there may be inadequate consultation along the way, production cannot implement the designs as received; design changes are then necessary which, in turn, cause ill feelings, delays and cost overruns. By the end of such a project, the word 'team' is hardly descriptive of the warring individuals and departments involved. There is a whole history of such catastrophes. The Ford Edsel car, intended to be the star in the company's crown, was later to be described as a 'camel produced by a committee setting out to design a horse'. One major chemical company's multimillion investment in a novel chemical process achieved only 75 per cent of design output and was 18 months late in coming on line, as well as being grossly overspent.

The lessons about the management of uncertainty in complex organizations, which emerged through experiments with matrix structures and the like, led to further experiments in temporary organizational structures. TRW Systems in the United States pioneered the way with project teams, which were set up as temporary, multidisciplinary teams, with clear-cut and shared goals. It is claimed that the success of one of the most complex tasks of all time, getting a man on the moon, was entirely due to such a team approach. The chemical company referred to above, having adopted such an approach, got its next major plant up to 100 per cent of designed production rates a month before the scheduled completion date and well within the cost target.

As with matrix organizations, such temporary teams need a lot of nurturing to make them work. It is important that people from the various functions who will be involved are both very competent and able to get on well with others. It is not enough to throw them together and hope for the best. Effort needs to be invested in team building activities, perhaps using management games initially to simulate stressful situations and to show how behavioural patterns help, or disrupt, team working. Team members can begin to develop norms for working together, such as openness in confronting issues, thus removing the blocks to smooth teamwork and developing mutual trust in, and respect for, each other. Once they have reached this point, they can then turn their minds to the major task which they have been given and to the way in which they will organize it, allocate responsibilities and define their mutual roles.

People working in such ways quickly lose their narrow functional viewpoints. Their concern becomes one of driving the project towards a quick and successful conclusion, and the multidisciplinary approach leads to much better problem solving and more elegant solutions. Incidentally, it also leads to the growth of a great deal of broad-based management potential, which is a considerable bonus. As with matrix organizations, it is important that rewards structures reinforce desired behavioural patterns.

Whilst much of the foregoing is particularly relevant to the larger firm, the smaller company can also benefit from such experiences. There is a need to have regular coordination and interaction between different specialists and it is useful to establish project teams with clear remits, even if the size of projects does not warrant full-time involvement.

Style of Management

Some reference has been made to different mini-cultures; it is important to stress the critical importance of organization culture and management style on effective innovation and product development. One of the main reasons why some matrix organizations, and indeed project teams, have not worked as successfully as hoped, is that the culture and style have been inappropriate. It is impossible to achieve the necessary degrees of openness, trust, team working and shared sense of responsibility in an organization which is bureaucratic and autocratic and has a highly controlling style of management. There is a need to move from the bureaucratic kinds of behaviours outlined in table 14.1 towards those at the more progressive end of the spectrum. Such a change in style may be perceived as highly threatening by long established managers who have been developed, promoted and rewarded for behaving in the autocratic and controlling ways characteristic of traditional organizations.

At a recent conference of directors from a wide range of large companies, two of the top human resource problems identified were the need to change the style of management to a more open and involving one, and within that,

Table 14.1   Impact of management style on how an organization is run

Area	Bureaucratic	Progressive
Management style, organizational culture	Controlling Decision taking at top	Open communications Joint, cross-functional problem solving Delegated decision taking
Objectives	Quantified Imposed	Objectives about the development of people's potential as well as quantified ones Agreed objectives and standards
Structure	Functionalized Hierarchical	Matrix Coordination at lower levels
Roles	Tightly defined Relatively narrow	Broadly defined Emphasis on team as well as on individual
Systems	Controlling Information to top for decision takers	Supportive Providing the information required for lower level decision taking
Work organization	Fragmented jobs Use of measurement for control Strong supervision	Whole jobs Greater autonomy and decision taking within work group Self-control within agreed standards

to deal in a more understanding way with the blocks to such a movement represented by the long established middle managers. Sir John Harvey-Jones (1988), the recently retired chairman of ICI, sees such a change in style as being essential to, as he calls it, 'switching people on, not switching them off'. It is always easier to change when an organization is feeling threatened. Faced with tough international competition, many companies have had to examine their traditional ways of doing things and implement changes which will improve their competitive edge and enable them to survive and grow. The introduction of the kinds of changes referred to here is therefore likely to be seen as consistent with, and an important contribution to, necessary wider changes and overall success.

Factors Critical for Success

The first and undoubtedly most important factor is the behaviour of senior managers. There is little point in them coming out with fine statements about a more open style if they continue to operate in their former autocratic and controlling ways.

The second and almost as important factor is to ensure that the new required behaviours are properly rewarded. Actions speak louder than words. If people are collaborating more, are achieving standards of excellence and showing imagination and inventiveness in their work, they must be given recognition through the rewards structure. There must be clear differentiation between performers and non-performers. If the company as a whole is doing better, so should its employees and particularly the high performers.

How the members of project teams are rewarded is particularly important. One of the reasons for the failure of some matrix organizations is that the reward of team members is left entirely in the hands of their functional managers. If members of a product development team are doing their jobs properly, their broader views of the area of business with which they are concerned will almost inevitably cause them to question what the department is doing and why. Because of this, individuals may be perceived sometimes as 'pains in the neck' and, if their reward is left entirely to their functional boss, their performance may be assessed unfairly. It is therefore most important that the project manager should have a big say in any performance rating.

The third factor, related to the first, is the need to appoint a design or innovation director to the board. This is a good outward sign that new product development is being taken seriously but, more importantly, it means the board will constantly have to face the key issues of effective product development.

Conclusion

The key to success is in the hands of senior managers. If they can establish open, problem solving cultures and structure their organizations in ways which will encourage interdisciplinary working and the release of energy, imagination, talent and ideas; if they can point their organizations in the right directions and establish and reward standards of excellence; then they will be managing the innovation process in such a way that their company has every chance, not only of surviving, but of beating the competition, growing and enjoying the fruits of success.

References and Further Reading

Harvey-Jones, J.: *Making it Happen – Reflections on Leadership.* London: Collins, 1988.

Lawrence, P. R. and Lorsch, J. W.: *Organization and Environment.* Boston: Division of Research, Harvard Business School, 1967.

Wilson, B.: Stimulating creativity, innovation, collaboration and productivity, through progressive organisations and cultures. Manchester Business School: *Creativity and Innovation Network.* Vol. 9, no. 4, October–December 1983.

15 How Ford Broke the Detroit Mould

CHRISTOPHER LORENZ
Financial Times, UK

Introduction

Ford's cars used to look characterless and utterly nondescript. If they were distinguishable at all from other makes, it was only for their anonymity. Today the story is very different. In both the US and Europe the company's sleek range of aerodynamic cars stand out from the crowd.

Behind this change in appearance lies Ford's transformation from design dullard, and commercial also-ran, into leader. Starting in Europe in the late 1970s, it completely broke away from its traditional strategy of providing worthy but unexciting products on a narrow sales platform of value for money. And in the early 1980s the parent company in Dearborn, Michigan, abandoned its standard approach of plodding along steadily in the wake of General Motors with lacklustre, unimaginative vehicles.

Instead, the entire Ford organization seized on a nascent European trend towards aerodynamics and 'drivability' to leapfrog its competitors, not only in Europe but around the world. In one risky bound it scrapped the time-honoured Detroit tradition of evolution at a snail's pace, and catapulted itself forward by a generation. As one of its top executives said at the time: 'We clearly identified the need to re-establish our image in the marketplace. You can't do that if you're timid.'

Despite tricky initial problems in a few European markets, Ford's adventurousness was quickly rewarded. In Europe it fought its way to market leadership for the first time, and in the United States its 'jellybeans', as the popular press dubbed its new range of aerodynamic cars, not only restored several precious points of market share that it had lost to GM and the Japanese, but completely transformed its standing with the customer.

The revolution in Ford's design was far more than styling. Its new image reflected a Galileo-like revolution on several fronts: in marketing strategy, where it moved sharply away from its traditional production and sales-led thinking towards marketing proper; in design itself, where skin-deep styling was replaced by an integration of form and function; and in terms of organization, with the company's product planners, industrial designers and engineers being brought together really effectively for the first time. The whole transformation constitutes a classic case study in the management of product development, and the changes that are necessary in executive attitudes, corporate structures and market research procedures, if a company really wants to use adventurous design to create products of meaningful distinction from the competition.

Europe Shows the Way

The shift in Ford's competitive strategy began in the mid 1970s, in its main market stronghold of Western Europe. Faced with an overwhelming Japanese challenge for its traditional platform of cheap and cheerful cars, Ford of Europe's top management in Britain and Germany decided

to get Ford cars out there that people desperately want, rather than cars they will buy because they are the lowest priced on the market. You can't do that any more because the Japanese have taken that part of the market away from us.

The words are those of Robert Lutz, the former BMW executive who ran Ford of Europe between 1977 and 1982, first as president and then as chairman, and again for a short period from 1984. In a frank interview in 1982 he admitted that

The Japanese have taken over the no-nonsense, no-frills, high value for money, reliable transportation part of the market. My goal is to be a mass producer of the type of cars BMW and Mercedes have a reputation for making. We are moving up in technology and credibility so we get the same price elasticity as they have.

Lutz might have added Volkswagen-Audi to the list of companies Ford was trying to emulate. But that would have been too near the knuckle, since it was precisely the upper end of the VW range, and the entire Audi line, at which Ford had most decided to pitch itself. This quest for quality was the first plank in its new European strategy.

The second was aerodynamics. By the late 1970s a number of Italian design consultancies had begun to display prototype 'ideas cars' at the leading annual motor shows, fusing together a growing obsession with aerodynamics, the mushrooming requirements of safety legislation, and unusually skilful packaging – the squeezing of as much interior space as possible into tightly restricted exterior dimensions. The Italian prototypes went well beyond the achievements of Citroen and Rover, which had been making aerodynamic-looking cars for some time.

Amid signs that Audi and a number of other companies were beginning to recognize the commercial advantages to be gained by offering remarkable improvements in fuel efficiency, together with extra passenger comfort and a futuristic new appearance, the then head of Ford of Europe, 'Red' Poling, and his new vice-president of design, Uwe Bahnsen, decided to apply some of this thinking to the new Escort model which was already under development. (A small to medium car in European terms, it was classed as a subcompact in the US.)

But winning the approval of Ford of Europe's all-important product planning and design committee was far from straightforward. Strange as it may seem in retrospect, the committee's heavily conservative membership of product planners, engineers and salesmen/marketers was initially sceptical of the notion that aerodynamics would produce fuel saving.

Still all too aware of Detroit's disastrous experiments with aerodynamic styling in the 1930s, they were nervous of the 'notchback' which Poling and Bahnsen proposed to give the Escort, in preference to the hatchback shape which VW and Renault had pioneered and which was then becoming standard in cars of the Escort's class.

By undertaking additional market research, and making more intelligent use of the results than in the past, it was eventually possible to persuade them that its unusual silhouette would not deter consumers from buying the car. And so it proved. Launched in 1980, the Escort was a stunning success in almost every European market.

For Bahnsen's team of designers, it was the big breakthrough. They had broken the ice of established traditions, and convinced the product planners and engineers that the use of aerodynamics required certain principles to be respected throughout the car's design. If this breakthrough had not been made, it must be doubtful whether Ford would have been able to move on to the near-revolutionary approach of deciding a car's shape early in the development process, as an integral part of its functional engineering.

At the same time as the Escort was under development, Ford of Europe was working on a replacement for its ageing line of light and medium trucks, the D series. With Mercedes and other competitors moving away from the traditional generation of harsh, aggressive-looking vehicles, it was obvious that Ford should proceed in the same direction. But its solution, christened Cargo, was all its own: a slim, elegant-looking cab with unparalleled all-round vision, which was so well designed the company was able to abandon its original plan of building a second type of cab for heavier trucks. Ford was also able to use it on the Brazilian-made 'world truck' which it launched in 1985, with the US and Brazil as prime target markets.

The two key design factors behind Cargo's appeal were driver convenience (which was sufficient even for long distance truckers) and the provision of enough airflow under the cab floor to cool a large engine. The latter is an extremely tricky design task if the cab floor is to be kept virtually flat, as town delivery operators demand.

The Cargo cost £125 million to design, develop and tool. Ford of Europe's next project, the Sierra medium saloon car ('sedan' in American terminology), required more than five times that financial commitment, including £95 million for design, development and engineering alone.

Cost was by no means the only measure of the importance to Ford of the Sierra, the replacement for the company's 20-year-old line of medium cars known in Britain as the Cortina and in Germany as the Taunus. It was the first vehicle to be developed from start to finish under the company's new marketing and design strategy.

The Sierra was adventurous indeed. In place of the blandness of the Cortina/Taunus came a vehicle which simply reeked of technology, drama and excitement. With an even more radically aerodynamic shape than Citroen's famous DS series (which some people nicknamed 'the platypus'), and its subsequent CX, it was not surprising that Lutz and Bahnsen again took time to win the support of the product planning and design committee, which initially favoured some of the less radical pre-programme alternatives. Together with the research engineers, Bahnsen also had to lobby for the adoption of some of the more ambitious aspects of the design, notably a large, injection moulded polycarbonate bumper which was crucial to the car's low nose and ambitiously aerodynamic airflow management: the Sierra's drag factor of 0.34 was a remarkable 24 per cent better than its predecessor's.

Fully realizing the risks of the dramatic step it was taking, Ford resorted to ingenious methods to soften up the market. During the 12 months before the Sierra's launch in late 1982, it put a very slightly altered version on display at several European motor shows, under the name of Probe III. Hints were then dropped to the motoring press that this futuristic vehicle, totally unlike anything Ford had ever produced, might – just might – be the replacement for the Cortina/Taunus.

All the same, the Sierra had a mixed reception in Britain, where much of the car market is dominated by conservative company fleet buyers. Though the car's packaging gave it a very spacious interior, its narrow front and curved lines made it look deceptively small. Some critics likened it to a cake that had gone flat at the edges and others said it looked as if someone needed to take a bicycle pump to it. In the face of unprecedented competition, not only from the Germans and Japanese but also from a revitalized Vauxhall, the local General Motors subsidiary, its success initially fell short of Ford's expectations.

But in the crucial German market it sharply increased Ford's share of the segment for medium cars. And in Western Europe as a whole it boosted the company's share of that category by half. At a time when the size of the total European car market was running at 15 per cent below the forecast level – on which Ford and every other manufacturer had based factory capacity levels – this was not bad going. Competition in the European car market had never been so intense, nor profits so poor.

All the same, the Sierra's difficult baptism demonstrated the risks Ford had run in being the first to lead the mass car market towards aerodynamic design. It is never easy or comfortable to champion a revolution. Another lesson was how difficult it can be to design one car to suit a set of regional markets, let alone the much-hyped global market; it was the very aspects of the car which appealed to German buyers that deterred the more conservative fleet operators in Britain.

Within two years of the Sierra's launch General Motors had begun to follow suit, first with a restyled version of its directly competitive model, the

Vauxhall Cavalier (Opel Ascona), and then with a smaller Sierra lookalike, the Vauxhall Astra (Opel Kadett). By the spring of 1985, when Ford launched a larger aerodynamic car to fill the top of its range, called Granada in Britain and Scorpio in other markets, it was clear that it had succeeded in setting the trend for future European car design.

Dearborn Follows Suit

Over in Dearborn, the parent company had also swung into line. In the late 1970s, when the European Escort programme was already under way, Dearborn had joined it in a belated attempt to take advantage of the growing US demand for compacts and subcompacts. But the top American management of the time had insufficient confidence in their European colleagues. They insisted on re-engineering the car from front to back, and top to toe. As a result it grew more costly to produce, heavier, less fuel efficient and less satisfactory to handle. Sold under the names of Ford Escort and Mercury Lynx, it was reasonably successful but made little money for the company.

Behind the scenes, however, the Dearborn designers had begun to press the case for aerodynamics. They wanted to adopt a proper industrial design approach, in which shape was an integral part of the car's function, rather than a last-minute wrapping-up of the working parts in a more or less stylish envelope.

So they were ready and waiting when Donald Peterson was named president in early 1980, at a time when Ford's market share and profits were plunging through the floor. In full knowledge of Ford of Europe's decision to launch the revolutionary Sierra, one of his first actions was to stimulate a last-minute rethink about the new US models under development at the time, which were scheduled for launch in late 1982 and early 1983.

As part of their newly revised strategy of abandoning full-frontal competition with GM and the Japanese, Peterson and his senior colleagues decided to target their attentions on particular market segments, especially the mushrooming hordes of young professionals ('yuppies'). Starting with the Ford Thunderbird and Mercury Cougar in late 1982, and even more dramatically with the Ford Tempo and Mercury Topaz compacts the following spring, Dearborn broke away from the 'boxy styling offered by our domestic competitors', as one of its senior executives put it – a description which applied perfectly to Ford's own previous line of models.

Though theoretically committed to the idea of global products, Ford was unable to use the Sierra itself as a centrepiece of its American strategy. This was partly because Ford of Europe had decided, for several reasons, to design the car with rear wheel drive. Thanks in particular to the promotional efforts of GM, American consumers had been successfully convinced that a car of the Sierra's size needed to have front wheel drive if it was to be frugal on fuel. So Dearborn had to 'stretch' the floorpan of the front wheel drive US Escort as the basis for the Tempo and Topaz models.

Appropriately enough for the two cars which marked Ford's phoenix-like revival from the depths of consumer disfavour (and from a mammoth loss of over $3 billion between 1980 and 1982), Tempo and Topaz were first launched into the public's gaze on the flight deck of an aircraft carrier which had survived the Pacific War with Japan – the USS *Intrepid*. The success of the two cars contributed heavily to Ford's remarkable feat in 1984 of lifting its US market share by a full 2 per cent to 19 per cent. By contrast, GM's share fell slightly, and Chrysler's rose by 0.3 per cent.

By then Ford had irrevocably committed itself to a strategy of integral design on its next two models. The ultra-aerodynamic Ford Taunus and Mercury Sable, mid-sized twins which went on sale in December 1985 after a record $3 billion development and tooling programme, were no reskinned affairs. This showed in their remarkable aerodynamic performance: a drag factor of only 0.29, the lowest of any US production car at the time, and in sharp contrast to the value of 0.37 for the boxy cars they replaced, the Ford LTD and the Mercury Marquis. Fuel economy, passenger room and road performance were improved dramatically; car magazine writers were quick to compare the cars with Audi's high status (and high cost) imports from Germany. But the general public was most struck by their appearance. As *Fortune* magazine commented: 'These newcomers carry the jellybean look all the way – the front grille has disappeared altogether, and the hood curves down to the front lights.' To fill out its range, Ford also began importing high performance, sporty versions of the Sierra and Scorpio.

As early as December 1985, a *Fortune* cover story remarked: 'The mainspring of Ford's US revival is its new boldness in design. Once a plodding follower of GM, Ford has moved aggressively to put its cars in the forefront.'

By the end of 1986 Ford was making higher profits than the much larger GM, the Taunus had become America's best selling car, and its design team had been catapulted from obscurity into media stardom. The *New York Times* commented that 'aero design' had not only put Ford back on the right track, but changed the look of the entire American car industry.

How Design Grew New Muscle

Behind this series of dramatic model changes in Europe and the US lay substantial alterations in organizational structure and product development procedures, all of them aimed at improving coordination between product planning, marketing, engineering and design.

As far as design was concerned, the most significant change in the US was the decision in 1980, as Peterson assumed the presidency, to move a large part of the corporate design staff into a line management role within Ford North America's product development structure. Under Jack Telnack, an American who had previously headed the European design team, the influence of the Dearborn designers grew apace.

In Europe, a similar step had been taken as far back as the late 1960s. Design was put on a par with product planning and engineering, with all three reporting to a vice-president of product development. This elevated position in the organization structure, rare in any company until the 1980s, gave the European vice-president of design more influence than his American counterpart of the time.

Strategy and structure apart, one of the main sources of the designers' growing muscle during the 1970s was the decision to recruit a number of design engineers into the design centre. For the first time, this enabled the industrial designers to talk to the engineers in their own language; in the past the 'styling' department (as it had been called) had often been 'raped' by the engineers, according to the evidence of some of the engineers involved.

But nowadays the design engineers carry out a full costing of the designers' work, and undertake engineering feasibility studies, to make sure it is viable in economic and technical terms. In certain other motor companies, the work of industrial designers is still confined to styling and the creation of concept models, which are completely changed once they are handed over to the engineers. By 1984, Bahnsen's 370-strong department contained as many design engineers as industrial designers.

Until the beginning of 1985, Ford's European product development structure was highly departmentalized. No one below the level of the vice-president of product development possessed clearly defined authority to coordinate the work of the different departments. In the early stages of development the product planners tended to 'hold the ring' between designers, development engineers and production staff (production planning itself represented sales and marketing). But their grip slackened as projects progressed, with the result that Ford experienced many of the classic problems of multiple handovers from department to department: details were changed, work had to be done again, costs escalated and valuable time was lost.

To overcome these difficulties, and create more cohesion, a matrix structure was introduced. The product planning function was absorbed into four new programme offices, one each for small, medium and large cars, and another for power trains (engines and transmission).

Reporting direct to the vice-president of product development, like the heads of design and engineering, the four programme directors were given clear responsibility for the coordination of development. In addition to their own direct staff of product planners and engineers, they assumed dotted-line control for the designers and engineers, though both groups still report direct to their respective vice-presidents.

Rather than indicating any lessening in the influence of design, the change was welcomed by industrial design as a necessary streamlining of the development process, with design now very definitely accepted as an equal partner. There was more resistance from some of the engineering fiefdoms, however.

A Market Research Revolution

Just like any other executive, the actual muscle of a design chief depends heavily on the informal power structure, and on his or her own powers of persuasion. A measure of the muscle which Uwe Bahnsen managed to amass soon after his appointment in 1976 is provided by the dramatic changes in market research which he advocated, and which first were applied during the later stages of the Escort's development.

In common with hallowed American practice, Ford of Europe had always placed heavy reliance on product clinics – sessions at which competitors' current models were disguised and displayed alongside existing and potential new Ford vehicles, in order to be assessed by a sample of potential customers. The standard motor industry approach had always been to ask the participants to rate all sorts of details of the various models against each other, right down to which had the best bumpers, rear lights, arm rests and so on, as well as the most effective overall design.

Not surprisingly, the participants – whether dealers, fleet operators or drivers – tended to judge everything against the features of current vehicles. The result, as in so many narrow market research exercises in other industries, was support for only slight design improvements to the previous model, and resistance to any dramatic change.

These tests used to be accorded an almost religious significance, and taken as a precise guide to the way a vehicle should be designed. As Bahnsen complained, 'We asked them for detailed reactions in a most ridiculous way, and then used the results as a substitute for decision making.'

Starting with the last of the three clinics held during the Escort programme, Ford went over to a more conceptual approach. The questions asked in the clinics were focused more on people's general perceptions of what a vehicle should look like four or five years into the future. Various methods were used to try to get the participants to forget about what they had just been driving. One line of approach was to precede the clinic session with a so-called conditioning group, in which the respondents were given an idea of how trends were moving in other areas of design, such as architecture, fashion and office equipment.

As subsequent experience has shown, this helps adjust the participants' eyes and minds to the fact that some of the vehicles they are about to see belong to the future, not the present. The subsequent list of questions concentrates more on their overall perceptions of the designs, though there are still some detailed inquiries about particular design features.

Most important of all, the research results are no longer viewed as the Holy Grail by Ford's marketing staff and programme directors. The company has dispensed with its dangerous faith in the overwhelming power of market research, and has learned the true nature of marketing and design-led strategy.

Conclusions

The recent history of Ford shows how design can have a powerful influence on overall company performance and profitability. A fundamental lesson of the Ford experience is that design must be integrated into the product development process right from the beginning. It must not be seen as something which is secondary to engineering and production considerations.

The case study also shows that critical organizational barriers may need to be overcome to achieve this recognition of design. Thus companies intending to increase their emphasis on design must be prepared to engage in organizational reform and to challenge the conventional beliefs of managers.

16 Managers and Designers: Two Tribes at War?

DAVID WALKER
The Open University, UK

Introduction

There are two main views of how design should be managed. On the one hand design is seen as special, even unique – and by that token it presents special difficulties for managers. On the other hand design is seen as an element of business – to be managed like anything else:

It is important to argue that design must be managed and that design can be managed. There is considerable misunderstanding on both points. Some managers believe that design is something outside normal business practice and does not benefit from being managed but due to creativity and other uncertainties is regrettably unmanageable. In fact design has to be managed just as much as anything else and the uncertainties that are involved are no more serious or disruptive than the uncertainties inherent in any other task within industry that has to be managed, for example, commissioning a new factory or exploiting a new market. (Geoffrey Constable, Head of Industrial Division, Design Council: letter to the author, 17 March 1987)

It is reasonable, of course, to argue that design as an activity and designers as professionals should operate within normal managerial constraints, within budgets, to time targets, within the limits of known resources – personnel, facilities, equipment and so on. But we all know that design by tradition woefully, and sometimes wilfully, fails to be constrained. This is not merely a problem of designers who have creatively invented new processes while devising new products, but more that there are some fundamental differences between managers and designers that we should explore.

Different Outlooks

These differences are below the level of organizational structures; they start with personal aptitudes and interpersonal skills. You might think, at the outset, that the division is simply between pushers and pullers: that is, on the one hand the designers enmeshed in the process of development, and on the other the managers with their better consciousness of demand, user needs and the context of design. But is it that simple? Are there not deep anxieties and even antagonisms between management and design?

It is still possible to witness a clash of cultures when managers and designers confront each other. The difficulties are like those between two tribes who have grown up in different parts of the jungle (to take an anthropological view). Language, dress, signs, sudden movements and underlying intentions are all open to misinterpretation or, at worst, blank mutual incomprehension. Only a few managers reach the confident understanding of Stuart Turner, Managing Director of Plasplugs: 'Many managers are frightened by creativity – as if you asked them to write a song – and they say, oh no that's too difficult . . . but someone writes the words, and someone writes the melody.' If a song-writing partnership is the appropriate analogy, where the manager writes the words and the designer provides the melody, why do many managers seem to be uneasy about design? The reasons could be, as Turner says, that design involves uncertainties to do with creativity and visual judgements. He also comments that design represents an extra expenditure that perhaps looks like extravagance. Several observers have characterized the problem as one of miscommunication: 'Designers think managers have their heads in the sand, while managers think designers have their heads in the clouds' (Daniel Englander, formerly of the Independent Designers Federation).

Designers themselves can be perceived as eccentric in manner and dress and, by implication, difficult to manage. Another commentator has described the personal antipathy as 'the suspicion of men in suits for those in yellow trousers'. Sartorial contempt aside, beneath this surface there are fundamental differences. The lack of understanding between designers and managers, their difficulties of communication and their personal suspicion, have their roots in different aims, different education and different styles of thought. Let us think about each in turn.

Different Aims

To most managers, design is just another resource. If used properly, it can add value to products and services. Just like money spent wisely on sales literature or advertising, or the streamlining of office systems, an investment in design can produce good returns and contribute to improving profitability and effectiveness. So from the outside it looks like any other component of the organization, something to be managed, and not treated particularly as a special case.

In contrast it looks quite different from the inside. Designers may resist attempts to quantify their efforts in financial and organizational terms. They stress other priorities in their work such as keeping up with their peer group of designers, details of the products, wanting to improve the environment or

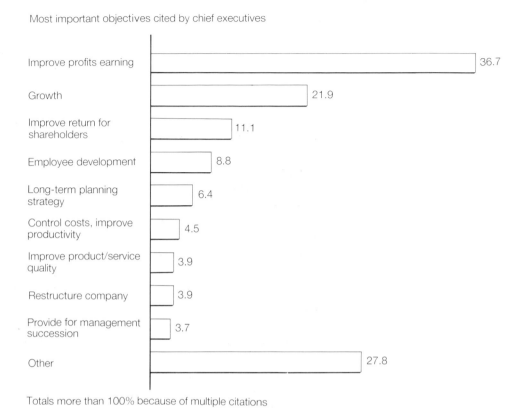

Most important objectives cited by chief executives

Improve profits earning	36.7
Growth	21.9
Improve return for shareholders	11.1
Employee development	8.8
Long-term planning strategy	6.4
Control costs, improve productivity	4.5
Improve product/service quality	3.9
Restructure company	3.9
Provide for management succession	3.7
Other	27.8

Totals more than 100% because of multiple citations

Figure 16.1 Priorities of managers as revealed by *Fortune* top 500 survey of April 1986 (percentages) *Source*: Walker (1989) and *Fortune* magazine (April 1986)

striving to elevate public taste. While managers might acknowledge these aims, a typical survey of US companies (figure 16.1) shows where the interests of top managers really lie – in profits, in growth, in long term planning, in employee development, in shareholders and so on. Design is not mentioned at all specifically, and even quality of products and services (which can be taken to subsume design) is given a fairly low priority.

On the other hand, designers are very preoccupied with novelty and the detail of products and services. Some of that commitment is to do with tangible qualities. Some of it is to do with career building. Designers can usually walk away from an organization with their portfolio of drawings and photographs. The opinions of their next employer, and their peer group, may turn out to be more important to them than long term loyalty to an organization. The careers of designers are constructed on different criteria from the careers of managers.

There is also a split in social abilities. Managers have developed both personnel and personal skills, because a large part of their job is keeping people happy – their peers, their subordinates, their bosses. Designers have developed technical abilities which concentrate on problems to do with things – artefacts, products, models, drawings and prototypes. Does this suggest that designers, typically, are bad at dealing with people? Perhaps it does. A recent survey on engineering designers reports:

A number of employers said that designers need to develop their communication skills and their ability to work effectively in project teams. One employer stated that some designers were unable to express opinions in a constructive way, and that their inability to communicate with production engineers had lost revenue for the company. (Coopers & Lybrand 1988)

We should not draw too wide a conclusion from the behaviour of one or two surly engineers. However, the natural medium of designers is not words but visual communication, and perhaps they are happier dealing with things rather than people. So, to put it in terms of caricature: the manager is interested in people, in profits, planning and growth, and the long term durability of the organization; the designer is interested in things, in experiment and the impact of products and services in the shorter term.

Different Education

Many managers experience great difficulty and unease when making judgements about 'things':

I perfectly understand the importance of design. But that is not to say that my education, experience or natural instincts lead me with confidence to produce it myself, or back others to produce it on my behalf. I believe that a large number of managers have exactly the same problem. (Sir Christopher Hogg, Chairman of Courtaulds, speaking at the SIAD Design Management Seminar, 1985)

This uncertainty can be traced to an educational gap. The education and training of managers tend to be based on analytical studies (such as accountancy and finance); therefore they are not very well equipped to deal with projects which involve unfamiliar concepts, predominantly visual (rather than written) information, fuzzy problems, high levels of ambiguity, and assessments which are, variously, subjective, personal, emotional and outside quantification. Without making design sound mystical, as if it relied solely on intuition and higher non-rational insights, it does have the quality of many complex human skills in resisting analysis.

Is this so surprising? Consider another kind of human skill like playing cricket or golf. The rule books and the routines that can be studied are only useful as a very crude delimitation of territory. The skill itself is locked into the activity and can only be really understood when felt, when experienced through constant rehearsal. This is also the case with design skills.

The management unease about design does seem to be related to a traditional cultural divide in UK education. Numeracy and literacy are rated highly, while the abilities which revolve around materials – manipulation, construction, tactile skills and visual literacy – do not even have a name. (Thankfully two suggestions, the horrors 'viseracy' and 'graphicacy', have never been given wide currency; Edward de Bono has coined the marginally better word 'operacy'.)

This cultural division runs through higher education into industry and is visible in many companies. Even some which depend on innovative, well-designed products for survival are managed, not by the visually literate, but by the highly numerate. In some industries this has led to a counter-reaction against this kind of cold neutrality, against managers who claim to be able to manage anything from cans of beans to motor cars. Accountants and systems analysts have given way, in some areas, to managers who are deeply involved with the products they manage. Hence there is a resurgence of the concept of product championship.

So these different aims and attitudes, fostered at all levels of education, have their effect on industry when students move into their jobs in the world beyond school. Mutual suspicion is rife:

Everyone knows that the trouble with UK industry is that it doesn't make any use of design – everyone that is except those who know that the trouble with UK industry is that industrial designers are of no use to practical businessmen. The upshot is that we have all the makings of one of those institutions so beloved of the British (what someone once called the sound of vintage British whine), the essence of which is that everything would be fine if it wasn't for the other lot. (Gunz 1986)

Pioneers such as Hugh Gunz and Peter Gorb in the UK, and Colin Clipson in the USA, have initiated projects in higher education which have brought design and business students together to work on real projects. These have had an uneven response from the students. Typically the coalition was felt to be valuable but was too little too late. The professional divisions and self-images were already deeply entrenched. Even a widespread campaign of educational reform following these pioneering efforts might not entirely overcome all the difficulties of bridging the gap. What if there are differences not only in aims and in education but also in intrinsic personality traits which engender different modes of thinking?

Different Styles of Thought

Do not managers by temperament like things to be orderly and precise? Is this not what we mean when we talk of someone being businesslike? Are not designers inclined to go off at a tangent and experiment with oblique ideas and unconventional procedures? These two thinking styles have been labelled 'serialistic' and 'holistic' by Gordon Pask, and the distinction owes something to a philosophical division between rationalists and empiricists, and to the established traditions of science and art (or at least popular versions of them; maybe exploratory science and genuine art have more in common than we think!).

A serialist, like a caricature scientist, prefers to work by proceeding in logical small steps, and tries to grasp every point before moving on. A holist, like a caricature artist, proceeds by picking up disconnected pieces of information and by approaching ideas from different viewpoints. This is not to say that one is better than the other. For example a serialist, when faced with a new urban scene, say as a tourist in a complex Italian town, would study the map and go by the most direct route to the recommended places of interest. A holist in the same situation would meander about, stop in a café, miss the cathedral, but find many interesting corners away from the main routes.

Over time the serialist would build up an exact structured knowledge of the town and do so very efficiently, even discovering errors in the map. The holist, given time, would similarly come to a comprehensive understanding of the whole town, which may lack structure at first but would be rich in vivid and significant details. It is unwise to categorize individuals exclusively as one or the other. How people operate depends on the task and on the circumstances. Yet it could be argued that managers tend to be serialists and designers tend to be holists. Let us take the argument a stage further.

Edward de Bono is well known for fostering creativity or, rather, what he terms lateral thinking. The popularity of de Bono's work to managers, at weekend courses and conferences, suggests that he has detected some widely recognized deficiency. Lateral thinking means a willingness to suspend judgement on what is conventionally regarded as true or factual, a willingness to approach problems sideways or in a spirit of provocation ('car wheels should be square'). This style of thought is counterposed against the anxious, severe, judgemental process of adult thinking. The mind has to be shocked into new perceptions and new territories, by chance events, humour or seeming contradictions.

De Bono has recognized that designers as a professional group are inclined to experiment, to seek novelty and to think laterally. Most managers are not. So once again we have a polarity. The work of Michael Kirton, a psychologist with a research interest in management, helps us consolidate this view. His studies, far from saying that people can be divided into creative and

Table 16.1   Adaptors and innovators

The adaptor	The innovator
Characterized by precision, reliability, efficiency, method, prudence, discipline, conformity	Seen as undisciplined, thinking tangentially, approaching tasks from unsuspected angles
Concerned with resolving residual problems thrown up by current paradigm	Could be said to search for problems and alternative avenues of solution, cutting across current paradigms
Seeks solutions to problems in tried and understood ways	Queries problems' concomitant assumptions: manipulates problems
Reduces problems by improvement and greater efficiency, with maximum of continuity and stability	Is catalyst to settled groups, irreverent of their consensual views; seen as abrasive, creating dissonance
Seen as sound, conforming, safe, dependable	Seen as unsound, impractical; often shocks others
Liable to make goals of means	In pursuit of goals, treats accepted means with little regard
Seems impervious to boredom, able to maintain high accuracy in long spells of detailed work	Capable of detailed routine (system maintenance) work for only short bursts
Is an authority within given structures	Tends to take control in unstructured situations
Challenges rules rarely, cautiously, when assured of strong support	Often challenges rules, has little respect for past custom
Tends to high self-doubt; reacts to criticism by closer outward conformity; vulnerable to social pressure and authority; compliant	Appears to have low self-doubt when generating ideas, not needing consensus to maintain certitude in face of opposition
Is essential to the functioning of the institution all the time, but occasionally needs to be 'dug out' of the system	In the institution is ideal in unscheduled crises, or better still to help to avoid them, if he or she can be controlled
When collaborating with innovators	*When collaborating with adaptors*
Supplies stability, order and continuity to the partnership	Supplies the task orientations, the break with the past and accepted theory
Is sensitive to people, maintains group cohesion and cooperation	Appears insensitive to people, often threatens group cohesion and cooperation
Provides a safe base for the innovator's riskier operations	Provides the dynamics to bring about periodic radical change, without which institutions tend to ossify

Source: Kirton (1980)

uncreative groupings, show that their creativity, including a capacity for problem solving and ability to make decisions, operates in two main characteristic ways. He labels the two groups 'adaptors' and 'innovators' (table 16.1).

He also argues that both types are necessary to organizations. Adaptors are at home in the smooth efficient operation of an existing system, creatively refining, improving and extending the thinking that underlies it. Without them the institution will soon fare badly. However, although adaptors possess these essential qualities, they should not be the only ones to be rewarded by promotion. Despite the fact that innovators have weaknesses – they are erratic, insensitive to the needs of others, impatient and inveterate risk takers, all weaknesses which are potentially dangerous to an organization – they too also have other essential qualities (Kirton 1980).

This, then, is an issue not about creativity but more about a split in basic aptitudes. Because of the visual techniques designers habitually use, they are good at manipulating models or analogues, and at depicting things in a concrete detailed way. Managers are more comfortable with general rules and abstractions. Managers have a critical, analytical frame of reference, whereas designers stick things together to see what results; they are better at synthesis than analysis.

Managers are good at dissection, cutting through irrelevancies, getting to hard facts and the basic structure of problems. They are very problem oriented. The whole of management indeed is sometimes portrayed as a series of problems to be solved. Designers by contrast are good at assembling, bringing unlikely things together. They work by leaping to detailed end results. They are solution led. This means that designers see opportunities and managers see difficulties. Designers can be wildly optimistic and therefore rash. Managers can be pessimistic and therefore inert.

Conclusions

The divergence between managers and designers can be detected in personality traits, in habits of thought and work, as well as in educational background. The familiar media of communication are split between verbal and visual languages; thinking styles are linear or lateral, convergent or divergent; personality traits are towards adaptation or innovation; decision making divides between analysis of problems and solution-led leaps. Extending Kirton, a list of broader polarities can be drawn up as in table 16.2.

Table 16.2 Manager–designer polarities

Characteristics	Managers	Designers
Aims	Long term	Short term
	Profits/returns	Product/service quality
	Survival	Reform
	Growth	Prestige
	Organizational durability	Career building
Focus	People	Things
	Systems	Environments
Education	Accountancy	Crafts
	Engineering	Art
	Verbal	Visual
	Numerical	Geometric
Thinking styles	Serialist	Holist
	Linear	Lateral
	Analysis	Synthesis
	Problem oriented	Solution led
Behaviour	Pessimistic	Optimistic
	Adaptive	Innovative
Culture	Conformity	Diversity
	Cautious	Experimental

The table might seem to present a divisive and stereotyped picture. It may indeed seem as if the two tribes are at war. If managers and designers have different aims, different educations, different ways of communicating and, to cap it all, have different styles of thought, how can they possibly deal with each other? Table 16.2, however, represents two extremes and is not a description of all managers and designers.

Many designers, naturally enough, would resent a portrait of them as product obsessive, short term careerists who, moreover, are insensitive to the needs of others. Similarly many managers would not recognize themselves as convergent, cautious thinkers, the dull plodders who form the backbone of the organization. Of course, managers can be found who are highly innovative, lateral thinkers – just as there are designers who are systematic and conformist. But this writer would still argue there *are* two cultures, derived from different educations and different personal abilities. Whatever we might want, traditionally the two camps do deal with each other with great difficulty.

More than that, it seems unlikely that one person can span the full range of aptitudes. Therefore, if it is unrealistic to expect all of these characteristics in one person, it may be better if a working partnership is formed between two or more people whose skills are synergistic; whether we call them managers or designers (or adaptors and innovators) is not so important as having the right mix of skills.

And what about the question posed at the beginning of this chapter: is design uniquely eccentric or something that should conform to conventional organizational practice? Is it possible that both views could be right? From a managerial perspective, from a little distance off, design looks like anything else: it employs skilled people, it consumes resources, it has to fit in with other parts of the organization. So if designers see themselves as constraint-free, somehow outside the limits of the organization, then they are in some fundamental sense 'bad' designers, wasting most of their time and that of their clients.

But the closer one gets to the design process, the more striking are its special characteristics: the divergent patterns of thoughts, the willingness to experiment, the leaps to concrete particulars, the visual means of conceptualization and communication. If managers of design do not understand these unique particularities, and not merely make allowance for them but value them and nurture them, then there is no chance that they will use designers and design techniques properly.

The very differences in perception, aptitudes, processes and skills are what make managers and designers important to one another at both a personal and an organizational level. They need each other. The one can scarcely function without the other.

References

Coopers & Lybrand: *Fit to Design? The Training Needs of Engineering Designers.* Report to the Design Council and the DTI, 1988.

Gunz, H.: Managers and designers: building bridges by accident. *Management Education and Development.* Vol. 17, no. 4, 1986.

Kirton, M. J.: The way people approach problems. *Planned Innovation.* Vol. 3, 1980.

Walker, D. J. et al.: *Overview Issues*, P791 Managing Design, Open University Press, 1989.

17 Design: Organization and Measurement

KEITH GARDINER
Lehigh University, USA

Introduction

There are many tomes addressed to architects, engineers, industrial designers and others engaged in design activity; the literature abounds with discipline-specific ideas which 'exactly' define design. However, these are not persuasive; prestigious studies, task forces and learned workshops continue to question exactly what is the nature of the design process. The more abstract conceptual questions relating to the design activity have been posed most cogently by Rabins et al. (1986). Notwithstanding the many tables, organization boxes and linked sequential diagrams published, it would appear that design has the characteristics of a true art and is not readily amenable to thorough scientific study, analysis and then successful synthesis and replication.

This chapter discusses design activity as practised by industrial enterprises with the ultimate objective of generating revenues from the sale of successful products. Several examples will be explored and methods for the implementation of successful design procedures will be developed. Perhaps Victor Papanek (1984) has it right when he says that 'We are all designers.' How, where and when you read this; what you happen to be wearing; what you do next: these are all choices made within some defined bounds which produce a change of condition. You, yourself, change an existing condition to a preferred condition. Design can be regarded as the act of choosing from amongst options.

Selecting a route to solve a transportation problem provides a very fruitful analogy. If we wish to travel from London to New York there are immediate questions with regard to resources available, time requirements, comfort, cost, amount of luggage and other considerations. We undertake a somewhat ordered sorting to design our solution, we develop or impose a suitable organization and a concomitant measurements system. Likewise, in industry, design cannot occur without an organizational and measurement framework which is compatible with the needs of the total system.

The Environment

Design considered as an isolated independent activity is nothing less than masturbation; design must be implemented to have any meaning. The means of implementation are an unavoidably integral part of the design process, as is the associated organization. In the context of the production of goods and services, the words 'design' and 'manufacturing' can be regarded as being synonymous or interchangeable. A definition of manufacturing which is useful in developing ideas of manufacturing systems states: 'Manufacturing is that activity by which materials and information are transformed into goods or services for the satisfaction of human needs ... [furthermore] it is the purpose of manufacturing systems to generate wealth' (Gardiner 1984).

Today, design and manufacturing are seen as highly important activities. There is renewed emphasis; perhaps at no time since the late 1880s has so much attention been concentrated upon these areas of human endeavour. Bankers, bureaucrats, business executives, economists, journalists and politicians are all pressing for design improvements and higher productivity. This attention raises many issues – and a seemingly endless barrage of popular and fashionable clichés. The media hype advertises 'design for assembly', 'design for manufacturability' and other slogans of questionable syntax. In reality, all that is being sought is plain simple good design and its implementation.

The computer is a predominant new feature in all this churning with calls for lower costs, higher output and better quality. The computer and its acolyte microprocessors can, it is promised, bring us the industrial or manufacturing renaissance that all economies presently seek. All an enterprise needs to do in order to survive and be prosperous is to install computer aided design or manufacturing (CAD/CAM), computer integrated manufacturing (CIM) and flexible manufacturing systems (FMS), together with the application of automation and robotics. All these, when tied together successfully, will comprise a well-structured integrated manufacturing system (IMS). Amidst these debates, only belated attention is being paid to the fact that the essential requirement for the survival, prosperity and growth of any enterprise is the provision of goods, or services, which meet the needs of the customers and, beyond this, generate customer satisfaction and additional desires.

The organization of the sequence from idea to design to implementation, and then to customer satisfaction, is a matter of paramount importance to the success of the design. In fact, it can be argued that even so-called 'good' designs can fail if the organization is not correctly configured for their prosecution and promotion; the organization must develop and grow to service the needs of the design/implementation activity. The management structure must also develop to cater for the accompanying financial, scheduling, volume, distribution and marketing considerations – which must be regarded and included as aspects of the whole design strategy.

Any existing organization structure which commences searching for designs to ensure its own survival is doomed to a painful although perhaps lengthy demise. Mature organizations, like bureaucracies, do best that which they have always done; change, discovery and novelty are anathema. Design activities and the ultimate delivery of customer satisfaction must drive and integrate every facet of the organization. Designs must transcend structure, because an established structure inhibits innovation. The creation, nurture and implementation of new ideas requires an organization structure which encourages change, meets the needs of each design and also promotes a satisfactory relationship with the customer.

Measurement of design success is an extremely important parameter, and is indicated ultimately by degree of customer/user satisfaction. Revenue and the prosperity of the designing/producing entity are manifestations of design success. However, often the impedance of this measurement cycle is too great for satisfactory control or management of ongoing projects. Some alternative ways of developing organizations and measuring their performance will be discussed.

Design Requirements

Today, the requirements placed upon the design activity are for the delivery of a continuing stream of ideas which facilitate profitable successful products. Cycle time from concept to customer satisfaction must be held to an absolute minimum, because once there is a concept the cost clock accelerates inexorably. Many organizations with successful histories have failed to recognize these requirements; their designs are restricted to following formats and styles preordained by existing structures and measurements. In very large organizations with high intrinsic inertia, long pipelines or product cycles, and traditional customers, it may be possible to continue to satisfy the stockholders for many years with such procedures. However, the organizations which will grow and prosper into the twenty-first century are likely to demonstrate design-driven organic structures with an orchestral or team nature.

These structures must evolve from the needs of the designs and the customers. New holistic or team measurement systems should be developed which will serve to promote and integrate the objectives of the whole enterprise. The emphasis must be upon collaboration and output, somewhat equivalent to overall artistic, musical or sporting impact, rather than upon measuring individual or departmental excellence. In such an environment, there may be many people with many specialized skills and many people with broad integrative abilities. They must all be considered as people involved with the project; their activities must not be separated by titles which restrict and limit, such as development, manufacturing, quality, process, product, reliability, service or systems. All employees must be accountable for and measured by their contributions to some portion of the whole cycle from concept to customer satisfaction. Organization and measurement schemes developed on this basis should ensure the success of the enterprise.

These 'good' products which generate customer satisfaction must be produced both economically and expeditiously. They must be available to the marketplace in a timely manner at some appropriate rate to avoid dissatisfied potential customers. In the final analysis, products must also be safe, user-friendly and reliable; they must perform, at least, to customer expectations, and must meet all advertising claims. The overall quality must be adequate to impress the customer with a notion of value obtained. Competition ensures heavy pressure to meet all of these requirements; for most products today there is only slender pricing elasticity, so it is necessary that costs are tightly managed to ensure an adequate return on resources invested.

Hence the success, survival and future of the enterprise rest upon the ability to deliver satisfactory products. A heavy responsibility falls upon the innovators, customarily considered separately and often housed in some ivory tower design department. A substantial responsibility also falls upon the personnel or group traditionally labelled as manufacturing and involved with implementing the design. Successful design and manufacturing require the seamless integration of these groups and the establishment of collaborative relationships throughout the whole reconfigured organization. Former adversarial and hierarchical structures must be replaced.

The Implementation System

A successfully integrated manufacturing system exhibits an organizational continuum from design, through the manufacturing processes, out to the marketplace and the generation of ultimate customer satisfaction. The design activity cannot exist independently; it must be integrated within the whole system operation. The system should not exhibit spikes or disruptions in any of the output; it should consume minimum energy and material. Overall it will conserve entropy, and this alone could be a sound theoretical measure of system and design effectiveness. A well-integrated, smooth-flowing manufacturing system will conform ideally to the needs of the society in which it is created; if there is incompatibility there will be inefficiencies. Hence, the whole system and its *raison d'être* should be viewed in both a thermodynamic and a historic context (Nierynck 1984).

To achieve this integration, it is essential to eliminate organizational barriers and departmentalized behavioural and thought patterns. In a well-integrated enterprise there should be flexibility, responsiveness and operational smoothness. Such arrangements can be greatly facilitated by the use of appropriate electronic communication and data networks which make information pervasively available throughout the enterprise (Gardiner and Olden 1985). This rapid information flow changes the traditional tasks of management and can render archaic the customary multilayered hierarchical structures.

Design Successes

Kidder (1981) gives an excellent account of the trials and tribulations that accompanied the development of a new range of Data General computing systems. The lesson which can be extracted from this case is one of design excellence notwithstanding the existing organization. Essentially, the designers/developers of the new system created their own organization and operated as a pirate group somewhat sheltered by their immediate management. They had an implicit measurement system for their whole team which was united in a grand challenge to achieve better designs and system performance than the primary design group located elsewhere which had the official mission.

The pirate group was probably successful because it was very constrained for resources and time, was under severe competitive pressures to demonstrate its worth to the enterprise, and was compelled to stretch intellectually. The problems themselves offered sufficient challenge and group members were striving to demonstrate the stupidity of the enterprise which had neglected them. The group dynamic could not have been stronger and there was good leadership. Mismanagement by the old organization created an environment, by omission, which nucleated a powerfully motivated new team. This success did not continue when the team was recognized and reabsorbed into the primary organization structure.

The cases of the Commodore 64 (Perry and Wallich 1985) and the Apple Macintosh (Guterl 1984) also afford some useful parallels. The Commodore team were part of a rather small operation, vertically integrated and with excellent communications. The designers were able to range through the factory, fabricate prototype circuits and chips, and then discuss, redesign and improve. This showed in the additional attributes of the 64 when it was compared with competing offerings. Unfortunately, growth of both plans and volumes necessitated embedding the design team within a larger organization and there were difficulties in subsequent development.

The Macintosh has been discussed extensively; it has been around a long time, originating as a concept in 1979 and then suffering much modification and several management perturbations. The essential creative thrusts derived from the work of a small, virtually autonomous group which supplied all its own expertise. The group was forcefully integrated by the sharing of inadequate facilities over a filling station, nicknamed Texaco Towers.

Although the final prototype was available in late 1982, incorporation into the Apple mainstream for volume manufacturing experienced some difficulties and normal production did not start until January 1984. Nevertheless, the design is proving very robust and is a mainstay of the current Apple offerings. It displays many of the attributes which were being sought by the Texaco Towers group, and shows the virtue of design by teams which are isolated from structured organizations: they are compelled to develop the relationships and measurements that they feel are needed for the success of the design activity.

The automobile industry surrounds us with visions of design aimed at seducing the cash from our pockets. The advertising, particularly in the US, emphasizes shape, form, finish and durability; but design also involves the unseen engineering from which develops performance, economy, comfort, reliability, safety and ultimate customer satisfaction. Ford have achieved a major turnround and are now challenging General Motors in volume for certain segments of the American marketplace for the first time since 1959. This improvement is due to emphasis on design from the very highest management levels, thereby ensuring a degree of organizational compatibility. Additionally, international design teams were created from individuals with wide-ranging educational backgrounds comprising all the necessary disciplines; the result is a range of vehicles with characteristics genuinely suitable for the global marketplace. This has been described in some detail by Cortes-Comerer (1987) and is mentioned in a useful history of Ford by Lacey (1986) (see also chapter 15 by Lorenz in this book).

Another example of successful organizational manipulation which supposedly permits the more effective prosecution of new designs has been labelled early manufacturing involvement (EMI). In this system the traditional organization structure which manages development and manufacturing as separate entities is retained, but people from the manufacturing group are assigned to assist with the development activity (Cortes-Comerer 1987). Thereafter, these manufacturing personnel carry the project forward to completion. This system, as with most others, is very dependent upon the personalities and abilities of the personnel assigned across functional lines.

Sometimes it works excellently, but in other cases the transfer lapses into a bureaucratic game played out between the functions; even in the better examples that have been cited it would appear to be a type of 'band-aid' solution applied to organizations which are change resistant. An alternative methodology builds a design team which is augmented with the necessary skills as the project advances. The key people in this team stay with the project right out to the marketplace; eventually they are available for reassignment for the inception of another product cycle. This leap-frog system has many advantages in transferring experience from person to person and project to project, but it does have the dangers that result from the separate existences and outlooks of the different communities that the project passes through.

Design Methodology

Design should be an activity which occurs in an unbounded environment with the objective of solving a well-defined problem. The only boundaries to the creativity and imagination of the designers should be provided by the problem, or the proposed product itself, and not by the structure of the organization. In the ideal case, an organization should evolve to accommodate the requirements and technologies which develop as a result of the design cycle.

Ideally, the design activity should commence with a clean sheet. There should be no encumbrances, preconceptions or restrictions which may inhibit the initial discovery or innovation phase. A clear and unambiguous understanding of the problem to be addressed by the design must be developed, and from this a comprehensive set of functional requirements devised (Suh and Rinderle 1982). These should be assigned relative priorities. It is preferable for this activity to proceed almost without structure because the imposition of structure, or format, inhibits creativity and tends to pre-ordain results.

After some explorations of various design concepts it will usually be profitable to rethink the initial activity. This can add more precision and understanding to the problem definition, and also adjust ideas of the functional requirements. Only after this open minded and unstructured sequence should the notion of constraints be permitted to arise. Constraints, and discussion of them, are always associated with negative thought trends which are destructive of the creative process. Some of the constraints may have been dealt with as functional requirements, any distinction being somewhat arbitrary. Functional requirements can be viewed as desirable objectives for the design to accomplish, whereas constraints can be construed as unavoidable imperatives resulting from the environment, or from the goals, resources or limitations of the enterprise.

Format, organization and structure are inimical to creativity, innovation and the design activity. This is borne out by the studies done by Root-Bernstein (1984) into the mode of problem solving apparently favoured by some of the most renowned scientists. However, design/manufacturing cannot be planned to occur spontaneously whenever required, so some formalization and structure is necessary. It may be a myth fostered by iconoclastic, eccentric and innovative designers that good design is innovation and cannot be measured or managed. Notwithstanding these ideas, there is measurement by the eventual customer and, even before this, by the engineers associated with implementation.

It is nevertheless clear that large, mature or traditional organizational entities severely inhibit innovation and the emergence of superior new design. Enterprises which foster better designs learn continually from the competition and often employ semi-autonomous groups, almost as pirate ventures outside the existing organization. These groups comprise the essential mix of creative, implementing and marketing skills and, whilst maintaining informal connections with the traditional establishment, customary discipline divisions and measurements are modified substantially, or even suspended. In many organizations some of the fiercest threats to continuance of existing products arise from conflicting internal design projects. Internal competition may seem wasteful of personnel resources, but if the designers of an erstwhile world-beating product are not encouraged to challenge the system then the output of the enterprise will lose any innovative edge and ultimately market share.

Despite the apparently high success rate for anarchic design activities, a methodology must be identified for accomplishing the design and thereon ensuring its successful implementation. The methodology in the final analysis becomes a part of the design and manufacturing process, and affords a constraint which must be considered and either delimited or evaded. It is an obvious requirement that a design must be capable of being reproduced with adequate consistency in the necessary volumes using the specified capital assets and consumable resources. The latter considerations are of prime importance to the success of the whole product programme and, indeed, the viability of the sponsoring enterprise. If the tooling does not produce satisfactory quality, volume, rate or sufficient variety with the use of planned manpower, materials and other resources, then the critical costing factors rapidly go awry and market projections can shift with disastrous consequence.

The use of computer assisted design or drafting systems (CAD) can be a great advantage within an existing enterprise with a relatively stable product family. These systems greatly facilitate implementation of group technology (GT) techniques for design and manufacturing process commonality and parts rationalization. The use and creation of historic data bases in this manner has value for future learning, cost reduction, design release communication, and implementation of process ground rules. However, the continuance of established proven materials, methods and techniques is ensured by the structure and format imposed by the software and the data bases – and equally so by the equipment, experience and traditions enshrined within the existing factories and management teams. This can significantly inhibit novel developments and innovation. The issue of data base constraint to imagination has been considered by Pugh (1985).

In summary, it must be realized that there can be no rules which axiomatically ensure good design. It is important to recognize early in any prototype phase that good designs engender appreciable serendipity when they enter manufacturing. It is prudent to avoid 'band-aid' changes to alleviate problems because poor design will cause later plagues in terms of higher costs, customer returns or other problems of types which cannot be forecast. The best designs are readily implemented but, ultimately, can only be recognized by the tests of time and many customers.

It is essential that all aspects of implementation be considered from the start. Capability for manufacturing flexibility to accommodate future design ingenuities and growth in product complexity and/or volumes must also be a factor. It is important, however, not to drive heedlessly for objectives like automation; the requirements for quality, throughput and volume must be balanced against sound analysis of capital and resource utilization. This leads to the final cost figure which is tied to ultimate customer satisfaction with the product performance. The responsibility of the designer covers this whole cycle. A measurement of success is the ease or transparency of the design through the transformation processes.

Measurements and Utility

These concepts of a methodology for design, based upon a responsive measurement system and relying upon a structured set of functional requirements, have equal relevance not only for products and manufacturing systems but also for facilities, organization structures, educational activities and other human endeavours – all of which represent problem sets which can benefit from systematic analysis. It could be argued that no objective measures of design excellence are possible; the area is too subjective. However, we have few difficulties as a society in measuring the beauty, elegance and competitive virtues of Olympic divers, ice skaters and ski jumpers. Thus there is little to prevent changing the traditional practices of measuring output of individual employees and their departments and, instead, measuring the excellence of collaboratively generated design ideas and details. Once the problem is thoroughly defined and understood, the development of appropriate measures or indicators of the likely success of any solution is a small extra procedure.

With regard to implementation, systems-oriented measures of performance are required for effective decision making and planning. In the past, measurements for manufacturing have been based upon traditional accounting and industrial engineering principles. These practices were developed when single-digit interest rates were common and the capital costs of tooling were at least an order of magnitude less than today. Additionally, there was a more leisurely pace in the whole marketplace. In the former environment there was more latitude, greater system elasticity, and inertia; development cycles could be longer and product life cycles were correspondingly prolonged.

Now, there are competitive pressures for shorter development periods with very rapid introduction of new products. These requirements are in conflict with the increasing complexity and process sensitivity of the newer product families and their essential components. Huge investment is required, and there is a very long approval, construction and commissioning cycle before any new advanced technology manufacturing facility can be brought on stream. The cost of these investments, together with the resource development requirements including personnel, militates against rapid transitions of major design parameters and product types. New accounting procedures are essential in order to make true assessments of worth which can aid resolution of the many design, process and tooling options; Leontief's (1987) input–output studies could be valuable in this respect.

For the satisfactory implementation of the design, the enterprise may be better served by planning to utilize a new facility with new equipment and fresh personnel starting from scratch on a vacant lot. This may be preferable to being compelled to design both the product and the manufacturing system to suit existing facilities and workforces. However, even if the latter is a requirement, it should be handled either as a low priority functional requirement or as a constraint for consideration only after first-pass conceptual design studies. Additionally it is essential, if existing resources are to be utilized, that realistic financial, productivity or efficiency measurements are imposed as a basis for resolving options.

Clearly if the enterprise owns foundries, say, and has a workforce with parallel skills, it may make little sense to develop a new product family which employs injection moulded plastics. However, this option must be explored in any initial design cycle and eliminated by later application of the constraints or by priorities accorded to the functional requirements. If this procedure is not followed with some rigour a wholly unsuitable product may be designed. This will result in a temporarily loaded, well-balanced production facility – but one which may be producing items unsuited to satisfying the latest marketplace fashion and pricing regimes. Indeed, going out of business may be a preferable course of action if consistent with the long range objectives and priorities defined by the management or stockholders of the enterprise. Rigorous comparative cost and performance analyses of potentially competitive offerings are essential for the planning of an effective manufacturing operation.

It is now more important than ever to attribute financial or accounting allowances which provide for the long term strategic goals of the enterprise in the global and competitive context. These should not only cover improved methods for justification of capital equipment but also take account of such items as employee retraining, redeployment, re-education, or even relocation. If these measures of the financial aspects are not included, inappropriate designs may be introduced which develop the resources of the enterprise in a less than optimum fashion. In this case, unless any new product shows some very convincing lead or advantages over the competition which provides a fortunate buffer, then disaster will ultimately result. The design activity carries the responsibility for conformance and delivery within these constraints, and if they are not understood, measured and managed well there will be only slender prospects for the future of the enterprise.

Conclusions

The design cycle cannot commence without a thorough investigation and development of an understanding of the whole problem to be addressed or solved. This can only be undertaken in the context of a well-founded knowledge of the objectives of the enterprise, together with an appreciation of the whole environment and the development of suitable organizations and methods of measurement. The measurement methodology, be it accounting practices or manufacturing equipment efficiencies, must be oriented to produce measurements which relate directly to the objectives for the system. The system must be responsive to means of control adopted as a result of these measurements. This applies equally to the measurement of design effectiveness.

Indeed, the future and prosperity of the enterprise depend upon the effectiveness of design as it is implemented in manufacturing so as to deliver customer satisfaction in the marketplace. The measurement techniques must be sufficiently accurate and precise to provide a means of conflict resolution to guide design decisions. Customer satisfaction is clearly the supreme objective, both for the enterprise to ensure its future, and for the subset of the design activity itself. To achieve this, any design must be suitable for manufacturing efficiently for delivery to customers in timely manner at appropriate cost. This can only be satisfactorily accomplished when the whole activity from the design pad through to the customer accepting delivery is viewed holistically and managed as an integrated system.

The design activity carries responsibility for the success of the whole enterprise. The information interchange, the management structure and the way that design is organized and utilized within the enterprise must all give adequate acknowledgement to the functional requirements of the total system. The design of the product itself is the key for successful implementation. It must be basic, rugged, perhaps elegant, but certainly simple and non-fussy. The design must solve the specific problems addressed without adding others.

Acknowledgements

Some of these concepts were initially developed as a result of teaching engineering classes at Southern University, Louisiana, and at the University of Vermont. They were first published in a paper 'Design for manufacturing', Proceedings of the CAD/CAM, Robotics and Automation International Conference, ASME Design Automation Committee, February 1985, pp. 437–42. Recent refinements are the result of the enthusiasm, contributions and patience of both students and colleagues at the IBM Corporate Technical Institutes, and at Lehigh University.

References

Cortes-Comerer, N.: Motto for specialists: give some get some. *IEEE Spectrum.* Vol. 24, no. 5, pp. 41–6, May 1987. This issue also contains other valuable design case studies.

Gardiner, K. M.: *Characteristics of an Ideal Manufacturing System.* Proceedings, American Society of Mechanical Engineers Winter Annual Meeting, Computer-Integrated Manufacturing and Robotics, PED vol. 13, pp. 185–201, December 1984.

Gardiner, K. M. and Olden, R.: *The Place of Information in Manufacturing.* Society of Manufacturing Engineers Technical Report EE84-825. Edited version published in *Circuits Manufacturing.* Vol. 25, no. 2, pp. 42–50, February 1985.

Guterl, F.: Design case history: Apple's Macintosh. *IEEE Spectrum.* Vol. 21, no. 12, pp. 34–44, December 1984.

Kidder, T.: *The Soul of a New Machine.* Boston: Little, Brown, 1981.

Lacey, R.: *Ford – the Men and the Machine.* New York: Ballantine, 1986.

Leontief, W.: The ins and outs of input/output analysis. *Mechanical Engineering.* Vol. 109, no. 1, pp. 28–35, January 1987.

Nierynck, J.: Technical progress as a result of the entropy law. *IEEE Transactions on Components, Hybrids and Manufacturing Technology.* Vol. CHMT-7, no. 3, pp. 211–14, September 1984.

Papanek, V.: *Design for the Real World* (2nd edn). New York: Van Nostrand Reinhold, 1984.

Perry, T. S. and Wallich, P.: Design case history: the Commodore 64. *IEEE Spectrum.* Vol. 22, no. 3, pp. 48–58, March 1985.

Pugh, S.: *CAD/CAM – its Effect on Design Understanding and Progress.* Proceedings CAD/CAM, Robotics and Automation International Conference, Tucson, AZ, pp. 385–9, February 1985.

Rabins, M. et al.: Design theory and methodology – a new discipline. *Mechanical Engineering.* Vol. 108, no. 8, pp. 23–7, August 1986. Based upon an ASME/NSF Conference Report, *Goals and Priorities for Research on Design Theory and Methodology,* September 1985.

Root-Bernstein, R. S.: Creative process as a unifying theme of human cultures. *Daedalus: Journal of the American Academy of Arts and Sciences.* Vol. 113, no. 3, pp. 197–219, Summer 1984.

Suh, N. P. and Rinderle, J. R.: Qualitative and quantitative use of design and manufacturing axioms. *Annals of the CIRP.* pp. 333–8, January 1982.

18 Managing across Organizational Boundaries

HUGH GUNZ

University of Toronto, Canada

Introduction

In this chapter I shall look at design as an organizational process. Design, as I use the term here, is about realizing an invention. The designers who produce the specification for the invention may well draw on the work of many other creative people. The design process means getting these people, and a great many more, to collaborate in turning the invention into a practical business outcome.

Each step along the route from idea to business outcome is complex and fraught with difficulties. It is a bit like the problems faced by baby sea turtles. Getting born is the easiest stage; even then, predators swoop to pick off ideas as they emerge hesitantly from their eggs and totter around looking for a safe place to grow. Unlike a sea turtle, though, an idea finds organizational depredation more lethal as it develops, because as this happens it needs more resources and becomes more and more of a threat to established ways of doing things (Kanter 1983).

The design process can be modelled as a series of steps of internal technology transfer, of handing the idea from one part of the organization to the next. Groups find it hard enough to work together when they all come from the same department. But people who work in different organizations or parts of the same organization have different bosses, reward systems, career structures, timescales for their work, and so on. Integrating their efforts can be a major problem. Every boundary is a potential obstacle which can slow the design process down, and speed is increasingly of the essence in designing competitive products (Pilditch 1987).

Two things make designing especially vulnerable to this problem. First, the process concerns something that does not yet exist, so that people in each unit can have quite different visions of what is being designed. Secondly, because designing often calls on skills which the firm does not have, outside organizations may be called in to help.

Something That Does Not Yet Exist

Design is about something which exists only in people's minds. A group of people collaborating in a design process has to solve the basic problem of making sure that everyone agrees on what is being designed. This is not trivial even amongst a group of people who work closely together. It means, for instance, that people have to learn each other's languages. Designers use visual means of communicating, while their non-designer colleagues are likely to be the products of educational systems which place a great premium on words and numbers.

But managers can be taught to draw, and designers to understand business concepts. Language differences are only part of the difficulty. It is a well-known upshot of organizing that perspectives vary according to functional speciality. This is called structural perspective: where you stand depends on where you sit. As soon as designs cross organizational boundaries the problems multiply, because everyone is likely to see the thing being designed from their own perspective. The problem is recognizing that this has happened; unless the group takes great care to uncover just what its members understand themselves to be designing, it is very likely that they will use the same words for quite different things. This problem is not created by working across organizational boundaries; it is exacerbated by it.

For example, a group of business teachers became embroiled in the design of a new educational software product, a major computer simulation of a company intended to let course participants experience the way managerial problems are interlinked. They convened a multidisciplinary team to coordinate the work. But although everyone ostensibly agreed on the shape of the product and was very enthusiastic about it, they each had quite different views of what it was to be. A finance specialist saw it as a multiperiod forecasting model, a demonstration of how a company should be run if his prescriptions were followed. A marketing specialist viewed it as a chance to show how companies are marketing led, and a production specialist thought of it as a very complex demonstration of the need to get materials requirements planning right. The software specialist managing the programmers was looking to produce something that broke new ground in simulation models. The behavioural scientists saw the core of the product as a complex set of communication pathways, and anything quantitative was an unnecessary complication. Yet everyone called the product a 'company model'.

There are a number of ways in which this trap can be avoided. Perhaps the most useful involve creative problem solving techniques (e.g. Rickards 1980), because of the way they take the group through an essential stage of group formation. Perceptions of the problem are brought into the open so that everyone can understand where the differences lie. The techniques then help the group come to a shared understanding of what they are trying to do, so that they can find ways of tackling it.

The alternative is for the group to discover the differences only when the product or process has taken shape. Not only is it much more expensive to resolve differences at this stage, but by now people will have become committed to their particular solutions. The engineers will resist to the death any alteration to their elegant engineering, the materials scientists will be most reluctant to see all their specification going for nothing, and so on. These problems are greatly exacerbated when the design work is done outside the firm.

Using Outside Organizations

Perhaps more than with most other functional specialities, firms regard design as a service that can be bought in. There are many reasons why this is a good idea, but it also creates a problem. If managing the design process involves working across organizational boundaries, working with a design consultancy involves a boundary of greater significance than many.

Consultants quite naturally have different interests from those of their clients. Their objective tends to be to sustain and enlarge their business and build a reputation for high quality, innovative design solutions. Being detached from the firms for which they consult, they are not part of the firm's pool of organizational know-how. These attributes have their advantages when the client wants to find distinctive new products or distinctive new solutions for existing products, but they have their dangers too.

For example, a research institute developed a novel design for a textile production machine. After testing the prototype exhaustively the institute arranged for a manufacturer to put it into production, and the first half-dozen were sold to a foreign company. They then hit a snag: subsequent machines kept overheating and breaking the fabric, and no solution could be found to the problem. The machines were withdrawn from the market. Strangely, no complaints were ever received from the first, foreign, customer. Some years later the research institute discovered by accident that the customer had hit the problem, tracked it down to a particular material used in a component, changed the material, and had no further trouble. It turned out that the material had been specified for the prototype by a freelance consultant who had kept the reason for the specification to himself in the expectation that he would be called in to consult for the production scale machine. But he had not been, so no one in the design team knew that the material in question was critical to the overall design.

There are other problems which are exaggerated by working with outside consultants. For instance, shooting the messenger is a typical reaction to unwelcome news, and it is easier to shoot a messenger who does not work for you than one who does.

For example, a small group of consultants was commissioned to research a novel product idea which would exploit the client's manufacturing skills and take the firm into a new market. The client had won a national award for innovative design and was very receptive to new ideas. Everyone – client and consultants – was very enthusiastic about the product, but it became evident as the concept was developed that it was beyond the client's capacity to produce unaided. In order to succeed it would have to be made in partnership with another firm, or perhaps be licensed. At this stage in the process the client stopped talking to the consultants. Financial commitments were met as agreed and a written report submitted, but the consultants were never able to discuss their findings with the client, who, from being most helpful, became totally 'unavailable'.

Identifying Problematic Organizational Boundaries

The issues highlighted so far illustrate my point that a key problem in design management is working effectively across organizational boundaries. The first step in tackling this problem is to find a way of identifying which boundaries are likely to be the most troublesome. Having identified them, one can then find ways of redesigning them so as to reduce the problem. One simple way of identifying problematic boundaries is to consider the interest each group has in the design process. Although it is common to ascribe problems in organizations to personality clashes, structural perspective can explain a great deal of why people behave as they do.

People's behaviour is strongly conditioned by being put inside organizational boxes called departments, each with its own job to do. Knowing the role of each group, it is possible to make surprisingly accurate predictions about the attitudes the groups' members will have to something they are presented with, such as a design project. One can make a great deal of progress with analysing the problems managers face in working across organizational boundaries through knowing no more than each manager's interest in the situation, arising from the demands of his or her job.

It is not difficult to put oneself in the shoes of each group along the line and to reflect on the project's significance to it. For instance, let us take two managers, for each of whom a particular design project may be of either high or low significance. The 'significance' can be assessed intuitively. Alternatively it can be measured crudely, for instance by looking at the proportion of manager A's time spent collaborating with manager B on the design project, and vice versa. The range of possible outcomes is shown in table 18.1.

Any boundaries of types 1, 2 and 3 need careful handling, because to either or both of the managers involved the contact is potentially not high enough up their priority list to get the attention it will need for the design process to be a success. It is possible, of course, to enrich the analysis with other information one might have about the people involved, the detail of their workloads, and the exact nature of the design project. But managers

Table 18.1   Assessing cross-boundary cooperation problems

		Significance of the project to manager B	
		Low	High
Significance of the project to manager A	Low	1 Probably not a difficult boundary to manage, unless close collaboration is needed	2 B depends on A to make the contact work; to A, B is something of an interruption
	High	3 A depends on B to make the contact work; to B, A is something of an interruption	4 Probably the least problematic boundary: both need each other equally and have few other distractions

are usually surprised by the amount of useful information they can get from the very simple approach summarized in table 18.1.

Having built up a picture of where the difficult organizational boundaries might be along the path of the design project, one can choose an organizational design which addresses the potential problems that have been identified.

Redesigning Organizational Boundaries

I have assumed so far that firms are organized in a traditional way, based on functional departments. But there has been a great deal of research reported over the past 20 years which has established that such structures are neither suitable for all applications nor universally used. Alternatives exist, many intended to cope with the problems raised by the need to work effectively across organizational boundaries. Managing the design process, in other words, may involve designing organizations as well as products.

One can think of an organization as a system for processing information. For a small firm, the traditional organization chart acts as a crude map of information flows for control purposes: crude, because things hardly ever happen like that in practice. As the firm grows and the world gets complicated the communication channels get overloaded. A simple departmental structure assumes omniscient management and instantaneous, total communication between everyone who needs to communicate.

The world is not, on the whole, filled with omniscient, total communicators. In a classic analysis of the situation, Galbraith (1973) argues that firms can choose, explicitly or implicitly, between four organizational design strategies:

1   Rely on slack in the system to cope with the problem (e.g. stocks of raw

materials, work in progress, finished goods; under-utilized production facilities; queues of customers).

2 Invest in information technology to make information flow more efficiently.

3 Split the organization into self-contained parts (for instance setting up a completely separate unit to design and build a new car or aircraft).

4 Off-load the information processing to the level in the organization which needs it; in other words, build lateral links into the structure.

It is vital to realize that none of these strategies is cost-free. The organizational design problem is to choose the strategy (or combination of strategies) which is least expensive. The cost of strategy 1 is the slack itself. Strategy 2 needs capital for the technology and is typically more expensive to implement than people planned; also, it is not always suitable for one-off situations such as design projects. Strategy 3 is fine for design projects if they are large, but if they are not, the duplication of resources that it requires can be unacceptable. It is one thing to build a new computer like this (Kidder 1982), but if one has a series of projects in hand concerning a range of small domestic appliances it is most unlikely that one can afford a separate design organization and its necessary support services for each appliance. That leaves strategy 4 which is, in many ways, the most appropriate for R&D in general and design in particular. But it, too, has a cost.

Lateral links can take many forms (a classic account can be found in Lawrence and Lorsch 1967: 137–40). Examples include people or groups given the job of managing links between departments, and temporary and permanent cross-functional teams. Cross-functional teams have attracted the label 'matrix organizations', and they come in many shapes and sizes. The idea is simple: one defines a two-dimensional structure in which people belong both to their functional department and to a cross-functional team with its own management. But it is not an easy arrangement to make work because of the ambiguities it builds in. For instance, it seems to violate a classic tenet of organizing, namely 'one person, one boss'. And it involves endless formal and informal meetings, which Galbraith specifies as the cost of strategy 4 (crudely, the hassle involved in making the arrangement work).

Space does not allow a full description of matrix organizations and their variants (see Knight 1977; Davis and Lawrence 1977; most standard texts on organizational behaviour or organizational design discuss the issues). The key to understanding the contribution they can make to design management is to see the design manager as a project leader. Two basic models of project leadership provide two alternative rationales for the matrix (Gunz and Pearson 1977).

The first model is more suitable when the work of many different groups has to be coordinated and the boundaries between them are of the type shown in cell 1 of table 18.1. Their major contributions probably will not all be needed at the same time. For instance, there is no point in getting the production engineers to design detailed production layouts until the product design is settled on. But that is not to say that their advice should not be sought until then, because they may have important suggestions for the design which will make the product easier or cheaper to manufacture. So in this model of matrix management the design manager is a coordinator, with the most complete information about the state of the project's progress. He or she effectively contracts work out to the various contributing groups, including outside design consultants if necessary, and informs each group when their contribution will be needed. The project group is not so much a cohesive team as a group of colleagues who meet regularly to discuss progress.

The second model is preferred when closer cooperation is needed, and the boundaries are of the kind shown in cell 4 of table 18.1. It is not always possible to parcel work out to groups to work on in comparative isolation. For instance, the design of successful artificial hip joints needed close collaboration between various medical and engineering specialities. Time may be too short for the project to be coordinated by means of regular, routine review meetings. Here, the design manager becomes more of a leader than a coordinator, whose role it is to weld the team together and spur it to action. The team is now much more identifiable as such, even though each member may have other duties apart from the design project itself.

Most matrix organizations have aspects of both coordination and leadership styles. Both involve potential clashes of authority between the design manager and the line managers of the various project group members. The more the matrix moves to the leadership model, the more the line managers feel that they are losing control over subordinates who are members of the design group. On the other hand, the more it moves to the coordination model, the greater the design manager's difficulties in making sure that the team members are allowed by their bosses to give the design project the priority it needs.

It is quite possible for a project to start with the leadership model and to move to the coordination model as it matures and as more easily defined packages of work are identified. In pharmaceutical development, for example, the later stages of testing drugs before they can be released for clinical trials involve highly specialized groups who use standard procedures examining, for instance, the drug's toxicity. The tests simply have to be allowed to run their course.

Alternatively, when boundaries are of the types shown in cells 2 or 3 of table 18.1, a mixed form of matrix might be appropriate. A small, close-knit leadership group may need the help of other groups who provide a similar service to many other clients. Internally, examples of these could include analytical or materials science laboratories, and externally, design consultants. So the design manager may involve these other groups in the rather more arm's-length manner of the coordination matrix.

The design manager acts as the focus of the project, whichever model is chosen. This means that this person needs the authority to get action. Authority has many sources (see Handy 1985 for examples). Design managers have control over information: they know most about what is happening. They may also have control over resources if they have a budget with which to commission work. Charisma helps, although it is an elusive and rare quality. All organizations have so-called opinion leaders – respected figures to whom others come for advice. An opinion leader who also has some managerial skills and is not too identified with only one part of the firm may be a good choice as a design manager. But probably the most important source of support the design manager can call upon is that of the chief executive of the business unit everyone in the group works for. If the chief executive is known to be deeply interested in the progress of the design project, and is reasonably accessible to the design manager, then the team members and their line managers will find it hard to ignore the design manager's requests.

Conclusion

My starting point for this chapter was to consider design as an organizational process, that of taking a design from concept to commercial realization. Managing design involves ensuring that the groups of people involved in this collaborate efficiently and effectively. A key problem therefore becomes that of working effectively across organizational boundaries.

Working across boundaries is never easy, but design management introduces two particular problems. First, designing involves something which does not exist yet, so a major problem is establishing agreement on what is the object of the process. Second, design often involves using outside organizations such as consultants. I outlined a four-step approach to working across organizational boundaries:

1   Consider each group of people in turn who are to be involved in the design process. What is their interest in the design project?
2   Review the boundaries between the groups who need to work together. In the light of the conclusions from step 1, which boundaries are likely to prove troublesome to manage? Look particularly for those where one or other party may not be able to give the project high enough priority.
3   Focusing on the troublesome boundaries, consider alternative organizational forms which may reduce the difficulties. There are many possible approaches, each with its costs, and the task is to choose the one which is least costly. Some form of matrix structure may be helpful, and the more the project needs close cooperation between groups, or urgent action, the more firms should consider the use of close-knit project teams.
4   Identify the role of design manager, who is given the responsibility for managing the design project and the contributions of anyone, including external design consultants, associated with the project. It well may be

possible for the design manager to combine the role with other duties, but he or she must have the support of top management in order to be effective.

It is important to emphasize that these steps are not intended to provide a universal guide to the problems of design management. Successful design management is about designing effective organizations as much as it is about designing good products. In many ways it is a process of managing change, perhaps one of the most challenging fields of managerial activity. My aim has been to draw out the aspects of design management which are explicitly to do with organizational barriers to innovation, in order to suggest organizational solutions that can be brought to bear on them. As with most managerial problems, there are a number of simple diagnostic procedures that can be used to identify problem areas and suggest possible solutions. Implementing the solutions is not so easy, however, and I have tried to highlight some of the better known pitfalls.

It will be evident from this account that redesigning the firm so that the design process can be managed more effectively is a time-consuming and difficult process. But the costs of doing nothing may be yet higher.

References

Davis, S. M. and Lawrence, P. R.: *Matrix*. Reading, Mass.: Addison-Wesley, 1977.

Galbraith, J.: *Designing Complex Organizations*. New York: McGraw-Hill, 1973.

Gunz, H. P. and Pearson, A. W.: *Matrix Organisation in Research and Development*. In Knight (1977).

Handy, C. B.: *Understanding Organisations* (3rd edn). Harmondsworth, Middlesex: Penguin, 1985.

Kanter, R. M.: *The Change Masters: Corporate Entrepreneurs at Work*. Allen and Unwin, 1983.

Kidder, T.: *The Soul of a New Machine*. Harmondsworth, Middlesex: Penguin, 1982.

Knight, K.: *Matrix Management*. Aldershot, Hampshire: Gower, 1977.

Lawrence, P. R. and Lorsch, J. W.: *Organization and Environment: Managing Differentiation and Integration*. Boston: Harvard University Press, 1967.

Pilditch, J.: *Winning Ways*. London: Harper and Row, 1987.

Rickards, T.: Design for creativity: a state of the art review. *Design Studies*. Vol. 1, no. 5, July 1980.

19 Corporate Identity

WALLY OLINS
Wolff Olins, UK

What is Identity?

Identity is a profound, simple and basic manifestation of the human condition. We all, as individuals, want to belong. What is more, we all want to be seen to belong. In every country in the world, human beings demonstrate their affiliations visually. Descriptions like laid back, yuppie, redneck, white collar, blue collar, Catalan, Basque or Corsican all demonstrate and underline this. People live and sometimes die for their identity. Look at the Palestinians, the Israelis, the various factions in Ulster and elsewhere.

The outward and visible manifestations of belonging are immediately recognizable, both to other people who belong to the group – and to outsiders. In our ordinary lives, we drop visual hints about who we are and what our affiliations are all the time. What we look like demonstrates what we believe in. It is our style.

What is interesting and significant is that these signs are more or less natural. What emerges on the outside is not so different from what is on the inside. The signs we make are largely visual. So identity emerges primarily, but not exclusively, through design.

Expressing identity is natural in organizations as well as amongst individuals. Interestingly, organizations that are, for the most part, in the same line of activity tend to emulate each other. Schools, universities, sporting clubs and, of course, business organizations all demonstrate their identity, usually without knowing it. So they often end up, without particularly meaning to, looking alike. Advertising agencies all over the world look the same. Their reception areas all seem to project the same slickness. In visual terms, design terms, they have white walls and black leather chairs with chrome arms. And agencies even tend to use the same kinds of names – a series of initial letters, such as GGK, WRCS, FCB, BBDO. Government departments around the world also resemble each other. They all have a similar air of punitive neglect.

But it is not only in the bigger things that identity emerges; it also manifests itself in a multiplicity of very small things. It emerges through some things that are apparently so insignificant that they hardly seem to exist at all – even through objects which are apparently anonymous, like coat hangers or signs.

So what can we learn from all this? Three things:

1   That every organization has an identity. Usually the identity emerges naturally in a rather haphazard, unmapped sort of way. Often organizations doing the same thing have the same kind of identity.
2   That identity emerges primarily through what you see; it is mostly visual. Design is the filter through which identity emerges.
3   That design involves not only a few rather big things, but also a multiplicity of minor things. It manifests itself in a holistic sort of way, taking into account everything from the very large to the almost insignificant.

In short, what we call corporate identity is the process of explicit management of some or all of the ways by which an organization is perceived.

Identity in the Organization

In business organizations – more or less any business organization – identity largely emerges through three areas that you can see, that are designed:

Products or services   What you make or sell.
Environments   Where you make or sell it.
Communications   How you present and promote what you do and how you do it.

Coordinated design is the key by which these can work together. Identity also emerges through one area that you cannot see, although it is just as important – and that is behaviour.

In smallish companies run by one person, the identity is an expression of the personality of the founder, his or her obsessions, dreams, ambitions. At this stage in an organization's life, things can be, and usually are, relatively simple. If the boss likes blue, the shop will be blue, the stationery will be blue. If it is a retail business, the identity is largely expressed through the shop. If it is a manufacturing company, the way the product looks and feels will bear the special stamp of the person at the top.

In larger companies, though, things are different. The relative significance of each of the various manifestations of identity varies according to the nature of the organization.

For example in a product-based company, say an automobile company, the product is the most significant way by which the company's identity emerges. It is primarily the way the car looks and feels, how the doors open and shut, how big it is, what kind of engine it has, how it performs, what it costs, that makes people feel the way they do about the product and therefore about the company. To paraphrase an ancient McLuhanism: the product is the message. Environments and literature affect the issue, but the product encapsulates the company's identity. Here, therefore, the product designer is responsible not just for the product but very largely, and possibly by default, for the company's identity as well.

However, there are many situations in which an identity is not product led but environment led. In retailing, in leisure businesses, hotels and theme parks, it is the environment that dominates the identity mix. Here, the environmental designer, the architect, the space planner and the interior designer carry the main burden of projecting the company's identity. In Bloomingdales or Harrods, for example, it is the environmental experience that makes them special places to visit. It is not, funnily enough, the products they stock because, after all, there is not much that you can get at these stores that you cannot get elsewhere. The products and communications are not unimportant; it is just that they are less important in the identity mix than the place itself.

Then there is, of course, a vast range of products and of services which are, in identity terms, communication led. Few consumer products, for example, have much character of their own, have a strong personality. A personality is bestowed upon them by the way in which they are packaged and advertised, that is promoted.

Advertising and Identity

Consumer products are given life through advertising. Advertising is a prism through which many products that we use in everyday life – fizzy drinks, soaps, toothpaste, breakfast cereals – are projected. Almost inevitably, therefore, many people have come to associate advertising with identity – or with image. This is a misleading and potentially dangerous idea, because it devalues the real power of product, environment and behaviour in the identity mix, at the cost of overvaluing communication.

Perhaps even more significantly, the idea that identity is somehow inextricably associated with conventional communication techniques (and particularly with advertising) can inhibit real and genuine cooperation between all those people within an organization – product designers, graphic designers, architects, management development people and communications people – who are collectively responsible for identity. Organizations which fail to see clearly their own special mix usually put identity under the communications umbrella, frequently with mournful results.

Style and Structure

Identity can clearly present the style of an organization, whether it is traditional, modern, regional, international, aggressive or whatever. The style of Holiday Inns is, for example, very different from the style of Ciga Hotels, although both are in the same business. Equally, the style of Apple is very different from the style of IBM. But identity is also used to convey a complementary and equally significant set of messages, which are nothing to do with style at all; they are to do with corporate structure.

In the old days, most companies were very simple monolithic structures. They normally used only one name and one identity. Most little companies are like that today; so are a few very big ones, like Sony and Shell. In fact, in the first companies which consciously used identity programmes, the railways of nineteenth century industrializing Britain, those programmes were developed as a homogenizing tool to hold together organizations that were geographically widespread. It was because they were monolothic that identity was so valuable. Since those days, however, things have become much more complex. It is now possible to classify identity structures into three broad types.

Monolithic identities   These forms are always used by companies that have grown organically. The fundamental strength of the monolothic identity is that each product and service launched by the organization has the same name, style and character as all the others. Everything that any part of the organization does is therefore supported by the rest of the organization.

This concept of mutual support and reinforcement makes promotions and new launches more economical and enables relations with staff, suppliers and the outside world to be consistent, predictable and comparatively easy to control. Companies with monolithic identities have high visibility which can be advantageous in the marketplace.

Endorsed identities   In today's world, most companies grow not organically but by acquisition. They buy competitors, customers, suppliers; they expand vertically and horizontally. Each of these acquired organizations has its own name, reputation, tradition and culture. Each has its own network of audiences, and each its own goodwill. The company which makes the acquisitions usually ends up by keeping some, if not all, of the names of the companies which it has acquired – but then it has to deal with the complexities which this situation inevitably creates.

The intention of an endorsed identity is to show how organizations forming a group can retain their original identities and at the same time become part of the group as a whole, sharing the group's values. Companies which seek to create a corporate identity involving a group of subsidiary organizations with complementary but sometimes competitive backgrounds have a difficult task.

On the one hand, certainly at corporate level and for corporate audiences, for shareholders, investment analysts, recruits at various levels and so on, they want to create the idea of a single but multifaceted organization that has a sense of purpose. On the other hand, they want to allow the identities of the numerous companies and brands they have acquired, often at considerable expense, to continue to flourish in order to retain goodwill in the marketplace – particularly, of course, for their customers.

This requires a balancing act. These aims are frequently conflicting. They can only be achieved simultaneously if the greatest sensitivity is used. Most commercial companies try it. Some do it well; most do it badly.

Branded identities The third kind of identity is the one in which the company operates through a series of brands which are apparently unrelated both to each other and to the organization as a whole.

Companies which operate in this way are often in the food and drink or other fast moving consumer goods business. At the corporate level, these companies may reach out to all of the audiences of the monolithic or endorsed company, but they do not present any kind of corporate face to the customer. As far as the final customer is concerned, the corporation does not exist; what the customer sees is only the brand.

Companies pursue this branded policy for several reasons. First, brands are thought to have a life cycle of their own, quite distinct from the company. Secondly, the system allows competitive brands from the same company to appear on the same supermarket shelves. Thirdly, it enables brands to present specific identities of their own, appropriate to both the nature of the product and the consumer for whom they are intended. In some cases this symbolism may be simple and even naive; often it is more direct and focused than the visual symbolism of a corporation.

Marketing and Identity

It is tempting to assume that identity is some kind of marketing tool – that its purpose, at least its prime purpose, is to project the ideas of the organization and its products to its customers. While brand identities are aimed almost exclusively at customers, corporate identity is not the province of the corporate marketing people. To think otherwise is to misunderstand, to underestimate its power and its purpose. The relationship between the audiences for all identities is subtle and overlapping. The customer audience is of great importance, but so are others.

The Audiences

Every organization has a wide variety of audiences, of which the most immediately significant is itself – that is, its own staff. The organization's own people have to know what it is, what it does and how it does it. If the organization is effectively going to articulate an idea of itself, its own staff must know and understand what this is; they, after all, will be a prime channel by which the idea is communicated to the world outside.

The organization has many internal audiences – at different levels, in different plants and offices, in different countries in the world, doing different things. All of these are prime targets. It also has what you might call quasi-internal audiences: people who own its stock, pensioners who once used to work for it, employees' families, and so on.

And then it has outside audiences; these are the people with whom in one way or another it does its business. Some of these are close and have a special relationship with it, like direct customers, dealers and final consumers. Others are equally close but have a different relationship with it: competitors, suppliers, the local community, trade unions. Some of the organization's audiences are, of course, not so close. These audiences are often called opinion formers: legislators, journalists, investment analysts, educational institutions and trade and industry associations, for example.

All are in some senses separate from each other and, in particular, each has a somewhat different relationship with the organization. Inevitably the organization will want to emphasize different facets of itself to different audiences. However, the organization's audiences are also overlapping. Some customers may live in the local community. Other customers may be pensioners. Still more customers may be stockholders or parents of children who are thinking of coming into the company. Customers will be married to stockholders – and so on.

It is unrealistic to put all audiences into separate compartments and assume they have nothing to do with each other. It is, therefore, unrealistic to attempt to convey different, even contradictory, messages to different groups of people. The fact of the matter is that each of the different audiences of an organization will form a view of it based on the totality of the impressions that it makes on them. And people will inevitably pick up their ideas about an organization from more than one source: from both advertising and editorial in the media, from personal experience, from other people's experience, from rumour and gossip.

Where these ideas are conflicting, where messages received in one place are different from those received somewhere else, the overall impression will be negative and confusing. Which is why an organization needs to control its identity and plan its identity programme.

Planning an Identity Programme

Most large companies today are trying to do three things:

Diversify
Decentralize
Internationalize.

Managers in distant countries, working in recently acquired companies, in technologies unfamiliar to the main company, want to be left alone to get on with it. And they often are. They claim, quite reasonably, that if they are given profit responsibility, they should also be given the means to achieve it. This means setting up separate cost and profit centres around the world. Inevitably this attitude creates powerful centrifugal forces which, if left alone, will simply tear the company to pieces.

In order effectively to control a diverse, international, decentralized company, it is essential for a central management to create a countervailing set of coordinating forces. In a sense, this is another version of the old battle between centralization and decentralization. The coordinated resources which a management retains at the centre, in order effectively to continue to control the business, usually embrace

Finance
Investment
R&D
Management development
Product quality
Technology.

Identity is also a central resource. It must be recognized and coordinated from the centre like all other central resources.

Two Models

A major difficulty which any organization inevitably encounters when it attempts to manage its identity as a central resource is that, if it is to be managed properly, identity will cut across traditional departmental territories – marketing, personnel, purchasing, properties and so on. Inevitably, effective management of identity will create, at least in the early days, some dislocation within the organization.

In order effectively to manage identity as a corporate resource, it is important to examine how other corporate resources are managed and to use these, where appropriate, as models. Generally there are two models we can use: financial management and the management of information.

Financial management   This is the prime management resource. Every company knows that unless it manages its finances effectively it will not survive. Finance affects every decision and every department. It is as important in investment as in purchasing, as important in marketing as in staffing.

In every enterprise, finance is represented both at board level and throughout the organization. Normally, however complex and far-flung the organization, the financial reporting system is standardized throughout. All this also applies in identity management. Identity affects every department of the organization. It needs to be represented at board level and below and applied ubiquitously.

Information management   While nobody questions the necessity to manage finance as a coordinated resource, there will be many people inside a company who will find the analogy between finance and identity a bit far-fetched. That is why it is useful to find another analogy, a resource which is also new and had some difficulty in establishing itself.

Twenty years ago or so, the concept that any organization could successfully coordinate and harness its information resources across a broad range of corporate activities was not just unknown – it was simply unimaginable. For example, typing was typing, it had nothing to do with accounts. Today, however, word processing and payroll go without comment on the same computer system.

It took some years and an enormous amount of effort from hardware and software suppliers to break through the barriers of tradition. But information technology is accepted today as an integral and essential corporate resource. Identity management has now reached the same stage of acceptance as information technology a few years ago. It will take a few more years before it is absorbed into the corporate bloodstream.

Creating an Identity Programme

There will be many organizations where there is no formal structure for either introducing or maintaining an identity programme. In such companies it is necessary to create a corporate identity formally, so that it can be introduced and disseminated throughout the organization. The work of creating an identity can be conveniently divided into the following phases:

1   Investigation and analysis leading to development of design brief.
2   Creation of a new visual identity incorporating, where appropriate, name changes and visual ideas based on the organization's agreed personality, strategy and structure.
3   Launch and introduction of the programme.
4   Implementation of identity across all areas of the organization – product, environment and communications.

Since, in the nature of things, most organizations are unfamiliar with the methodology involved in creating an identity, and in addition are likely to be too close and emotionally involved with themselves to see their own company objectively, it is desirable for them to go about the task of creating a new corporate identity in conjunction with design consultants.

Today, there is a profusion of choice. Design consultancies vary in size from one-person practices up to 300 or 400 people. Their skills, experience, dedication and structures are as varied as their size. The client company should pick a consultancy in which the individuals with whom they will work have experience and with whom they can get on well as working partners. Creating an identity effectively involves the closest collaboration between client and consultancy over a period which can vary between three months and five years. Sympathy between the parties is therefore essential.

Identity Management

In the end, every company will manage its identity in a way particular to itself. So although guidelines can be laid down, each company will have to create its own structure.

A board member should be made responsible for identity/design matters and must be supported by his or her colleagues. A commitment to identity/design matters must be demonstrated through a proper financial budget, for which individuals must be accountable. Someone must be appointed to manage and coordinate design throughout the organization, and will be responsible for creating and coordinating task forces by both function and geographic area. A major part of this task will be not only to introduce but continually to promulgate and promote the identity idea.

This identity manager should be responsible for codifying the design rules by way of a manual which will be monitored and updated on a regular basis. The identity must be introduced and developed across all functions and geographical areas. The identity manager must pay particular attention to coordinating work in all of the areas in which the identity manifests itself – the visible ones like product, environment and communications, and the less palpable ones like behaviour.

The identity must be managed in such a way as to recognize the company's real priorities. In a product dominated company, it is vital that appropriate attention is paid to product design and identity – and so on.

Identity management will inevitably take somewhat different forms in different organizations. It may even have different names – identity management, design management, communications management – but its essential task will be the same: to manage the company's identity across every single thing it does so as to project a coherent idea.

Identity management is a permanent task. Like finance management and the management of information technology, it must never stop.

PART IV

THE NATURE OF DESIGN PROCESSES

20 Preparing for a Design Project

MARK OAKLEY

Aston Business School, UK

Introduction

Design projects may come about for all kinds of reasons, not always logical ones. Sometimes, the urge to create a product which is better or new completely obscures the possibility of considering any other course of action. In older firms – where products or parts of products have always been designed in-house, from scratch, drawing only on the skills and experience available internally – it may never occur to managers that it might be better if products were designed by other means. Indeed no designing may be necessary, perhaps because any problems really lie elsewhere or because it is now much more cost effective to source this type of product from other manufacturers.

Ideally, senior managers will have considered these kinds of issues when formulating their company's design and general business strategies (part III). If so, a framework may exist within which decisions about embarking on design projects can be taken. This chapter indicates the types of specific questions that should be asked as part of the process of deciding whether to embark upon a design exercise or to select some other course of action.

What is the Problem? What Result must be Achieved?

The fact that a design project is being proposed implies that there is some problem which needs to be tackled. The problem may relate to an undesirable turn of events; perhaps sales of the product are falling, or selling prices are being squeezed so much that an adequate financial return can no longer be made. On the other hand, the problem may be of a more positive kind such as how to take advantage of a new opportunity which has arisen in the marketplace.

Hence, the first questions to be asked must accurately pin-point the problem or opportunity which is thought to exist. Just as important, but often more difficult to do, is to specify the outcome or series of outcomes which will be evidence that an adequate solution to the problem has been achieved.

Is Designing Really the Answer?

Once it is agreed that a problem does exist and its characteristics are fully understood, it is time to decide on the best course of practical action. There should be scope for considerable lateral thinking at this point. Although there may be great pressure to move quickly into a design project, a much wider range of options may be worth considering than might appear at first sight.

Sometimes, it may be possible to avoid the time and expense of designing either completely or to a large extent by, for example, reconsidering the wisdom of commitments to existing markets. It is always possible that a product which is struggling in a current market might prosper in a different one with little or no modification necessary. In other words, rather than redesign the product, can we find a new market for it?

Similarly, even if a dated or over-complex product design is judged to be the basic cause of high manufacturing costs, it may be possible to restore profit margins by reducing total manufacturing costs in other ways. Suggestions might include modernizing the manufacturing system, subcontracting some or all of the work (perhaps to regions or countries with lower wage costs), buying more components rather than making them, and so on. There may even be real marketing advantages to such a course of action; customers may be very well disposed towards the existing design of the product if only it can be made available to them at a competitive price.

A more extreme possibility, but one which should always be considered, is to discontinue the product in question and concentrate on other opportunities instead. Except for companies which are dependent on a single product (and they ought to regularly question such a strategy), most companies will benefit by adjusting the balance of their efforts to reflect changes in markets and technologies. Even the largest firms may find they are able to achieve better results by limiting the variety of products which they sell.

What are the Alternatives to a Design Project?

If, after asking these questions, it is concluded that a new or improved product really is necessary,a further set of questions remains as to whether or not a design project needs to be set up. There may be other ways in which a suitable product can be obtained and these should be seriously considered, especially by managers of those firms which have not enjoyed great success with design projects in the past.

The purchase of rights to make and/or sell an already designed product is one possibility. In many fields these days, this is the only realistic choice – for instance, where design costs are prohibitive because of the technologies involved or where there are already numerous well-proven designs for which licences are readily available. To try and produce yet another design could be a great waste of resources which would be much better used to support a marketing campaign or to improve the distribution operation.

On a grander scale, suitable new products might be acquired by taking over, or merging with, another company. Clearly, such action is likely to involve a wide range of considerations but, in many cases, access to new or better products is a major motivating factor.

None of these possibilities may completely eliminate the need for design work; at the very least, products may need to be relabelled or repackaged. In fact, quite a lot of additional work may be needed if styling changes are required to bring the product into line with the rest of the company's range or to conform to a corporate identity programme. Nevertheless, the time and cost involved should still be much less than with a full 'from scratch' design project.

Something else which requires less resources than a full project is what might be termed an adaptive or imitative approach. No one admits to copying the designs of other companies' products but it is clear that the practice is widespread. Novel products and distinctive designs are rarely free from imitation for very long. Any direct copying is, of course, an infringement of the originator's legal rights – but the deriving of inspiration is a different matter!

Managers should never be afraid to challenge designers to explain why a competitors's successful design cannot be used as the basis for a new product. There *may* be valid reasons why this is not a wise course to follow – but designers' attempts to preserve the mystery or purity of their art should not be part of them.

Setting Up the Project: Recognizing and Addressing Constraints

After the decision to embark on a design project is finally made, it will then be necessary to start planning the practical details. The two most important things to get right are the staffing, organization and management of the project (part III) and the preparation of product and project briefs. In order to make the correct decisions about these matters, comprehensive information must be available, especially concerning the various *constraints* which will shape the project.

External constraints   External constraints relate chiefly to two groups of people – customers and competitors. However, it is possible that account may also have to be taken of the needs and views of many other groups or individuals such as material suppliers, subcontractors, retailers and service agents.

Customers' concerns and requirements are likely to be paramount. As well as identifying as precisely as possible the major needs which the design project must address as top priority, the company may need to review other less obvious issues. These could include many disparate matters ranging from the extent to which customers display brand loyalty to details of the kinds of sales outlets from which purchases are likely to be made. Information such as this may be very important in correctly setting the limits to the design project.

For example, if the product is sold and installed entirely by specialist dealers, its design may not need to take account of possible installation by unqualified personnel. On the other hand, if the product is sold through do-it-yourself outlets, provision may have to be made to ensure success can be achieved by even the most inexperienced installers.

Competitors' activities and plans may also be significant, particularly where these indicate that new design or business directions are being followed. Care must be taken not to blindly pursue other producers on the assumption that they have identified the best route to customers' wallets; it may well be that there is room in the market for a variety of design solutions. However, many products are highly influenced by fashion and other trends, so it is vital to at least ask why competitors are moving as they appear to be and to consider what the long term results may be.

Internal constraints   Internal constraints on design projects may be many and varied. Almost every function or department in the company may seek to influence, at some stage, either the management of the project or the design decisions being made – or both. Managers of design projects must anticipate these interventions, many of which will be justifiable and necessary if a result is to be achieved which is generally *compatible* with the rest of the company's activities.

Manufacturing constraints are often highly significant although, as elsewhere, efforts must be made to distinguish between genuinely important constraints and those which are being put forward merely for the sake of making a point or even to antagonize. The kinds of issues that may need to be taken into account could include utilization and capabilities of equipment, material procurement and handling, quality norms and methods of checking – and many more. Similar lists may be drawn up for other areas of activity and influence such as finance, marketing, distribution and general administration.

Drawing Up a Design Brief

No design project of any substance should be started before a competent brief has been prepared, considered and approved. The brief should cover two main areas:

1   The new or modified product or service which is to be the outcome of the design exercise (its features, customers, costs, etc.) – the *product brief*.

2   The design project which will be set up to do the work (time, resources, etc.) – the *project brief*.

The form and size of the brief should be appropriate to the nature of the product, firm and industry; that is, a 20-volume document should not be needed for a new range of saucepans but, equally, a single sheet of A4 would be inadequate for a new passenger aircraft.

Exactly whose job it is to draw up the brief will vary from firm to firm. In some, the work might be done by top management. In others it could be the responsibility of marketing, design or manufacturing management – or a combined effort.

In all cases, the manager who will be ultimately responsible for the project ought to be involved. Too often, such design managers find themselves presented with quite inadequate instructions (sometimes none at all). They ought not to proceed regardless. Their professional response should be to insist on the preparation of an adequate document. Sometimes, they may well have to be prepared to do the job themselves or advise and badger others about what is needed.

The Product Brief

The first thing that needs to be done is to check whether the proposed new or redesigned product falls within the general strategic plan of the company. Assuming the company knows which markets it is aiming at and what kind of products it wants to offer – does this current proposal make sense?

If *not*, go no further until top management has reconsidered and confirmed that the implications of what is being proposed are understood (e.g. possible change of company direction, new markets, new technologies). When satisfied on this point, check that all the key factors relating to the market have been identified, including:

1   What is the most significant part of the market? Do we know the precise target we are aiming at? The markets for most products are varied and extensive; rarely is one version of a product right for the whole market.
2   What are the characteristics/needs/wants of our target market? They *must* be stated, otherwise there is little hope of correctly achieving the right design.

What product features are dictated by these charactistics, needs or wants? There may be two extremes:

1   Features the market already recognizes and knows it wants. These are likely to involve low design costs, low risk and low returns for the company. Examples include many domestic appliances, electric typewriters and basic office furniture.
2   Features the market might have a need for but does not yet recognize because the features are unknown to it at present. These are likely to

involve high design costs, high risk and high returns. Recent examples include fax machines, computer systems, video recorders and the Sinclair C5.

Senior management, marketing, etc. may not have this information (or may have customer information but not know how to translate it into product features). Often it is the designers or design managers who are best equipped to do this. This is especially so when it comes to looking ahead and predicting trends, rather than just reading the market as it stands at the moment.

When all the essential product features have been identified, their design implications need to be quantified:

1   What combination of design skills and project time will be required?
2   Do we have these skills? If not, can we get them? At what cost? From where? If skills are lacking, we might want to ask why. Is it because management has failed to recruit enough people, or because too little attention is paid to training? Can we influence a change in attitudes for the future?
3   What range of skills is required? Just engineering design and styling design? Or other specialisms?
4   How much will this cost? Is the likely cost within the limits that the company can afford? Does the company *know* what it can afford?

At this point, the product brief can be drawn up. In the light of the foregoing, it will be possible to specify the product parameters and features. Typically, these may be presented in two or three categories:

Mandatory   Examples are dimensions, performance levels, operational features, compliance with standards and legal requirements, and costs, all of which must be achieved.

Highly desirable   Examples might include distinctive appearance, target fuel economy and robust design (i.e. capable of future modification).

Desirable but not essential   Examples are difficult to give as such features will vary from company to company. Great care must be taken to double check before any feature is deemed not essential.

All items recorded in the brief should be quantified in some way, either in absolute terms (e.g. 'Dimension *X* must be between 27.5 and 29.5 mm') or by means of targets (e.g. 'Operating speed must be a maximum of 30 seconds per A4 page; 15 seconds would set a new industry standard and is the target').

Features which are more subjective – typically concerned with product appearance, but which cannot easily be quantified – must still be specified in some way. Comparison might be made with competitors' products as a means of setting the standard. Or reference may be made to company 'house styles' (if available). Or the company may seek the collective judgement of particular consumers, design critics, etc., perhaps through participation in design clinics. Additionally, there may well be aspirations which cannot be specified in terms of features or parameters (because they will not be known until the design team has determined them) and can only be expressed in terms of desired functions.

Finally, price and cost breakdowns must be provided. Too many design projects are undertaken without adequate regard to product price and cost – the attitude being that costs can be sorted out later. This is usually a misguided approach. The product part of the brief should record:

1   Target selling price(s) of the product.
2   Target manufacturing costs of the product – broken down as appropriate into target costs for components and subassemblies. A design to cost approach can then be attempted: at review points during the project, any failure to meet target costs can be investigated and action taken, e.g. respecify subassembly/component, or seek alternative design solutions.

The Project Brief

Basically, there are three issues that need to be considered and specified: resource availability, project cost and project duration.

Resources   This also relates to the next item (project cost). If funds are unlimited, resources are unlikely to be a problem since any skills, equipment, etc. needed can just be bought in. However, for most companies operating in the real world, funds are strictly limited. So:

1   Design skills must be used to best effect; put another way, design projects may need to be selected and organized to use the skills realistically available. Hence, the brief must indicate which skills will be used and how they will be provided if not already available to the firm.
2   Similarly, design management skills must be assessed. What level and quantity of people management, budgeting, coordination and other skills are necessary?
3   Non-human resources have to be considered – accommodation, equipment, computers, etc. What is needed? Is it available? What are the alternatives?

Project costs   There are two main determinants: the costs of the resources demanded by the project, and the duration of the project. In predicting each of these, an indication of the degree of certainty is needed. If either is likely to be exceeded, the possible consequences must be considered.

For example, if the design project were to take twice as long as first estimated and costs were roughly doubled, but expected returns from sales were still the same (or even less because a competitor was able to 'cream' the market), would the venture still be worthwhile? This should be spelt out in the project proposal; the *risk* associated with cost, time and resources can and must be evaluated before designing starts.

Project duration   In addition to the points just discussed, duration needs to be considered in relation to market windows, e.g. an annual trade show or similar unalterable deadline. Also, regard must be paid to other projects within the company which might be competing at certain times to use skills or accommodation. The *phases* or milestones of the project are important and should be recorded – including, for example, when the review points will occur and when funding instalments will need to be made available.

Obtaining Approval for the Brief

The design manager or project manager can save a lot of trouble during later stages of the project if a little time is spent before the start of the project making sure that all who need to be involved know what is proposed and approve of it.

The best way of obtaining approval is by making sure that people feel they have been personally involved in formulating the brief (even if they only contribute a very small part) rather than presenting the document out of the blue for approval. The temptation needs to be resisted to leave out awkward or nitpicking colleagues who really ought to be involved; eventually they will find out what is happening and insist on taking revenge at the most inconvenient time!

Especially in the case of major projects, approval and authorization may be needed from the highest levels of the company. The involvement approach while drawing up the brief may not be feasible – but it is important to get formal acknowledgement that the document has been seen by, and is supported by, top management. Also, make sure that any subsequent revisions after the project are communicated and agreed. The real problem is in those firms where the top managers just do not want to play a part in design decision making. Here the design manager's persistence may need to be great indeed.

Conclusion

The decision to set up a design project should not be lightly taken. Alternatives to designing include the purchase of a new product from another manufacturer or the search for new markets for an existing product. If, after considering these options, the go-ahead is given for a project, a proper brief should be drawn up. This should detail the full implications in terms of product features and costs, staffing and skill requirements, and project costs and duration.

21 The Nature of the Design Process

KNUT HOLT
University of Trondheim, Norway

Introduction

The design process varies from the design of an atom to the design of the universe, from the design of a new corporate strategy to the layout of an advertisement, from the design of a fashion hat to the design of a nuclear power plant, from the – you name it. In such a situation it is difficult, if possible at all, to find a common denominator or a useful general model. The perception of the design process depends on the project as well as on the person performing the task. This chapter views the design process from four different angles.

View 1: Design as a Problem Solving Process

The problem solving process can be presented in numerous ways. For analytic purposes it is practical to identify three models.

The analytic design process The basic steps are shown as a checklist in table 21.1. This should be considered as a thinking guide. One may start by going quickly through the five basic steps to get an overview and then work through each step in the list.

The first step is often neglected by practitioners. This may be due to the nature of technical education, where the emphasis mostly is on the analysis of the problem as given. In the literature on problem solving, great attention is paid to the development of alternatives, and the selection and implementation of solutions. The importance of defining the problem by asking the right question is often mentioned but little information is given about how to do it. This is unfortunate. The definition of the problem is the most important step, as it determines the scope and direction of the following steps. It includes the assessment of the problems and the needs of those involved – users, employees, owners, the community and society at large. Next comes determination of internal constraints in terms of time as well as human, physical and financial resources, and external constraints such as laws and regulations. By matching needs and resources one determines objectives for the project. If necessary, the project is divided into subproblems, each with an assigned priority.

Table 21.1   Checklist for the analytic design process

Definition of the problem
Description   Make a preliminary description of the situation including problems and needs of those involved
Controllable factors   Indicate factors which hamper and promote the solution
Simplification   Make simplifying assumptions
Constraints   Indicate internal and external constraints
Character   Indicate type of problem (technical, economic, social, etc.), importance, depth, time for development of solution, reversibility, consequences, and influence on other departments/persons
Definition   Indicate objectives that must be reached
Priority   Divide the problem into subproblems with priorities

Collection of data
Facts   Get all facts that can be controlled in a practical manner
Assumptions   Develop forecasts or make assumptions where it is not possible to provide reliable information
Reliability   Evaluate facts, feelings and attitudes

Analysis of data
Models   Clarify relationships by means of mathematical, graphical or conceptual models
Principles   Apply those that are relevant

Development of alternatives
Requirements of the solution   See to it that it satisfies the objectives and can be implemented within existing constraints
Description   Make a description of each alternative
Quantitative factors   Indicate the size of factors that can be measured or calculated
Qualitative factors   Indicate and evaluate importance of factors that cannot be measured
Risk   Evaluate technical and non-technical risk

Selection of solution
Range   Range alternatives after the most important success criteria
Comparison   Evaluate alternatives by means of relevant success criteria
Choice   Make a decision
Check   Check the solution and each step in the process
Implementation   Make arrangements for implementation

Source: Holt (1987)

The problems and needs of those involved, especially the users, can be expressed in a needs specification. The challenge is to determine how important the various needs are and what it will cost to satisfy them. Often there will have to be a tradeoff between needs and costs, so the specification will usually give less than 100 per cent need fulfilment. The needs must be classified in groups related to performance, ease of use, safety, maintenance, status, appearance, service, etc. Another alternative is to organize the needs in hierarchical order, for example primary needs, secondary needs and so on.

The next step is to evaluate and rank the various groups according to their importance. This may be done by questioning potential users. Another approach is to undertake the evaluation by means of pooled judgement by experts; ideally they should have a good understanding of the user situation and of the needs of others influenced by the design of the product. The results may be expressed in a ranking of the various needs according to their importance. One may for example use the following classification: absolutely necessary, very important, important, less important, unnecessary.

This analytic model assumes full knowledge of the alternatives and their consequences. It is a useful approach for design projects with low uncertainty. Such projects are need initiated and characterized by imitations and minor changes.

The iterative design process

This is a modification of the analytic design process. In most situations the designer does not have all the relevant facts. Part of the information may be obtained from other sources, for example by talking to colleagues, reading relevant literature, using consultants or undertaking research and experiments. This may help, but still the designer may lack sufficient relevant knowledge. It may take too much time and money to get the information – even if it is known what knowledge is required at the start of the project. The designer therefore has to make assumptions that will range from educated guesses to well-founded forecasts. The basic steps of the analytic model are followed, but there has to be iteration to provide new data and assumptions, analyse them and undertake corrections and modifications to previous conclusions. The model is best suited to medium risk projects such as radical improvements and adopted innovations.

The visionary design process

This model fits situations where it is not possible to define the problem properly, such as when an entirely new product, process or system has to be created and the designer must break new ground and develop an original solution. In such cases there is not much help from analytic or iterative problem solving models. It is not possible to start the process by defining the problem in terms of the needs of those involved.

On the contrary, the problem will exist only in a vague form or will have the character of a dream, a vision or an intuitive sense of what must be done in order to reach a desired future state. The solution concept and its consequences are unknown or only perceived in an abstract manner; gradually, they emerge through a complex iterative and interactive learning process. The focus is upon knowledge development, as emphasized by Norman (1977) who sees planning under uncertainty as a step-by-step process that starts with a vision. When the consequences of a solution concept have been studied and the conclusions evaluated, more information is acquired and analysed, and the vision is adjusted.

A person with vision has the capability of anticipating future conditions, of perceiving opportunities ahead of anybody else, and of communicating the vision in a simple and easily understood form. A successful inventor, Rabinow (1977), claims: 'When you invent something really new, you create a need. The big trick is not to invent to satisfy a need – anybody can do that. The trick is to recognize a need that people don't realize is a need. You create the need when you put the product on the market.'

This statement is somewhat exaggerated. There are many examples of innovations based on well-recognized user needs. However, tremendous benefits may be obtained if one is able to sense something that will satisfy a need that is not yet perceived as such by those concerned. Some designers have been so successful with their innovative effort that they have created not only new products but whole new industries. Such a result is most likely to happen in companies that attempt to obtain technological leadership by developing new products with superior functional performance. As soon as a solution that works is found, it is quickly brought to the market in order to reap the benefits by being first.

However, many fail, often due to lack of market understanding. This is particularly true for those high tech companies which operate like nineteenth century inventors rather than twentieth century innovators (Drucker 1986) and apparently believe in the dictum 'If you invent a better mousetrap, the world will beat a path to your door.' It does not occur to them to ask what makes a mousetrap better. They are infatuated with their own technology, believing that quality means technical sophistication rather than value for the user.

This often gives the company with a 'follow-the-leader' strategy an opportunity. By looking at the product from the point of view of the user, it can exploit the results of the pioneer, for example by adding new features or by providing different versions for different segments of the market. It is a risky venture to base an innovation on a vision. In addition to lack of technical and market knowledge, one may run into human problems and find different and even conflicting views on what is desirable (Checkland 1979). The ranking of needs is a value judgement based on changing criteria. A typical example is the increased emphasis given to the quality of working life and to environmental and ecological problems. In such situations, one should make different views explicit, work out implications and test them against other views which may be equally valid within other frames of reference. The visionary model of the design process is relevant for high risk innovations of a basic or incremental nature.

View 2: Design as a Creative Process

The creative process is a mental activity that can be divided into four stages:

Preparation   Includes recognition of a need or the wish to accomplish something; thinking and deep involvement; provision of information by experiments, by reading or by talking to people; definition and reformulation of the problem.

Incubation   Characterized by a number of subconscious processes of an intuitive nature; the problem is dropped for a while from the conscious level.

Illumination   Characterized by synthesis and creation of an idea which is brought from the subconscious level to the conscious level; this is the ah-ha or eureka moment.

Verification   Includes evaluation, elaboration and refinement of the idea.

Creative behaviour is needed at all steps of the design process. It is most important at the first steps, which are concerned with generation of the basic idea or concept. The result of this will determine the direction of the subsequent steps which cover the processing of the idea into a useful product, process, system or method and the practical application of the solution. The creative design process is needed at all levels; the higher up in the hierarchy, the more important it is. The strategic far-reaching decisions are taken at the highest management level. Although creativity is highly connected with innovation, it is also needed in minor changes and improvements. Here is the area where most employees have the opportunity to be creative.

Creative ability varies between people. Of particular importance are the highly creative designers. They should be well taken care of. This requires that one accepts their lack of ability to follow rules and procedures, and shows understanding for the eccentric form of behaviour that many of them demonstrate. If one is able to do this – and that can often be difficult – considerable benefits will result for both the company and the individual.

The need for creative behaviour is so strong in most companies that it is not possible to rely only on a few specially gifted individuals. A determined effort must be made to utilize all creative resources. It is increasingly being recognized that most employees can contribute with valuable ideas. The best results are obtained through participation in groups based on the systematic application of creative techniques such as brainstorming, morphological analysis, forced relationships or synectics.

Computer assisted creativity   By means of creative techniques one is able to generate a large number of ideas; 50 to 100 ideas are often the output of a half-hour session. The next step is to screen and evaluate these ideas and select the best one among them. A computer can be of help in both cases. It contributes to systematizing the idea generating activity through its ability to record, store, manipulate, structure, edit, communicate and present information. Computer aided creativity (CAC) provides more time and opportunity for creative thinking by removing much of the routine work. By computer support the evaluation process is more systematic, more consistent and less time-consuming.

The computer can be used in support of individual techniques or as an integrated tool combining several techniques. An example of such a tool is IDEGEN developed by Virkkala (1985) which can be used for creative problem solving by both individuals and groups. The program is by and large self-explanatory and consists of eight activities which can be selected from a menu. When used on real problems it mixes in a practical way advice, personal experience of the problem solver, and theory. During an idea session the user makes short, sketchy notes via the keyboard or by hand. The editing is done after the session by means of a separate text handling program.

Another example of a CAC program is the Idea Generator distributed by the Centre for Commercial Innovation, Duiven, Holland (Trost 1987).

Organizational climate

In order to realize the creative potential of employees, with or without the support of computers, an organizational climate that stimulates creative behaviour is required. It can be described and measured with a set of factors

Table 21.2 Form for measurement of the organizational climate

---

Measurement of the organizational climate   Date:

Indicate how the situation is within your department/group for each of the factors below by marking one of the numbers on each of the scales. Managers indicate how they feel that the situation is for their subordinates. Employees indicate how they perceive the climate in their work situation.

Time for creative activity   Have sufficient time to think, read, study, discuss, experiment, invent, etc.

| Agree 100% | 5 | 4 | 3 | 2 | 1 | Disagree 100% |

Reception of new ideas   Ideas and proposals are received positively and evaluated objectively independent of status and age of the persons submitting the idea.

| Agree 100% | 5 | 4 | 3 | 2 | 1 | Disagree 100% |

Attitude of supervisor   Stimulates creative behaviour by giving encouragement as well as professional advice and support.

| Agree 100% | 5 | 4 | 3 | 2 | 1 | Disagree 100% |

Recognition of creativity   Good ideas and proposals are given proper recognition.

| Agree 100% | 5 | 4 | 3 | 2 | 1 | Disagree 100% |

Interaction with others   Have freedom to establish contacts inside and outside the firm.

| Agree 100% | 5 | 4 | 3 | 2 | 1 | Disagree 100% |

Composition of staff   Have frequent contact with persons of different professional backgrounds.

| Agree 100% | 5 | 4 | 3 | 2 | 1 | Disagree 100% |

Other factors   Indicate on the back of the form other factors that might be of importance for creative behaviour and characterize the situation by means of a five-point scale.

Importance   Indicate with a cross in the margin the three factors which you consider to be most important for a stimulating organizational climate.

Source: Holt (1987)

of the type shown in table 21.2. Studies of the climate may include the whole company, one or several divisions, departments, sections or groups. The results of the measurements can be presented as in figure 21.1.

Studies in both large and small companies indicate that the most important climate factors are 'time for creative activity', 'attitude of supervisor', and 'recognition of creativity'. Studies of the organizational climate are useful in stimulating the creative process. The mere fact that measurements

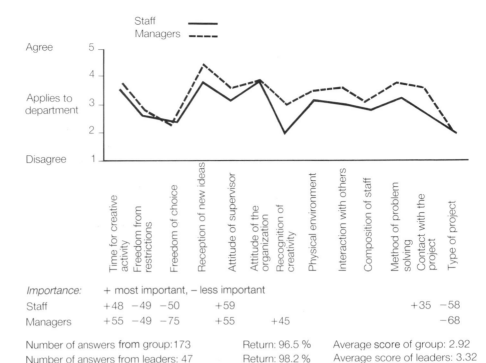

Figure 21.1 Example of
climate profile
Source: Holt (1987)

are made motivates study and analysis; they spotlight weak points in the climate and give a quantitative base for discussions of ways and means of improving it.

View 3: Design as a Need Fulfilling Process

In many situations, quality is an important aspect of the design process. The term *quality* here refers to the achievement of a specific amount of need satisfaction. In many situations one is faced with a quality gap – a discrepancy between needs and product characteristics.

The needs of the user   With increasing demand for product quality, influenced by greater user awareness, consumerism and legislation, the provision of proper information about user needs is an important aspect of the design process. This is now commonly recognized. Millions of words are written each year about the importance of market orientation, of bringing the user's problems and needs to the centre of the planning process. However, it does not appear that the many words have had much impact on industrial practice.

One of the major conclusions of a study of successful American companies (Peters and Waterman 1982) is that in too many companies the users are considered a nuisance despite all the lip service given to market orientation.

In the 'excellent' companies the situation is different. They are characterized by a remarkable user orientation; staying close to users is of vital importance in order to satisfy their needs and anticipate their wants. The situation in Europe is not much different. One of the major findings of a study of German, Italian and Norwegian companies (Holt et al. 1984) is that information about user needs in most cases results much more by chance, or from the informal approaches and intuition of gifted individuals, than from the application of specific methods and well-defined procedures. Few companies apply need assessment methods systematically; those doing so often show a great deal of creativity in detecting the needs of present and potential users.

The chief executive of Brown Boveri (Hummel 1983) claims that an exact assessment of the problems and needs of the user is an essential task. Producers of capital goods for industrial markets cannot create new needs. Their products are determined by the market. Even if this triviality is obvious, it is surprising how often it is neglected. In many cases the interpretation of the needs is wrong. Instead of finding out and basing the product on the real problems of the user, too often the starting point of the design process is the technical solution perceived by the designer. In this way the design effort is headed in the wrong direction and the firm continues to allocate resources to the perfection of the wrong technology. For every company it is of fundamental importance to identify long term problems in the market that require new solutions. Choice of strategy is nothing else than defining the market problems that the company wants to solve in the long term.

In order to properly assess the needs and problems of the market a systematic approach is required. Through the cooperation of researchers in several countries, 27 methods for assessing user needs have been identified and presented (Holt et al. 1984). Seven of them (customer information, staff information, government information, competitor information, trade fairs, literature and expert information) are based on utilization of existing information. They are best suited for revealing rational needs such as those related to function, performance and practical use. These needs determine the basic design and can often be expressed in quantitative terms.

The other methods (user questioning, user employment, user projects, multivariate models, dealer questioning, user observation, active need experience, simulation, brainstorming, confrontation, morphological analysis, progressive abstraction, value analysis, Delphi method, scenario writing, systems analysis, product safety analysis, ecological analysis and resource analysis) require generation of new information. Most of them are concerned with rational needs but several also reveal emotional needs related to status, novelty, form, style, colour, appearance and other characteristics of an aesthetic nature. The creative techniques and the forecasting techniques may help to indicate future problems and needs.

The needs of other stakeholders   In addition to the needs of the user, marketing must consider the needs of the various links in the distribution chain. Then there are the needs of those making the product; the negative aspects of many production systems are now so strongly felt that the quality of working life has become a key issue. Finally, there are societal needs related to resource depletion, energy conservation and protection of the environment. Those responsible for design should take into consideration the needs of all stakeholders, inside and outside the company, that will be influenced by the solution of the problem (Holt 1988). The weight given to the needs of the various groups depends on the situation, the power of those concerned and the attitude of those involved in the design process.

View 4: Design as a Human Activity Process

There is no design process without the designer; indeed, the designer remains the most important part of the process. If we look at the behaviour of the designer, we find many different types (Holt 1987). At one extreme we find the perfectionist, strongly motivated by the problem and the work itself, deriving great satisfaction from inventing and creating something new. But, obsessed with a particular solution, this designer may not be able to see other alternatives – or even want to hear about them. At the other extreme is the standard practitioner who prefers security and whose contribution may be limited to minor improvements. Designing is generally considered to be interesting work, but this may not always be true. It is important that managers should be aware of the factors that may cause frustration amongst designers – and the steps that can be taken to stimulate and encourage outstanding design results.

Frustrated designers Particularly in large firms one often finds designers in a frustrating work situation. They have routine jobs of a highly specialized nature such as upgrading existing solutions. Their work is like an intellectual assembly line controlled by fixed procedures, standards and rules (Cooley 1985). The frustrating situation of many designers is reflected in a study at IBM by Ritti (1971). A comparison with salesmen, field designers, foremen, managers and customer service employees showed that frustration was highest amongst designers. They were seldom responsible for entire projects, unless they were also project managers, and tended to work only on small parts of systems.

Satisfied designers Some designers have challenging jobs with plenty of opportunity for creative behaviour. In small companies they are often responsible for the major part of the design process – from idea generation to final design. Similarly, in large companies we also find designers with interesting work – such as those responsible for the system solution or the basic technological concept.

Routinized work is often dreary but it need not exclude creative behaviour. Even a simple component can be made in many different ways. By applying creative techniques, the designer has a good chance of finding a better design than by following standard practice. The engagement of the designer in the design process depends to a large extent upon the individual, but is also influenced by the organizational climate. Special care should be taken of highly creative designers. One approach is to let designers work on personal projects selected by themselves and approved through an informal procedure with a minimum of bureaucracy involved. In order to encourage such projects in large companies where there may be highly formalized procedures for evaluation and approval of projects, special representatives must be appointed and authorized by the chief executive.

Another approach to stimulate designers to creative behaviour is to give them free time, say 15 to 20 per cent of working time, to spend on their own projects. Many firms have obtained positive results by the free time approach. Others have negative experiences, perhaps because the designers get time for their own projects only after they have finished other urgent tasks. An interesting variant is the concept of illegal research, also called smuggled research, moonlighting or bootlegging. This refers to hidden projects which do not follow the formal evaluation and approval procedures – 'the art of pirating unauthorized time and resources from ongoing operations'. At IBM, moonlighting has been widely used both in the USA and in Europe, and several centres of competence have grown out of such projects (Foy 1974).

The illegal approach depends on innovation-oriented supervisors who are willing to take a personal risk by supporting the creative designer. This has often led to good results but it puts the supervisor in a delicate position; the attendant problems are knowing which designers deserve support and learning when to look the other way to allow for sanctioned but not unofficial projects. One advantage is that nobody gets hurt by a failure. On the other hand, if a project succeeds, it is a surprise and all are happy.

Conclusions

It is not possible to give a definitive description of one standard design process applicable to all situations. However, by the adoption of different viewpoints, including those of customers and of designers, some of the main elements of design processes have been explored and special concerns highlighted.

References

Checkland, P. B.: The problem of problem formulation in the application of a system approach. In Bayraktar, B. A. (ed.): *Education in System Science*. London: Taylor and Francis, 1979.

Cooley, M.: *Architect or Bee?* London: Hogarth, 1985.
Drucker, P.: *Innovation and Entrepreneurship*. New York: Harper and Row, 1986.
Foy, N.: *The IBM World*. London: Eyre Methuen, 1974.

Holt, K.: *Innovation – a Challenge to the Engineer*. Amsterdam: Elsevier, 1987.

Holt, K.: *Product Innovation Management* (3rd edn). London: Butterworths, 1988.

Holt, K., Geschka, H. and Peterlongo, G.: *Need Assessment – a Key to User-Oriented Product Innovation*. Chichester: Wiley, 1984.

Hummel, P.: Interview in *Teknisk Ukeblad*. No. 40, pp. 6–7, 1983.

Norman, R.: *Management for Growth*. Chichester: Wiley, 1977.

Peters, T. J. and Waterman, R. H.: *In Search of Excellence*. New York: Harper and Row, 1982.

Rabinow, J. In Wolff, M. F. (ed.): *Managing the Creative Engineer. IEEE Spectrum*. pp. 2–57, August 1977.

Ritti, R. R.: *The Engineer in the Industrial Corporation*. New York: Columbia, 1971.

Trost, R.: Computer assisted creative thinking and problem solving with the personal computer. Paper presented at the Research Conference of the International Society for Product Innovation Management (ISPIM), Brighton, UK, August 1987.

Virkkala, V.: *IDEGEN – a Computer Programme for Stimulating and Guiding Creative Thinking*. Helsinki: Kone, 1985.

22 A Systematic Approach to Engineering Design

KEN WALLACE

Cambridge University, UK

Introduction

The aim of every manufacturing company is to create products which will sell profitably. These products have to be designed. In a broad sense, design can be considered simply as foreseeing a desirable future situation and planning to achieve that situation. In this sense, every employee in the company is part of the overall design activity, which extends from identifying a market need to supplying a product to meet that need (figure 22.1). The overall creation of a new product is often referred to as product development (Andreasen and Hein 1987) or total design (Pugh 1986). In a more restricted sense, engineering design is the process by which a technical requirement is converted into the detailed information from which a product can be made. This chapter concentrates on a systematic approach to engineering design; it also identifies some of the methods and tools which are available.

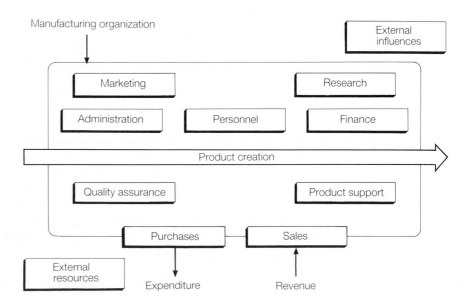

Figure 22.1 Product creation in a manufacturing organization. In most companies there will be a number of product development programmes in progress at any time

In the technological and commercial environment of today, the design process has to be carefully managed (Engineering Council 1986; Design Council 1983). Intense international competition and rapidly changing technology are forcing manufacturing organizations to regularly introduce new products and update existing ones. For a complex product, a large multidisciplinary team is needed and product liability legislation requires accountability from every member of that team.

The engineering design process involves converting information from the general to the particular, from the abstract to the concrete. After many iterations, the resulting information, for example in the form of drawings, software, assembly and test instructions, represents a description of a product which will, if all the predictions are correct, function as required when it is built. The aim of design managers is therefore to produce the best possible product description by maximizing the effectiveness and efficiency with which the information is processed.

Product Life Cycles

All products, whether a power station (large) or a motor car (medium) or a paint brush (small), follow a broadly similar chronological sequence, often referred to as the product life cycle.

A typical life cycle for a medium-sized product, such as a motor car, is shown in figure 22.2. It is split into a number of main stages with an appropriate input and output for each. The sequence starts with a product idea, often the result of careful market research. Once the need is established, the product must be planned and its feasibility determined. In the next stages, the product is created by being designed, developed and manufactured. Only seldom is it possible to get things right first time, and costly prototype development is required. For example, motor car manufacturers build and test many prototypes before a new model goes into full-scale production – a point that is often referred to in their advertising campaigns. The result is a new product which can be sold and earn revenue for the company. The product is then delivered to the customer, which can involve packaging, transport, installation and commissioning. During use, maintenance and repair may be required. Over a period of years, the design is continually refined as information about the performance of the product is fed back to the design team. Eventually the product becomes worn out or obsolete and has to be disposed of. As many elements and materials as possible are recycled, and the rest are scrapped. The information and experience gained often triggers off fresh product ideas and the cycle repeats itself. For a larger or a smaller product the overall sequence is essentially the same, expanded or contracted appropriately.

All the stages of the life cycle must be carefully considered during engineering design. The earlier stages determine the input to the design team and

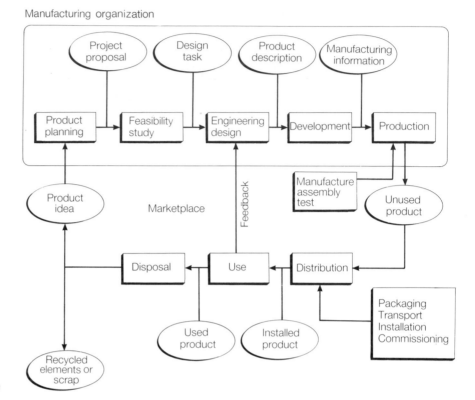

Figure 22.2 Product life cycle

the success of the subsequent ones is largely determined by the output from the team.

Engineering Design

There have been various attempts to describe the design process and to prescribe a general approach for tackling design tasks (Verein Deutscher Ingenieure 1987; Pahl and Beitz 1984; Hubka 1982). In common with most strategies for tackling complex problems, the design process is broken down into a number of main phases or stages and these in turn are broken down into an appropriate number of steps, depending on the complexity of the task. Many methods (Jones 1980) and tools are available. Some are general and can be used at any time and at any level of resolution; some are more specific and can only be used for one particular stage or step. The aim of a systematic approach is to make the design process more logical, visible, transparent and comprehensible. These qualities are assuming increasing importance as the design process becomes more complex and as computers are being used more extensively.

A systematic approach is intended not to replace creativity, intuition, experience or insight, but to support and enhance these essential qualities. It must be applied flexibly, and adapted to suit the particular task and context.

A description of the technical requirement provides the input to the engineering design process (activity). The output is a description of the product or, in more general terms, a description of the technical system to

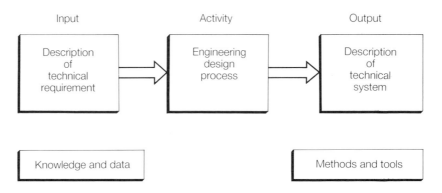

Figure 22.3 Engineering design activity

be manufactured (see figure 22.3). The design process draws upon the available knowledge and data and employs numerous methods and tools. It is essentially an information processing activity. So before considering the more detailed stages in the engineering design process, it is useful to consider information processing and technical systems in general.

Information Processing

At all levels of resolution the basic information processing activities are directed towards problem solving, and these activities require careful management as summarized in figure 22.4.

The recommended procedure for solving problems of all types is first to break the problem down into appropriately sized subproblems and then to apply the following steps to each subproblem:

1 State the problem clearly.
2 Establish the criteria.
3 Generate options.
4 Evaluate the options using the criteria.
5 Decide on the best option.

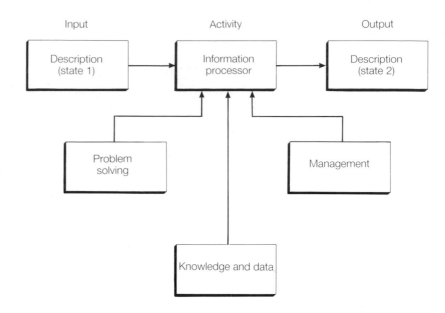

Figure 22.4 Information processing

In practice, considerable feedback and iteration will be involved. Finally the various subsolutions are combined to obtain an overall solution to the problem. The management of the problem solving activity involves the following steps:

1   Plan the programme and resources.
2   Communicate the plan.
3   Monitor and control progress.
4   Review the outcome.

Again the process is iterative. Effective communication is one of the most important factors in achieving a successful result.

Technical Systems

The physical structure of every technical system can be split into appropriately sized units as in figure 22.5. The overall product, for example a motor car, can be split into a number of major assemblies (engine, transmission, suspension, etc.), and these can be further divided into subassemblies (carburettor, gearbox, shock absorbers, etc.). Each subassembly is made up of standard components (fasteners, bearings, seals, etc.) and individual parts

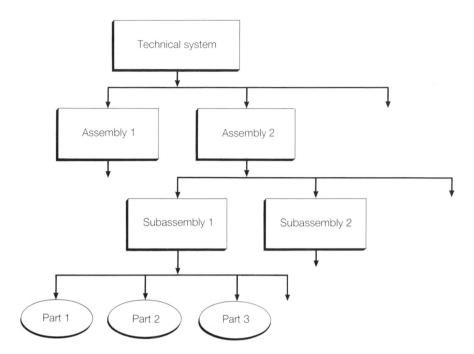

Figure 22.5 Physical structure of a technical system

designed and manufactured specifically for the particular application (actuating lever, gear wheel, casing, etc.).

The design process determines the physical structure. Sufficient information must be provided to manufacture every part and purchase every standard component. Further information must be provided to define how all these items are to be assembled into the final product. Any faults which occur during manufacture, assembly, testing or use will have to be rectified. The higher the level and the earlier in the process at which an error occurs and the longer it takes to discover, the more expensive it will be to rectify. This is not to suggest that faults at the detail level are unimportant. The majority of designs are incremental improvements to existing ones, so it is often the quality of the detail design which distinguishes one product from another. For example, the underlying concept of most washing machines is the same. It is the detail design which largely determines the differences in performance, appearance, ease of use, reliability and maintainability, and these are the factors which influence purchasing decisions. In addition, through careful attention to detail, the manufacturing costs can be substantially reduced, thus providing the manufacturing organization with the ability to sell at a competitive price and make a satisfactory return.

Design Process

It is usual to break down the design process into a number of main phases or stages. These phases can then be further divided into any number of detailed steps. Precisely where the divisions occur and how many detailed steps are chosen depend on the size and complexity of the proposed product and the type of manufacturing company involved. With a complex project requiring a large team, for example a power station, the design process may be formally split into many phases and steps with reviews held at important milestones. For a smaller project, such as the design of a new paint brush, the total design process might be the sole responsibility of one person. In this case, although a similar sequence of activities takes place, the design process will not be split into so many detailed steps and the management of the whole process can be less formal.

Many models of the design process can be found in the literature (Andreasen and Hein 1987; Pugh 1986; Verein Deutscher Ingenieure 1987; Pahl and Beitz 1984; Hubka 1982). Despite differences in their visual appearance, their division of the phases and their terminology, they all agree on the main sequence of activities. The purpose of the models is to represent an idealized approach, based on observations of best practice over many years, and thus assist the management of the design process. It is clearly understood, however, that conditions are seldom ideal; so the models, and the methods associated with them, must be applied in a flexible way and adapted to suit the particular situation and its context.

The model selected is taken from VDI 2221 (Verein Deutscher Ingenieure 1987) and is shown in figure 22.6. This model breaks down the design process into seven stages and the description of the stages which follows is adapted from that given in VDI 2221.

Stage 1: clarify and define the task

It is necessary to define and clarify the requirements of the task, which may have been expressed in vague and contradictory terms, by collecting all the information available and discovering where there are gaps; checking and supplementing external requirements; adding specific company requirements; and defining and structuring the task from the point of view of the design team.

During this stage, it is worth spending some time identifying the true need to be fulfilled. This should be expressed as a short statement of the task which gives no indication of any means by which it might be solved – in other words, a solution-neutral problem statement. A useful technique is to consciously raise the level of abstraction using the following steps:

1   Eliminate all requirements that have no direct bearing on the proposed function.
2   Change quantitative statements into essential qualitative ones.

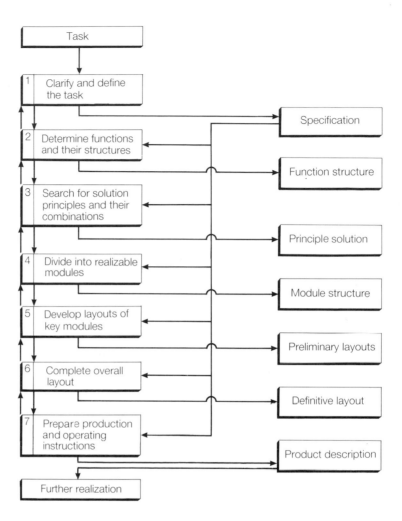

Figure 22.6 Stages of the design process
Source: Verein Deutscher Ingenieure (1987)

3   Formulate the task in solution-neutral terms at the appropriate level of generality.

For example, consider the following design task:

Design a 1 kW lawnmower to cut grass. It must be possible to use it in both wet and dry conditions.

This statement is certainly not solution-neutral; it suggests that a particular type (lawnmower) and size (1 kW) of device should be designed and that this device should operate in a particular manner (cut) under certain conditions (wet and dry). An improved statement, at a higher level of abstraction, would be:

Devise a means of keeping grass short.

This immediately increases the solution search space to include chemical or biological retardants or, in an extreme case, developing a synthetic grass.

The output from this stage is a specification, or a requirements list, which can be established independently of any solution. The specification is an important working document which should be constantly reviewed and kept up to date. Important findings in the course of the design process can lead to existing requirements being modified and new requirements being added. All modifications should be undertaken formally and on a regular basis.

Stage 2: determine functions and their structures

The functional description of any product relies on the fact that every technical system can be modelled in terms of the flows and conversions of material, energy and information which take place within its system bound-

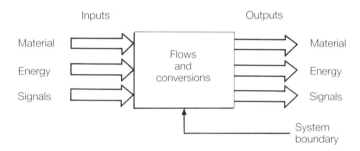

Figure 22.7 Flows and conversions in a technical system
Source: Pahl and Beitz (1984)

ary (figure 22.7). For example, in a motor car engine the main flows and conversions are those of energy, in a concrete mixer those of material, and in a television set those of information (signals).

First the overall function and then the most important subfunctions (main functions) to be fulfilled by the product being designed are determined. The classification and combination of these subfunctions into structures form a basis for the search for solutions for the overall product.

The output is one or several function structures. These are usually presented as formal diagrams but, in some cases, simple descriptions suffice. Figure 22.8 shows an example from Pahl and Beitz (1984) of the overall function and a possible function structure for a machine to pack carpet squares into lots of a specified size.

Stage 3: search for solution principles and their combinations

A search is made for solution principles for all subfunctions, or initially for the most important subfunctions of the function structure. Physical, chemical and other effects need to be selected for this purpose, and these must be realizable in principle by embodiment features. In the case of mechanical systems such embodiment features include, for example, the geometry, the

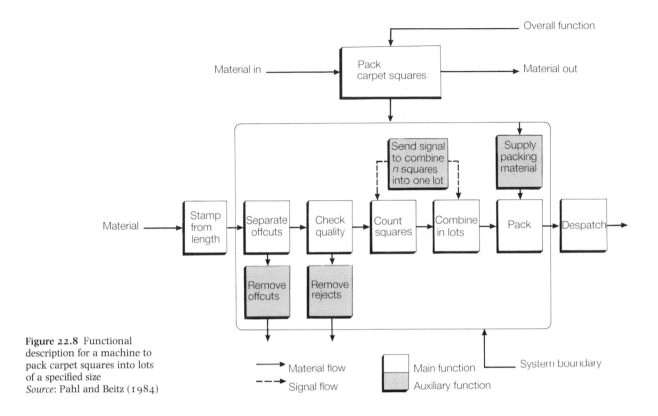

Figure 22.8 Functional description for a machine to pack carpet squares into lots of a specified size
Source: Pahl and Beitz (1984)

motion and the type of material. The solution principles discovered for subfunctions must subsequently be combined in accordance with the function structure. In so doing, further subfunctions (auxiliary functions) may become apparent, and these in turn make possible the realization of certain effects or solution principles.

The output is a principle solution which represents the best combination of physical effects and preliminary embodiment features to fulfil the function structure. It may be documented as a sketch, a diagram, a circuit or even a description.

Stage 4: divide into realizable modules

The principle solution is divided into realizable modules, before starting the complex and time-consuming process of defining these modules in more concrete terms.

The output is a module structure which, in contrast to the function structure or principle solution, provides a preliminary indication of the breakdown of the solution into the realizable groups and elements (assemblies, subassemblies and parts) which, together with their links (interfaces), are essential for its implementation. This can be represented in the form of layout drawings, process flow charts or circuit diagrams. A module structure is particularly important in the case of complex products as it facilitates the efficient distribution of design effort. It also helps with the identification and solution of embodiment design problems. Thus, for example, a distinction is made between: design modules limited according to working principle; assembly modules, allowing easy assembly; maintenance modules, allowing easy maintenance; recycling modules; and basic and variation modules, allowing a modular product system.

Stage 5: develop layouts of key modules

The key modules are further developed. The level of refinement of the geometry, materials and other details should only be pursued as far as to allow the optimum design to be selected.

The output is a set of preliminary layouts for the key modules, which can be represented as scale drawings, circuit diagrams, etc.

Stage 6: complete overall layout

The preliminary layouts of the modules are completed by the addition of further detailed information about the assemblies, standard components and parts not previously included, and about their combination into a complete product. Often it is possible to define those modules not included in stage 5 by selecting standard or commercially available ones.

The output is a definitive layout containing all the essential configuration information for the realization of the product. The main forms of representation are scale layout drawings, preliminary parts lists, instrumentation flow charts, etc.

Stage 7: prepare production and operating instructions

All the final production and operating instructions for which the design team is responsible are prepared, and the quality (British Standards Institution 1987) and production aspects of the design must be thoroughly reviewed and checked (British Standards Institution 1981; Institution of Production Engineers 1984). This stage thus overlaps the preceding one.

The output is a complete product description, in the form of detail and assembly drawings; parts lists; and manufacturing, assembly, testing, transport and operating instructions.

In all these stages, several solution variants are analysed, often tested in the form of models or prototypes, and then evaluated. The activities of problem solving and management take place in all the stages, but they have not been shown in figure 22.6. It must be emphasized that the stages do not necessarily follow rigidly one after the other. They are often carried out iteratively, returning to preceding ones, thus achieving a step-by-step optimization. For design management purposes, it is often convenient to consider the design process split into just four main phases (Pahl and Beitz 1984):

Clarification of the task (stage 1)
Conceptual design (stages 2 and 3)
Embodiment design (stages 4, 5 and 6)
Detail design (stage 7).

If the amount of time spent on each phase is recorded for a number of design projects, then a profile for the type of products designed by a particular organization can be built up; this will greatly improve future resource planning. A smoothed project profile based on data recorded during the design of a materials test facility is shown in figure 22.9 (Hales 1987).

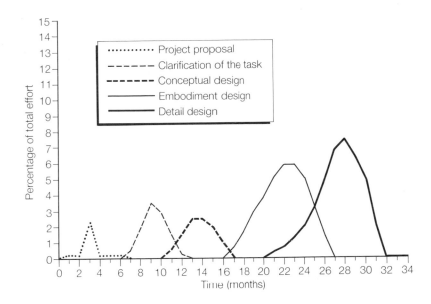

Figure 22.9 Project profile for the design of a materials test facility
Source: Hales (1987)

To assist with the various stages of the design process, there are numerous methods available such as brainstorming, morphological analysis, cost–benefit analysis and fault-free analysis. There are also numerous tools increasingly associated with an interactive computing environment, such as modellers, simulators, optimizers and data bases. A useful list of methods and tools, classified according to their applicability in each of the seven design stages, is given in VDI 2221 (Verein Deutscher Ingenieure 1987). A more detailed description of many of them is given by Pahl and Beitz (1984).

Conclusions

Designing technical products is an extremely complex activity. The aim during the design process is to get the design as near to being right first time as possible. Alterations and corrections to the design are relatively easy and cheap to make while it is still on paper, but much more difficult and expensive once the product has been manufactured and is possibly in use. However, because of the difficulty of accurately predicting the performance of a product at the design stage, it is extremely rare to get all the predictions right. The expensive development programmes common in many industries, during which prototypes are built and tested and the design refined, highlight this point.

The management advantages of adopting a systematic approach to engineering design, particularly where large teams are involved, can be summarized as:

Improved planning, monitoring and control
Clearer role identification of the team members
Better control of the information flows and decisions
More effective use of computers.

The more effectively and efficiently the design process is managed, with the help of all the available methods and tools, the greater are the chances of realizing a successful product.

References

Andreasen, M. M. and Hein, L.: *Integrated Product Development.* IFS/Springer, 1987.

Besant, C. B.: *Computer-Aided Design and Manufacture* (2nd edn). Ellis Horwood, 1983.

British Standards Institution: *The Management of Design for Economic Production.* PD 6470. London, 1981.

British Standards Institution: *Quality Assurance.* BSI Handbook 22. London, 1987.

Design Council: *Design and the Economy.* London, 1983.

Engineering Council: *Managing Design for Competitive Advantage.* London, 1986.

Hales, C.: *Analysis of the Engineering Design Process in an Industrial Context.* Eastleigh: Gants Hill, 1987.

Hubka, V.: *Principles of Engineering Design* (ed. Eder, W. E.). Guildford: Butterworths, 1982.

Institution of Production Engineers: *A Guide to Design for Production.* London, 1984.

Jones, J. C.: *Design Methods.* New York: Wiley, 1980.

Pahl, G. and Beitz, W.: *Engineering Design* (ed. Wallace, K. M.). London: Design Council, 1984.

Pugh, S.: Design activity models: worldwide emergence and convergence. *Design Studies.* Vol. 7, no. 3, pp. 167–73, July 1986.

Verein Deutscher Ingenieure: *Systematic Approach to the Design of Technical Systems and Products.* VDI 2221. Beuth: Berlin, 1987.

23   Conceptual Engineering Design

DAVID G. JANSSON
Texas A & M University, USA

Introduction

Engineering design is a complex, multidisciplinary process made up of a wide variety of activities. Its complexity is further enlarged because it unfolds within the dynamics of the larger process of technological innovation, spanning the spectrum of processes from the recognition of a need in the marketplace through to the establishment of a mature market for a new product generated to meet the recognized need. Hales (1986) has created a contextual model for the engineering design process, providing an informative and helpful view of the many influences and environments which complicate the practice of engineering design. Wallace, in chapter 22 of this book, presents an excellent systematic view of the complete engineering design process.

This chapter focuses on a particular part of the engineering design process – conceptual design. Conceptual design can be thought of as the front-end portion of design during which new ideas are generated or configurations are created or selected to meet the demands of an identified need. The goal of the conceptual design process is the establishment of the core technical concept about which the remainder of the design will be built. Thus, although it can be viewed as a small part of a very large picture, it is an extremely important segment of the whole. Downstream from the conceptual design process are many other activities which can be called simply design execution. Design execution may be thought of as the realization or transformation of the core technical concept into a product.

This chapter deals with a few fundamental questions about conceptual design and attempts to establish some principles with regard to the process. In fact, the principles can be viewed as much more widely applicable than just to the engineering design process for which they have been developed. The questions are as basic as 'Why is it so difficult?', 'Where do new ideas come from?' and 'Why didn't I think of that?' These questions seem almost trivial or trite but they are fundamental to conceptual design, and an attempt to answer them helps to elucidate the subject.

Generating Ideas

First, let us consider why conceptual design or the generation of new ideas is such a difficult task. Implied, of course, in this question is the task of generating very good ideas as opposed to just generating ideas. It is quite a simple matter to generate many, many ideas when the demands upon you do not include the quality or goodness of the ideas. As we all know, many more ideas are generated than are developed; many more are developed to some level of completion than are made into products; and many more products are made than are successful. The real issue comes down to generating ideas of high quality and of good value to others.

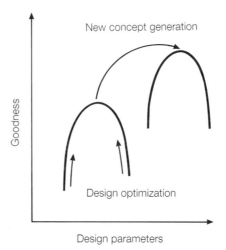

Figure 23.1 The relationship between conceptual design and design optimization

Figure 23.1 is a schematic view of why conceptual design is fundamentally more difficult than many other parts of the design process. The result of conceptual design is, hopefully, a new concept or configuration. It is this newness which makes the process so elusive. A new configuration which heretofore did not exist has been created, and the process of reaching that point is very different from the activity which takes place during the refinement of the concept downstream in the process. The horizontal axis in figure 23.1 is a one-dimensional representation of the many design parameters which can be varied within a configuration. The vertical axis is a one-dimensional representation of the quality or goodness of a concept. Such a measure may be imagined to be the ratio of performance to cost since, fundamentally, we would all like to pay less for more performance. In any event, such a measure is a convenient way to describe the desirability of concepts or configurations.

Consider now the curve on the left and imagine that this curve is actually a multidimensional surface in space and that the points on this surface represent various combinations of the design parameters which can be reached. All points on the surface are various realizations of a single configuration.

At this point, a simple example will help. Imagine that the curve on the left describes a fountain pen. That is, the core technical concept is described by the fact that the pen contains a nib through which ink flows from some sort of a reservoir to the paper. Points on this surface may describe pens that range from a quill pen to a sophisticated refillable fountain pen to a fountain pen which has a very wide tip used for calligraphy. These pens differ in the values of many design parameters but the core technical concept is still the same. Movements from one point on the surface to other points on the surface may come relatively easily by making minor changes to a design parameter. These might include changing the type or size of the reservoir of ink, changing the physical dimensions of the nib or making small changes to the properties of the ink. All of the realizations are still on the surface because they all share the same core technical concept. Each of the points represents one realization of the common attribute of 'being a fountain pen'. These movements may be described as design optimization – a selection of the volume of existing design parameters in order to meet a performance specification.

In contrast, conceptual design is represented by a leap from one surface in parameter space of design to a new surface, represented by the curve on the right. The new curve may not even be described by all of the same design parameters. Before the leap was made, the new core technical concept did not exist. To extend the example, imagine that the new curve represents a ballpoint pen. In fact, some of the design parameters that are pertinent to the fountain pen also are important to ballpoint pens. However, there are a number of design parameters which these two configurations do not share and there is a fundamental difference between the core technical concepts of a fountain pen and a ballpoint pen. All points on the curve to the left are fountain pens, all points on the curve to the right are ballpoint pens, and the generation of the ballpoint pen concept was not obvious in the consideration of the fountain pen designs. It is because this process is a leap to a non-existent curve that conceptual design is so difficult.

This discussion of the generation of new ideas has been only philosophical. However, since the preponderance of engineering practice and education is directed towards analytic modelling and refinement of configurations, represented by motions on the surfaces in figure 23.1, this view should help to clarify why the creation of truly new, high quality configurations or ideas is an event too rare in engineering practice today.

This chapter considers the nature of conceptual design based on very simple but profound ideas. The discussion draws on work which includes both observation of, and participation in, the conceptual engineering design process. A framework for a model of the process is described and then a conceptual design methodology based on this model is presented. Finally, some comments will be made on the implications for the management of engineering design.

Nature of the Process

Study of conceptual design allows us to make the following observations relevant to a description of the process:

1 Human creativity is more successful or productive when it is used in the search for solutions to simple rather than complex problems.
2 The best ideas are usually quite simple conceptually.

These notions need some explanation. The first deals primarily with the number of issues or factors which must be considered during the solution process. Thus low order problems, i.e. those which require consideration of very few parameters or issues, tend to be much more easily solved than higher order problems, i.e. those with a large number of variables or factors to be considered. That is why all of us try to break down problems into smaller pieces in an effort to make them more tractable. This simple notion comes as no surprise but has profound implications for the nature of successful creative processes.

The second observation relates to the simplicity of the results of the creative process. It is not that the resulting products based on these good ideas (core technical concepts) are simple but that the concepts themselves are most often based on simple but creatively new insights. Often these simple ideas can be described as resulting from transformations of the key issues in a problem into new relationships or sets of relationships uniquely identified by the creative designer. Thus we often ask ourselves, 'Why didn't I think of that?' In retrospect, these ideas are simple but the pathway to reach them may not be at all obvious.

These notions lead to parallel conclusions regarding a model for the conceptual design process. Figure 23.2 presents a simple schematic model. The output of the process is generally a physical object or group of objects which here are called configurations. Thus the various outputs as the process unfolds can be thought of as elements in configuration space. However, consideration of the above observations concerning problems and solutions within conceptual design indicates that movement from one point in configuration space to another point in configuration space is generally made at the level of concepts. Therefore, the model also contains another space, called concept space, in which are contained the ideas or concepts that are the basis for the elements of configuration space. Thus the process can be viewed as an iterative process of moving from configuration space to concept space, making changes or movements within concept space, and then

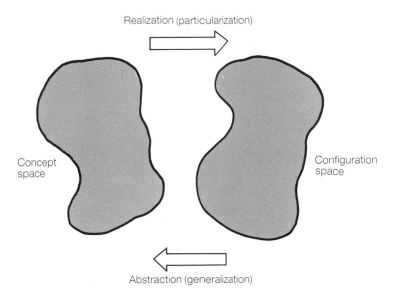

Realization (particularization)

Concept space

Configuration space

Abstraction (generalization)

Figure 23.2 A theoretical model of conceptual design

moving back to new points within configuration space corresponding to further creative generation of configurations based on these new ideas.

The process of moving from configuration space to concept space can be thought of as abstraction or generalization; that is, generalizing from the particulars of a physical configuration to a conceptual understanding in order to change it or improve upon it. Movement from concept space to configuration space can be thought of as realization or particularization – bringing to reality, in particular physical form, the technical concepts arrived at within concept space.

Readers should note that, in this process description, the elements of configuration space are not actually physical objects but only diagrams, sketches or other representations of physical objects motivated by the conceptual elements of concept space, which bring some real form to the thoughts created in concept space.

This model should be thought of as much more than a philosophical statement. Readers are encouraged to examine their own experiences in conceptual design to be further convinced of its significance. As ideas come to life, there are many points along the way which are either intermediate configurations (elements in configuration space) or conceptual statements or relationships (elements in concept space). Rarely within this process is a new configuration created from the previous one without an excursion to concept space, with new conceptual insight being the driving force for the new configurational result.

Methodology

This understanding of the nature of the conceptual design process leads us to generate a methodological approach to carrying out conceptual design. We must first state very clearly that this approach is not to be considered a method but, rather, a methodology. The importance of this understanding cannot be minimized since the word 'method' implies that a procedure that unfolds in an orderly and sometimes very predictable manner results from such an approach. As Rittel (1972) and Broadbent (1979) have observed, successful design methodologies must be much more open, iterative and flexible than a highly systematized method would allow. The methodology presented here, therefore, is a general approach based on the fundamental notions regarding the nature of conceptual design. A methodology gives one the general direction of how to approach problems rather than outlining a rigid procedure.

Readers should also appreciate that the structured framework in which the methodology is presented is only for the purpose of being able to describe the approach in an effective way. Thus, although the methodology appears in a very graphical and orderly form, this is only to illustrate the general principles upon which the methodology is based.

Finally, a general word about problem solving techiques is in order. As a way of describing various techniques to enhance creativity and, in particular, conceptual engineering design, one might describe various approaches as being environmental or methodological. What is meant by environmental here is that creativity is encouraged or stimulated by establishing an environment in which creativity is more likely to occur. Methodological approaches look at the enhancement of the creative process from the point of view more of active intervention in the unfolding of the process than of simply causing conditions to be met that will encourage creativity.

One methodology for conceptual design presented here is called parameter analysis (Jansson 1980, 1987) and is represented by the schematic of figure 23.3. Figure 23.4 indicates the correspondence between the theoretical description of the process and this methodological approach. The three types of activity within parameter analysis form an iterative loop. The first element, *parameter identification*, consists primarily of the recognition of the dominant parameters or issues in a problem. The word 'parameter' is used to describe in a very general way any issue, factor, concept or influence that plays an important part in developing an understanding of the problem and pointing to potential solutions.

The best descriptions of what takes place within parameter identification are simplification and transformation. The parameters within a problem are not fixed but rather those which are creatively recognized as the process moves forward. The recognition of interesting new parameters, based on

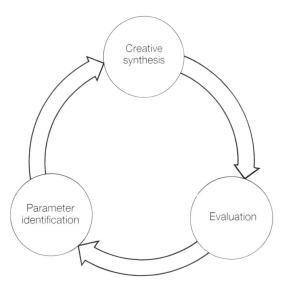

Figure 23.3 A schematic of the parameter analysis methodology

transformations of more commonly understood parameters, triggers new, unique solutions to problems. Also, temporary purposeful oversimplifications of a problem, to focus on only one of the key parameters which characterize the problem, help to concentrate creative activity so that it can produce

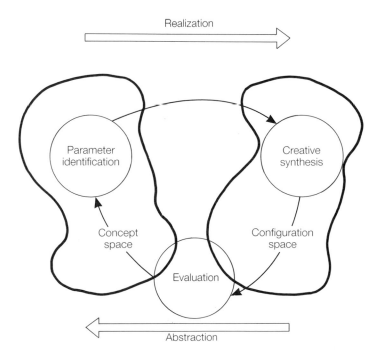

Figure 23.4 Correspondence between theory and the parameter analysis methodology

quality ideas. The identification of these key parameters through simplification and transformation does not yield results unique to a problem but is a highly creative process.

The second part of parameter analysis is the process of *creative synthesis*. In particular, this part of the process represents the generation of a physical configuration based on a concept recognized within the parameter identification process. Since the process is iterative, one should expect that many physical configurations are generated as the process unfolds, not all of which will be very interesting. However, the usefulness of the physical configurations is that they allow one to see new key parameters which will yet again stimulate a new direction for the process.

This process of moving from a physical realization back to parameters or concepts is facilitated by the third part of the parameter analysis methodology, the *evaluation process*. Evaluation is an important step since one must consider to what degree a physical realization is a possible solution to the entire problem. As parameter identification and creative synthesis proceed, one should expect many physical configurations to be generated which are far from being valid solutions to the problem but which, nevertheless, are important to the stimulation of the creative process. Evaluation should not contain analysis of physical configurations that is any deeper than required to create a fundamental understanding of its underlying elements. We might best call this process 'appropriate' analysis. It must be carried out in the light of the entire need, not just with respect to the particular simplified parameter of each iteration.

The major implication of the theoretical model which stands behind parameter analysis is that the burden of truly creative activity is shifted from what here is called creative synthesis – the generation of physically realizable configurations – to parameter identification, the creation of new conceptual relationships or simplified problem statements (all lumped under the term 'parameters') which will lead to the desired configurational results. Thus the task of creative synthesis along the way is only to generate configurations which, through evaluation, will enlighten the creative identification of the next interesting conceptual approach. Each new configuration does not have to be a good solution, only one which will further direct the discovery process.

An Example of Parameter Analysis

As an example of the principles of parameter analysis, we will discuss the creation of a clever mechanical device capable of measuring extremely small angles of tilt with respect to the local gravity vector. Figure 23.5 is a schematic representation of the tiltmeter with no input angle, and figure 23.6 shows the configuration of the device when a non-zero tilt angle is being measured. The circles are the pendulum weights, the solid dots are bearings, the cross-hatched areas represent connections to the base and the lines represent stiff members.

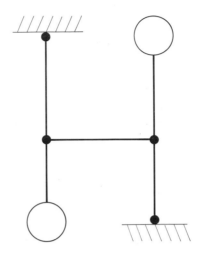

Figure 23.5 Schematic of the tiltmeter

It is fairly obvious, as it was to the inventor, Y. T. Li, a former colleague of the author, that a simple pendulum is a configuration which provides a measurement of tilt angle through a measurement of the lateral displacement of the pendulum weight. However, the size of the device presents a problem. The lateral displacement of the weight at the bottom of a simple pendulum is large enough to be measured at small angles of tilt only when the pendulum is extremely long. This is a relatively obvious concept. However, the inventor also realized that one can represent this displacement relationship as a simple

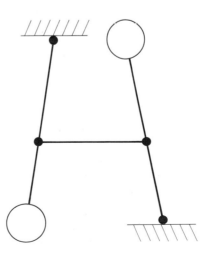

Figure 23.6 Tiltmeter in slightly tilted orientation

spring. A displaced spring pushes back with a force f which is proportional to the displacement $\triangle x$ ($f = k \triangle x$). Notice that the statement that the pendulum must be very long is the same as saying that k, the stiffness of the pendulum, must be very small. Continuing this logic, the inventor then recognized that there is another way to obtain a small spring constant in addition to extending the length of a simple pendulum. The difference between two large spring constants (short pendula) can yield a small spring constant (effectively long pendulum). This relationship is represented by $f = (k_1 - k_2) \triangle x$, but it requires a negative spring in order to obtain the $-k_2$ term. The inventor noted that an inverted pendulum is a negative spring. Thus, all that remained was the coupling between the two pendula at a point at which the resultant spring constant $(k_1 - k_2)$ is sufficiently small and positive to yield very high sensitivity (large lateral displacement) for small angles of tilt.

Discussion

This very brief description represents only a small portion of the process which took place in the creation of the new tiltmeter. However, it does describe the kernel event upon which the whole concept revolves. Of particular importance is the nature of the events which took place within concept space. First, there was simplification of the problem by considering only the pendulum configuration and ignoring, temporarily, many other parameters which obviously are important. Secondly, there was a transformation from the normal way of looking at a pendulum to viewing a pendulum as a spring. Thirdly, there was the creative step of recognizing the relevance of the difference between two large numbers to the situation at hand. Finally, we should not ignore the additional creative steps to generate the double pendulum configuration.

Figure 23.7 is a modification of figure 23.2 to include an important observation. On the right-hand side, within configuration space, are represented two realizations of pendulum devices: a simple pendulum and the new coupled-pendula tiltmeter. On the left-hand side, within concept space, are two concepts: the simple spring relationship and the spring relationship represented by the difference between two stiff springs. The transition from the simple pendulum to the complex pendulum within configuration space is a transition which is not likely to take place by itself. On the other hand, the transition from the concept of a simple spring, correlating with the properties of a pendulum, and the second concept, is motivated by recognizing the elements of these conceptual parameters. Thus the creative motion within configuration space is actually a movement driven by motion within concept space. In simple terms, the process unfolds with the movements as labelled numerically in the figure.

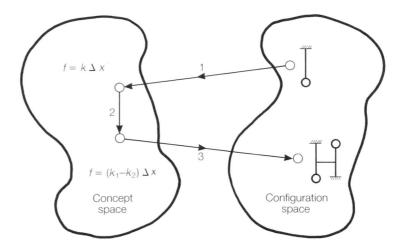

$f = k \, \Delta \, x$

$f = (k_1 - k_2) \, \Delta \, x$

Concept
space

Configuration
space

Figure 23.7 Simplified map of
the creation of the new
tiltmeter

Notice also that the movement labelled **1** was in itself a creative step. It represents a simplification and transformation of the typical understanding of pendula to a form which, downstream, led to a very interesting result.

Readers should be aware that this diagram of movements is extremely simplified. In this example of creative conceptual design, there were many more motions taking place back and forth between concept and configuration which have not been included in the diagram. Furthermore, the process at this stage of the description is far from over. There are many more conceptual issues which still remain, such as how to obtain bearings with the properties necessary to allow the new device to perform as required.

Of course, conceptual design problems have many solutions, as does this problem of measuring extremely small angles of tilt. The significance of the example is not the particular solution presented but the types of information which were used during the process and the relationship between the elements as the process moved forward.

Readers will remember instances when they were stuck in a rut. One may represent this situation as an effort to undergo creative movements within configuration space, when in fact it is more beneficial to make significant movement within concept space and between concept space and configuration space. Parameter analysis methodology is not merely equivalent to generating a clear problem definition. Often it is stated that if the problem is known it is nearly solved. However, good conceptual design goes much deeper than that, demonstrating the need for movement to drive the creative process.

Conclusion: Implications for Design Management

Some general conclusions can be drawn from this representation of the nature of conceptual design as it applies to the management of the engineering design process.

First, we should note that there is more to the stimulation of creativity than providing an environment. A deeper understanding of what the process is can enable one to take steps to improve the unfolding of the process.

Secondly, we see that there should be an increased emphasis on fundamental knowledge. Significant creative advances are almost always made at the level of fundamental concepts. Thus, our educational processes as well as continuing education within firms should acknowledge this property of the process and be designed accordingly.

Thirdly, as a result of the nature of the process, hiring policy should encourage breadth of background as well as depth of specialization. Often new concepts are generated by calling on material from disciplines outside one's own specialization.

Fourthly, individuals involved in creative design should be constantly observing, in depth, the parameters or triggers of the creative work of others and learn from this process. This will enable the motions described above to be more frequent and of higher quality.

Finally, so-called environmental approaches are not invalid and it should not be thought that this chapter implies such a position. When applied properly, techniques such as brainstorming are excellent tools for the stimulation of human creativity. Methodological approaches such as parameter analysis do not diminish the need for creative thinking. Rather, in parameter identification and creative synthesis, any approach which enhances the creative process should be applied. Furthermore, management strategy should include creating suitable environments to enhance conceptual design.

References

Broadbent, G.: The development of design methods. *Design Methods and Theories.* Vol. 13, no. 1, pp. 41–5, 1979.

Hales, C.: *Analysis of the Engineering Design Process in an Industrial Context.* Doctoral dissertation, University of Cambridge, December 1986.

Jansson, D. G.: Generating new ideas. In Li, Jansson and Cravalho: *Technological Innovation in Education and Industry.* New York: Van Nostrand Reinhold, 1980.

Jansson, D. G.: Creativity in engineering design: the partnership of analysis and synthesis. *ASEE Annual Conference Proceedings,* 1987.

Rittel, H. W. J.: *Second-generation Design Methods: the DMG 5th Anniversary Report.* DMG occasional paper no. 1, 1972.

24 Life Cycle Design

WOLT FABRYCKY and BEN BLANCHARD
Virginia Polytechnic Institute and State University, USA

Introduction

Design activities of analysis and synthesis are not an end in themselves but are a means for satisfying human wants. Design has two aspects, one concerned with the materials and forces of nature, the other with the needs of people. All products, systems and services which have utility (the capacity to satisfy human wants) are physically manifested. This is obvious with regard to a loaf of bread, an automobile or a school building. It is also true for intangible things. Music is enjoyed because sound waves strike the ear, pictures are seen via light waves, and even friendship is realized physically through the senses.

It follows that utilities are created by altering physical factors. The purpose of design is to determine how the physical environment may be altered to create the most utility for the least cost, in terms of product design cost, production cost and product service cost. Because design is practised in a resource constrained world, it must be closely associated with economics and economic feasibility (Blanchard and Fabrycky 1990; Thuesen and Fabrycky 1989).

Design for the Life Cycle

Products, systems and structures are designed and developed in accordance with a process which is not as well understood as it might be. Life cycle design is suggested as an integrative approach for bringing competitive products and systems into being in such a way as to minimize their deficiencies and life cycle costs (Fabrycky 1987). This integration involves design and development efforts to:

1 Transform an operational need into a description of performance parameters and a preferred product configuration through an iterative process which includes functional analysis, synthesis, optimization, definition, design, test, and evaluation.
2 Consider related technical parameters to ensure compatibility of physical, functional and project management interfaces in a manner that optimizes the total product definition and design.

3   Integrate performance, producibility, reliability, maintainability, man-
    ability, supportability and other 'ilities' into the overall design process
    (Blanchard and Fabrycky 1990).

The life cycle concept   Fundamental to the management of design is an understanding that the life
cycle of a product, system or structure begins with the identification of a
need and extends through conceptual and preliminary design, detail design
and development, production and/or construction, distribution, customer
use, support and then phaseout and disposal.

This life cycle process is universal in its applicability. It originates with
the perception of a need and terminates with product phaseout and disposal.
Between these points there are two major life cycle phases. The first is the
acquisition phase including the several iterative steps necessary to define the
need, perform the design, test and evaluate and, finally, produce and distribute the product. The utilization phase follows and involves activities
required to deal with the product in being. These include operating, maintaining, modifying, retiring and disposing of the product.

A life cycle design approach for bringing competitive products and
systems into being must go beyond consideration of the life cycle of the
product itself. It must simultaneously embrace the life cycle of the manufacturing system and the life cycle of the product service system. Accordingly,
three coordinated life cycles progress in parallel, as is illustrated in figure
24.1.

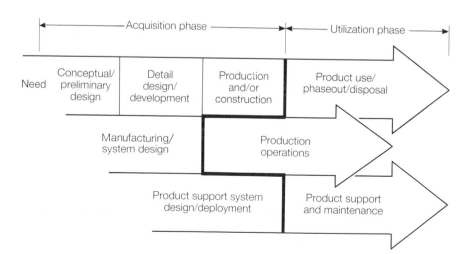

Figure 24.1 Product, process
and support life cycles
Source: Blanchard and
Fabrycky (1990)

The need for the product comes into focus first, initiating conceptual
design. Then, during preliminary design of the product, simultaneous consideration should be given to its production. This gives rise to a parallel life
cycle for the production process involving many related activities to prepare
for manufacturing (production planning, plant layout, equipment selection,
process planning, etc.).

Also shown in figure 24.1 is another important life cycle, It is for the logistic support activities needed to serve the product or system in use and to support the production facility during its duty cycle. Logistic and maintenance requirements planning should begin in a coordinated manner during product conceptual design.

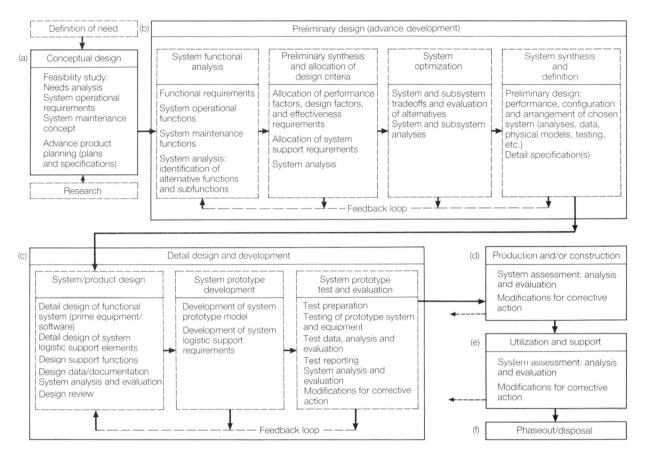

Figure 24.2 Life cycle design activities and interactions
Source: Blanchard and Fabrycky (1990)

A detailed presentation of the elaborate technological interactions which must be integrated over the coordinated life cycles is shown in figure 24.2. The progression is iterative from left to right, and not serial in nature as figure 24.1 might have implied.

Current design practice

In general, designers have focused mainly on the acquisition phase of the life cycle. The experience of recent decades indicates that a properly coordinated and functioning product or system, which is competitive in the marketplace, cannot be achieved through effort applied largely after it comes into being. Accordingly, it is essential that designers be sensitive to operational feasibility during the early stages of product development and that they assume the responsibility for life cycle design which has been largely neglected in the past.

Good design for a product's primary function often produces side effects in the form of operational problems. This is largely due to consideration of the primary function, rather than to the more challenging problem of designing in the face of the several 'ilities'. Enough specialized knowledge exists to solve this problem. The impediment to its solution is the integrated use of what is known in a systematic manner.

The communication and coordination needed to bring together the product, the process and the service systems is not easy to achieve. Progress in this will likely be facilitated by new technologies making possible the more timely acquisition and utilization of design information. CAD/CAM is only one of these technologies. Others are developing which can integrate relevant activities of the enterprise over the spectrum of life cycles. The most promising of these is computer aided life cycle engineering (CALCE) (Blanchard and Fabrycky 1990).

The Life Cycle Design Process

The design process comprises a logical sequence of activities and decisions associated with transforming an operational need into a description of product or system performance parameters and a preferred system configuration. The sections which follow present the life cycle phases of conceptual design, preliminary design and detail design shown in figure 24.2.

Conceptual design   Regardless of the product, system or structure, design begins with an identified need based on a want or desire arising out of a perceived deficiency (figure 24.2a). An individual and/or organization identifies a need or a function to be performed and a new (or modified) product is designed to perform that function.

Given a statement of need, a feasibility analysis must be performed, the scope of which will vary depending on the type and complexity of the requirement. It is necessary to identify all possible alternatives that will fulfil the requirement; to screen and evaluate the most likely candidates in terms of performance, effectiveness and economic criteria; and to select a preferred approach. The selected product or system configuration must be defined in terms of technical performance characteristics, effectiveness factors, supportability characteristics, economic goals and other criteria.

A feasibility analysis is conducted as part of, or an extension to, a preliminary market analysis to define product operational requirements, to develop a maintenance concept and to identify a system configuration that is feasible within the constraints of available technology and resources (money, people, equipment, material or a combination thereof). The output from a formal needs identification and feasibility analysis may be presented as a recommended preferred system configuration and design approach, or in the form of a decision not to proceed further because of the lack of available technology or resources. Technical parameters to be established for a product or system derive from an operational concept. This operational concept for a system includes but is not limited to the following:

Mission definition Identification of the prime operating mission of the system, along with alternative or secondary missions. What is the system to accomplish? How will the system accomplish its objectives? The mission may be defined through one or a set of scenarios or operational profiles.

Performance and physical parameters Definition of the operating characteristics or functions of the system (for example, size, weight, capacity, speed, accuracy, output rate). What are the critical system performance parameters?

Operational distribution Identification of the quality of equipment, personnel and facilities, and the expected geographical location including transportation and mobility requirements. How much equipment (and associated software) is to be distributed and where is it to be located? When is it required?

Operational life cycle Anticipated time that the system will be in operational use. What is the total inventory profile throughout the system life cycle? Who will be operating the system and for what period of time?

Utilization requirements Anticipated use of the system and its elements (for example, hours of operation per day, on–off sequences, operational cycles per month). How is the system to be used in its intended environment?

Effectiveness factors System requirements specified as figures of merit for cost/system effectiveness, operational availability, dependability, logistic support effectiveness, mean time between maintenance, failure rate, maintenance downtime, facility utilization (per cent), operator skill levels and tasks, personnel efficiency, etc. Given that the system will perform, how effective or efficient is it?

Environment Definition of the environment in which the system is expected to operate (for example, temperature, humidity, arctic or tropics, mountainous or flat terrain, airborne, ground or shipboard). This should include a range of values as applicable and should cover all transportation, handling and storage modes. How will the system be handled in transit? What will the system be subjected to during operational use, and for how long?

A maintenance concept evolves from the definition of operational requirements and responds to the question: how does the producer envisage that the product will be supported throughout its planned life cycle? It delineates levels of maintenance support, repair policies, maintenance responsibilities, logistic support requirements, effectiveness figures of merit (such as supply responsiveness, test equipment reliability and utilization, facility utilization, maintenance personnel effectiveness and cost constraints), maintenance environments and so on.

The maintenance concept serves two purposes. It provides a baseline for the establishment of supportability requirements (including reliability, maintainability and human factors characteristics) in system/equipment design. It also provides the basis for the establishment of requirements for total logistic support as illustrated in figure 24.1. Thus the producer must project how the product or system will be distributed and utilized by the consumer. The product must then be designed to meet the need under the conditions stated.

Preliminary design   Preliminary design begins with the technical baseline for the product, as defined in the feasibility analysis, and proceeds through the translation of established system-level requirements into detailed qualitative and quantitative design requirements (figure 24.2b). This includes the process of functional analysis and requirements allocation, the accomplishment of tradeoff studies and optimization, system synthesis and configuration definition in the form of detailed specifications as illustrated in the figure. Inherent in the activities identified in figure 24.2b are the aspects of planning, implementing and measuring, with feedback provisions.

A requirement is initially defined, a design approach is considered and the results are assessed in terms of specified effectiveness criteria. If the results are satisfactory, the proposed configuration enters the next phase of the life cycle. If the results are not satisfactory, alternatives are evaluated and changes are implemented as required. A continuous and iterative process must be followed.

An essential element of preliminary design is the utilization of a functional approach as a basis for identifying initial design requirements for each level of the product or system. A function constitutes a specific or discrete action required to achieve a given objective (for example, an operation that the system must perform to fulfil its intended mission). Such actions may be accomplished by equipment, personnel, facilities, software, data or a combination of these. The functional approach helps to ensure that all facets of system development, operation and support are adequately covered (including design, production/construction, test, deployment, transportation, training, operation and maintenance support); that all elements of the system (including prime equipment, software, test and support equipment facilities, personnel and data) are fully recognized and defined; and that a means of relating equipment packaging concepts and support requirements to given functions is provided (identifying the relationship between the need and the resources required to support that need).

The translation of system operational requirements and the maintenance concept into specific qualitative and quantitative design requirements begins with the identification of the major functions that the system is to perform, followed by development of functional flow diagrams. Functional flow diagrams are employed as a mechanism for portraying system design requirements in a pictorial manner, illustrating series and parallel relationships, the hierarchy of system functions and functional interfaces.

Functional flow diagrams are designated as top level, first level, second level and so on. The top-level diagram shows gross operational functions. The first-level and second-level diagrams represent progressive expansions of the individual functions of the preceding level. These diagrams are prepared down to the level necessary to establish the needs (hardware, software, facilities, personnel, data) of the system. Functions may be classified as independent or dependent and can be presented in a series format, a parallel format or a combination of both. Functions represented in each diagram should be numbered in a manner that preserves their continuity and provides traceability throughout the system to the functional origin.

The functions identified should not be limited to those necessary for operation of the system but must include those related to the possible impact of maintenance on design. Maintenance requirements (evolving from the maintenance concept) should be addressed to preclude the possibility of developing a technically feasible product from an operational viewpoint, without first determining whether or not it can be effectively and economically supported throughout its planned life cycle. The objective is to attain a proper balance of performance, effectiveness, support and economic factors as in figure 24.3.

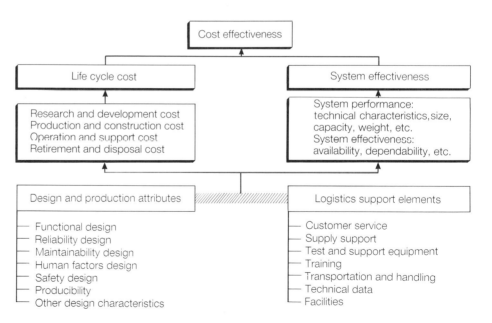

Figure 24.3 Cost and effectiveness factors to be balanced

These and other factors must be allocated to lower levels of the system to provide technical parameters and constraints, functional requirements and design criteria as needed. Otherwise, individual designers assigned to different elements of the system and working independently will establish their own goals. The results (when combined) may not comply with the initially established requirements for the overall system. Thus it is necessary first to establish requirements at the system level and then allocate these requirements to the depth necessary to provide guidance in the design process. Subsequently, as design progresses, it is necessary to check (on a continuing basis) that the design results at lower levels are, in turn, compatible with the overall system requirements.

In accomplishing allocation, the designer should consider all appropriate qualitative and quantitative criteria that will significantly influence the design process. These will vary from system to system and must directly support operational requirements and the maintenance concept. In some instances the allocation of design criteria to the subsystem and unit level may be sufficient, but in other cases it may be appropriate to assign goals down to the assembly and subassembly level. The depth of coverage depends on the design controls that the company may wish to establish. Controls that are too stringent may inhibit the decision making process relative to allowing for tradeoffs at the lower levels in order to meet system-level requirements. On the other hand, insufficient design control may not provide the desired results.

The allocation process establishes boundaries and constraints for system design (maximum and/or minimum values to which the design must conform). Within these bounds and constraints, any number of design configurations may be envisaged that will satisfy the specified requirements. The problem is to select the best approach possible through the iterative process of analysis and synthesis.

Detail design and development

The detail design phase (figure 24.2c) begins with the concept and configuration derived through the preliminary design activities identified in figure 24.2b. When an overall design configuration has been established, it is necessary to progress through further definition to the realization of hardware, software, trained personnel, data and items of support. This process includes:

1   The description of subsystems, units, assemblies and lower-level components and parts of the product and the elements of logistic support (test and support equipment, facilities, personnel and training, technical data, spare/repair parts, etc.).
2   The preparation of design documentation (including specifications, analysis results, tradeoff study reports, predictions and detail drawings) describing all elements of the product or system.
3   The definition and development of hardware and computer software.
4   The development of an electronic data base, an engineering model, a service test model and/or a prototype of the product and its elements for evaluation to verify design adequacy.
5   The test and evaluation of the physical model or prototype.

6 The redesign and retest of the system, or an element of the system, as necessary to correct any deficiencies noted during initial testing.

In detail design and development, the design process continues based on the foundation established in the conceptual and preliminary phases. The primary emphasis is that of verification and feedback.

Conclusions

Life cycle design is an organizing concept useful in the management of the process of bringing products and systems into being. To ensure the economic competitiveness of the end item, design must be pursued in the context of a coordinated set of life cycles. During life cycle design, an evaluation process incorporating iteration is very important. This may involve the accomplishment of specific tradeoff studies where two or more individual system parameters are reviewed independently in terms of the effects on each other. Individual tradeoffs may deal with different performance parameters in terms of cost and other factors. Ultimately, these individual tradeoffs are combined and reviewed in terms of higher-order system parameters such as system effectiveness and life cycle cost.

The system life cycle design process emphasizes aspects such as functional analysis, the allocation of requirements, synthesis, tradeoffs and optimization. These activities (by themselves) are not new; nor is the approach presented by any means novel. However, in many instances the system designer will start identifying hardware for a given system without first fully identifying requirements for such. Further, the designer will often ignore reliability and maintainability factors, life cycle cost factors, etc. The emphasis here relates to a design process discipline necessary for the orderly development of the product or system in an effective manner.

The life cycle design process is applicable to both small and large scale products, systems and structures. Further, the process is applicable to many different categories of systems – an aeroplane or missile system, a ship system, an electronic system, a manufacturing or production system, a structure or facility, etc. Although the nature of the requirement may vary from one application to the next, the process is essentially the same. Its implementation does not necessarily require the expenditure of additional resources – but it does require discipline on the part of the designer and those involved in design management.

References

Blanchard, B. S. and Fabrycky, W. J.: *Systems Engineering and Analysis* (2nd edn). Englewood Cliffs, NJ: Prentice-Hall, 1990.

Fabrycky, W. J.: Design for the life cycle. *Mechanical Engineering.* Vol. 109, no. 1, January 1987.

Thuesen, G. J. and Fabrycky, W. J.: *Engineering Economy* (7th edn). Englewood Cliffs, NJ: Prentice-Hall, 1989.

25 Creative Problem Solving Techniques

TUDOR RICKARDS
Manchester Business School, UK

Introduction

Designers use many and varied tools in their work to help them structure the design process. Creative problem solving techniques (CPSTs) are examples of such tools. Unlike more tangible aids such as electronic calculators, graph paper or design notebooks, a CPST can assume many different forms in use, involving the user in design activities at two differing levels.

First, there is the level of designing the problem solving exercise itself, selecting from a range of techniques and variations within techniques. Secondly, there is the level of designing within the exercise in order to achieve the output – a new and relevant product. Depending on the purpose of the exercise, the product may be one which we would all associate with design activities – a refinement to a manufacturing system, a computer game or a consumer fashion product. Or the product could be a strategy or a plan of action, both of which are often the outcomes of deliberate design activities. In what follows, we describe this process of designing the techniques and also look at the second level of design activities.

The core beliefs of CPST practitioners seem to be as follows:

1 Everyone has the potential for creative action, but for many people this potential is never fulfilled.
2 The potential for creativity can be tapped if the blocked individuals are helped to behave in ways that have become associated with other people who are widely acknowledged as being creative.
3 A powerful means of stimulating creativity is through the use of so-called structured aids to creativity or creative problem solving techniques (CPSTs).
4 Techniques and behaviour patterning can be introduced effectively through training programmes.
5 When this is carried out, the individuals and their organizations benefit considerably.

Taken together, these statements help to explain the motivations of advocates of creativity development and training and the importance of the CPSTs as a means to an end.

The Major Creative Problem Solving Techniques

Among group techniques, the brainstorming family remains of prime importance. Individual techniques of conceptual blockbusting (Adams 1980) are dominated by those popularized as lateral thinking by Edward de Bono (1971). A further important category of concept structuring methods deserves a mention, with versions of morphological analysis particularly widespread in a range of applications including engineering, design, education and industrial new product development (see Carson and Rickards 1979 for an industrial application).

Brainstorming variations

These all involve procedures which assist the proliferation of ideas under conditions of deferred evaluation. Typically brainstorming is a group technique, although any individual can attempt to brainstorm. 'Nominal groups' (Van der Ven 1980) are made up of individuals brainstorming, at least part of the time, without interacting. The process of brainstorming is enhanced by a process facilitator, an experienced group, a means of capturing the ideas (for example, a flip chart) and a relaxed environment. More sophisticated variations involve generating many differing problem statements, then producing ideas on selected definitions and, finally, brainstorming ways of gaining acceptance for the preferred ideas (Parnes et al. 1977).

Lateral thinking

This is a comprehensive approach to going beyond conventional thinking. Among the lateral thinking variations are reversing the obvious logic in a situation and introducing a fresh element in problem solving – the so-called random juxtaposition method. Yet another lateral thinking technique is to escape from a view of reality by considering a desirable but impossible idea. Sometimes this intermediate impossibility reveals a 'new' reality; the impossible provides a stepping stone to insightful thinking and an innovative outcome.

Morphological analysis

Finally, there is the modelling approach of morphological analysis. Many systems which the creative designer wishes to investigate are complex. There may be several dimensions, each containing many possible elements. Even to examine all potential combinations is a difficult task and so, without a way of structuring the system, the investigator tends to ignore the vast majority of permissible combinations. Morphological analysis is a means of systematically revealing the combinations within a system, thus avoiding subjective search biases and increasing the possibilities of discovering valuable combinations of ideas. A three-dimensional matrix which has played a significant part in new product searches is described in Carson and Rickards (1979). Recent evaluations show that well over 100 industrial new products have resulted from searches involving versions of that matrix, with its axes of raw materials, processes and markets.

Creative Problem Solving and the Design Process

In a series of penetrating writings, Jones (1981, 1984) has demonstrated a view of design as essentially a creative activity. However, with a few notable exceptions, practitioners of creative problem solving techniques and designers have developed their activities in mutual isolation. This is a pity as there is much that they could share. Consider the essentials of the design process. An individual (or team) is engaged in a process or task which has some identified purpose or goal. To achieve the goal, the designer has to solve problems in a process which is partially constrained by the requirements of the goal.

This overall process is one of problem solving under uncertainty. The designer will draw on experience and know-how so that any design will be strongly influenced by what has already been learned. While this process may be intuitive, the designer may recognize sequences of procedures or operational mechanisms. These in turn may relate to a broader strategy or design methodology (establish function criteria, establish constraints, consider optional approaches, etc.).

Similarly, creative problem solving is a process aimed at achieving a solution to a problem. In the majority of non-trivial activities, the process is partly constrained, so that there can be a great variety of outcomes. Again, the process can be intuitive and the problem solver can develop a conscious repertoire of procedures. These are the operational mechanisms of creative problem solving techniques.

Creative Analysis: a Methodology for Designing CPSTs

It is not immediately obvious why industrial applications of CPSTs involve an element of designing. It may be imagined that once the ground rules for operating a technique have been acquired, the technique can be plugged in when and where appropriate. Recently I have gone into the counter-argument in some detail (Rickards 1988), showing how the process requires a design-like ability to decide from a range of possibilities that have not yet been realized – that is, the intended technique sequences.

This creative analysis reminds me of a mental rehearsal of what it is hoped will happen, testing out various possibilities and making judgements from past experiences, in the light of what are imagined will be the circumstances of the creativity session. This is the first level of design, in which the plan or blueprint for action is created. It is creative because it produces something new and relevant from imagination. It is analytical in as much as it involves data manipulation, comparisons and decisions. Creative analysis requires the designer to be sensitive to the operational mechanisms of the techniques – and to their component subcategories. Furthermore, there must be an appreciation of the rationale – the 'why' of the operational mechanisms – so as to test their performance in practice.

There are analogies with the longer established role of the designer. Each project offers scope for habitual or routinized behaviours: short cuts, tricks of the trade and so on. However, there is always scope for departure from routine as no design will be exactly the same as an earlier one. Without analysis of the outcomes of design activities no development is possible. The greater the attention to subroutines, the greater the opportunity for constructing combinations of them appropriate to differing circumstances. The creative designer is one who avoids over-reliance on one or two strategies and who can be flexible regarding the combinatorial possibilities within the strategies.

Relating this to the process of creative analysis, we can imagine that the experienced practitioner follows a strategy in using CPSTs. This strategy is triggered by assorted stimuli, particularly needs for ideas perceived either by the designer or by would-be clients. Some form of decision taking (intuitive or formalized) leads to the application of one of the available families of techniques. Further design decisions are called for, and taken, so as to select in detail the various specific techniques and to assemble a planned set of operational procedures. In practice, it is found important to collect evidence about the problem, the client and the environment. These bits of information help in tailoring the design to the specifics of the situation.

Perhaps it should be stressed that the process is far from a totally rational one. Under day-to-day pressures, I may find myself designing on automatic pilot, arriving at a proposed meeting without a very clear idea beyond an intuition that the most appropriate approach will perhaps involve some form of brainstorming with a few blockbusting interventions. If the session leads to success (new and acceptable ideas for the clients involved) it is because the CPSTs are robust in use, and because there are always opportunities to make mid-session adjustments to plan. There is a fine line between leaving design options open as a matter of deliberate policy and because it is the easy thing to do. The following notions of best practice have served me well when I make the effort to follow them through. They are also a source of improvements for exercises which have gone badly wrong.

Plan Make it a habit to plan any attempt to generate ideas. This is when you have a conscious choice over the type of technique used, taking into account the context – objectives, who should be involved, timescales, locations and so on. The choice of technique system will then be more considered and less reliant on your first impulse or dominant mind-set. For more complex group exercises, I find it useful to carry out a mental dummy run of what might happen in the actual event.

Keep creative exercises and analysis apart Analysis before and after the session is vital. During a session, attention to the 'why' keeps the left brain too active and engaged on the process. Even the group leader, who is responsible for the mechanics of group process, should attend during the meeting to how the group is progressing, not why things are turning out well or badly.

Review   Immediately after the session it is excellent practice to check the feelings and observations of participants. Instructive suggestions for future approaches often emerge.

Keep records   Most practitioners of creative problem solving techniques do not like the discipline of careful recording of sessions. There are few reported comprehensive case examples out of the thousands of activities that take place every year in dozens of different countries. At a minimum, these sessions should be treated as seriously as any other project and descriptions written for future reference and analysis. Video and tape recordings can be used to assist in a creative analysis, although such efforts are time-consuming and are perhaps best left to the serious researcher. (Gordon 1961, for example, used tape recordings of sessions in progress to refine the operational mechanisms for stimulating creativity.)

Understand what you are doing and why   The 'what' comes from studying the practical activities and outcomes, that is, the operational mechanisms; the 'why' comes from the less tangible reasons or theoretical justifications we call precepts. In addition, the context must be taken into account in studying any outcomes.

Learn from the experiences of others   Become involved with other practitioners, through conferences and networking. In time you will see the rationale behind new techniques as you analyse the operational mechanisms and the thinking behind them.

Experiment   Creative problem solving is still an emerging subject, needing more experiments as well as closer links between theories and practice.

A Case Illustration of Creative Analysis

I will illustrate the process from a reconstructed example of a recent industrial exercise. I will concentrate on the design decisions which led to the format of the creative session as it actually took place and consider some of the infinite variations of unplayed melodies which might have occurred. The main events of the story concern a marketing manager with a fast growing consumer electronics organization and his efforts to solve some distribution difficulties.

The manager, David Guest, was responsible for developing the European market for a major new technology product. He had a track record of success with a competitor and had been head-hunted into his present position by Dirk Channon, the (UK) general manager of his multinational company. He had found the challenge exciting and had helped extend the market from direct sales to electronics enthusiasts to multiple sales via commercial buyers. However, rapid growth was introducing strains into the system. David had received training in creative thinking from his previous company and eventually contacted me 'to help us run a brainstorming to get some fresh thinking on marketing our new product'.

After his initial telephone call to me, I was invited to meet David and Dirk Channon. David conveyed the impression of a conscientious and competent executive who was trying to introduce some longer term thinking into a very fast changing situation. Dirk was a more explosive character with a hands-on style which resulted in an exhausting work schedule. One reason for the meeting was clearly to allow Dirk the chance to find out directly how brainstorming might help him solve his problems – which he described as coming from an overstretched operation, itself made worse by the company's very rapid growth and projected targets. He had several more meetings that day and rapidly agreed to the brainstorming, leaving David to 'sort out the details' over lunch.

My design thinking at that stage (as best as I can reconstruct it) went along the following lines. I had been asked to run a brainstorming session. But was brainstorming, in some form or other, what was really needed? This family of techniques was suited to the production of large numbers of ideas, rapidly. Successful results had been obtained for a wide range of topics, including marketing ones of the kind we had here. Something seemed wrong in this simplistic design diagnosis. This was an experienced team of market executives. They would have had experiences that suggested ideas for marketing the product. So what was wrong with the ideas that had already occurred to them? Under such circumstances, the barriers to progress may be due to failure to challenge assumptions about the problem. Thus, any brainstorming for new ideas should include a technique or a subroutine which would help redefine the problem.

The culture I had been briefly exposed to suggested that once some idea had been accepted by Dirk he would drive it through to completion. From David's implicit analysis of the situation and my own encounter with Dirk, I felt it important to structure the brainstorming so that it involved him in the implementation decisions. David could be relied on to put such proposals into action. In selecting ideas, their intuitive acceptance by the managers (especially Dirk) would be more important than analytical techniques.

Over lunch our discussion turned to the operational details of running the session. David felt strongly that the key people to be involved in the first instance were, as well as himself, Allen Evans the production manager and Winston March the financial controller. He asked whether such a group would be the right mix to get the sort of ideas needed. My own view was that the timescales made it difficult to bring in outsiders and, in any case, the subject was one of some commercial delicacy. Bringing in outsiders can broaden the range of possibilities being considered; conversely, they can inhibit the members of a closely knit team and prevent them coming to terms with organizational blind spots. If the need was to broaden thinking, we needed outsiders; if it was to challenge assumptions, we should stay with insiders.

We agreed to stick with David's team selection for the first meeting. This gave me further design constraints. Typically, six or seven people can make up an effective brainstorming team, although teams of just two people are very common in creative work in advertising and script-writing. A small team would gain in intimacy, compensating for any loss in the range of ideas which might have been brought in by outsiders. By the end of the lunch a meeting format had been agreed which was in essence a problem exploration stage, based on a structured discussion of the situation, followed by an idea generation session on whatever key themes emerged.

This creative analysis had helped us refine the broad technique strategy ('run a brainstorming') and we had been able to pin down some aspects of the meeting to a hybrid problem searching discussion, followed by some brainstorming. The meeting was scheduled to take place as quickly as possible, which turned out to be several weeks later. It is instructive to follow its course and compare our initial design with what actually happened.

At the brainstorming meeting, David arrived with the news that a major internal crisis had cropped up. Dirk and the others were 'sorting it out' and would be delayed. Dirk eventually arrived an hour late, having left Allen and Winston to follow through with the immediate problem. We established it would be at least six weeks before any further meeting could be held. We would be advised to make the best of the few hours left at our disposal. The omens were not good for creative problem solving. Dirk was preoccupied by the events back in the office (some miles down the road from the hotel we were using). I offered to act as a sounding board for the two managers if they wanted to review their situation and, also, to explain some of the thinking behind the creative problem solving approach. My intentions were to modify the original plan to meet these changed circumstances. Now, the warm-up session was to be an attempt to win Dirk's confidence and encourage him to open up about the nature of the problem.

Very rapidly Dirk began to outline the difficulties in the situation. I wrote these down as a series of numbered statements; it emerged that many of them reflected problems with the company's distribution system, the original focus of the meeting. The retailers were needed to support business to personal purchasers and to small firms, but were not needed for the proposed market extensions into large companies whose executives had indicated their preference for dealing direct. At the end of the meeting Dirk had come to the conclusion that they needed new ways of dealing with the distributors. A few new ideas had emerged for doing that. He requested a second meeting of a similar format to see where this line of thought was taking us. The client's orientation had moved away from brainstorming, towards a rather conventional meeting with an outside process consultant.

Nevertheless, at the next meeting (attended by Dirk, David and Allen, the production manager), I encouraged the listing of ideas without criticism on the topic 'how to deal with our distributors'. Some 20 ideas were listed – rather a smaller number than if we had followed a more typical brainstorming format. Most of them had been thought of before by the company but some issues emerged as fundamental, particularly those connected with winning over the distributors into accepting new relationships with the company for large institutional orders.

After the meeting, I felt that I had failed to carry out the agreed brainstorming and worried about the effect this would have on subsequent contacts with the client. The conclusions to the meeting seemed rather mundane and might have emerged from any internal meeting on the topic. To my great surprise, I later learned that the meetings had produced substantial changes in the company's policy towards their distributors and had, in a few months, led to renegotiations and a new agreement which incorporated some creative 'win-win' features suggested at our second meeting.

Design Issues Raised in the Case Study

This example shows clearly how plans are needed, if only to be a platform from which to depart in the light of new information. The modification to the original design was considerable but was no more than happens in many instances during the execution of CPSTs. The modifications in this example are rather extreme compared with most commissioned brainstorming sessions. Nevertheless, it is always necessary to proceed and to be willing to depart from the prearranged plan in the light of circumstances. The 'master plan' of my own personal approach to brainstorming, with considerable scope for such modifications, is shown in figure 25.1. More typically the modifications might involve greater or less time spent in 'warming up' the group or in finding interventions in response to a failure of the group to escape from mundane and conventional thinking.

For example, a lacklustre idea generation session might benefit from the introduction of a metaphorical excursion (Gordon 1961). In a recent exercise on organizational response to unclear directions from the board of management, the corporate planning team was extremely stuck during brainstorming. An excursion was suggested in which the team compared their situation with that of a drifting ship in a sea fog trying to decide what best to do. A range of good ideas emerged.

The Mechanics of Creativity within a CPST

In the case study, we concentrated on the process of designing the technique framework – the first-level creative design task. We will now take a closer

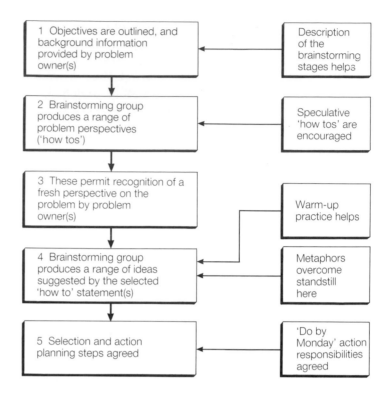

Figure 25.1 Brainstorming
flow chart, with notes
Source: © Tudor Rickards and
Associates Ltd 1987

look at the second-level process, in which the CPST contributes to the creation of a new and relevant product.

A recent exercise involved a management team within a well-known secretarial college. The nine-person team had achieved considerable savings in the areas of office equipment and supplies following the advice of an experienced CPST practitioner. The team had simply listed as many 'why don't we' statements as possible and narrowed down their selection on the basis of simplicity of introduction. These changes in office practice could be viewed as new, created designs assisted by brainstorming. However, an example of a more obviously invented and designed new product using these methods then emerged.

The group turned its attention to ways of motivating the students, and began to look for a 'best student' prize. The ideas produced by brainstorming were considered rather mundane. The facilitator (in first-level design mode) suspected that they would benefit from some individual thinking that challenged conventional logic, and suggested they spend five minutes looking for unusual ideas triggered off by the lateral thinking 'random word' approach. A set of nine words was produced, one for each group member.

Interestingly, two members generated the same idea from two quite different starting points, namely 'waterwheel' and 'octopus'. Each team member claimed to have seen from the random word the idea of a best student brooch in the design of a daisy-wheel. The basic concept was enthusiastically developed by the team and the personalized silver daisy-wheel is now established within the college's graduation award ceremonies.

It is interesting to speculate on how the idea arose. The team had been working previously on the problems caused by the theft of daisy-wheels from their electronic typewriters. Perhaps the technique achieved its eureka impact through a subconscious combining of the daisy-wheel and octopus images, 'a kind of floppy daisy-wheel' as it was described; and similarly for the waterwheel, with its associative and visual links to the daisy-wheel. What does seem likely is that the first-level creative analysis ('try lateral thinking and random words') assisted in the creation of the design of a new gift product. The second-level product (a daisy-wheel brooch) was produced with the help of the selected creative problem solving technique.

Conclusions

Creative problem solving techniques are design tools. However, because they are conceptual tools, they themselves can be designed for each specific application. The designing of the design tools I have called creative analysis, or first-level creative design. It requires a professional consideration of what is being done and a codification of experience. The output of creative analysis is a product characterized by novelty and relevance, the hallmarks of designed artefacts. We saw how CPSTs could produce not only these conventional products but also less tangible products such as marketing strategies and decisions. CPSTs should be a part of every designer's conceptual toolkit.

References and Further Reading

Adams, J. L.: *Conceptual Blockbusting: a Guide to Better Ideas* (2nd edn). New York: Norton, 1980.

Carson, J. W. and Rickards, T.: *Industrial New Product Development.* Farnborough: Gower, 1979.

de Bono, E.: *Lateral Thinking for Management.* New York: McGraw-Hill, 1971.

Gordon, W. J. J.: *Synectics: the Development of Creative Capacity.* New York: Harper and Row, 1961.

Jones, J. C.: *Design Methods.* Chichester: Wiley, 1981.

Jones, J. C.: *Essays in Design.* Chichester: Wiley, 1984.

Kuhn, T. S.: *The Structure of Scientific Revolutions.* Chicago: University of Chicago Press, 1970.

McPherson, J. H.: *Structured Approaches to Creativity.* California: Stanford Research Institute, 1969.

Parnes, S. J., Noller, R. B. and Biondi, A. M.: *Guide to Creative Action.* New York: Scribner, 1977.

Rickards, T.: Designing for creativity: a state of the art review. *Design Studies.* Vol. 1, no. 5, 1980.

Rickards, T.: *Stimulating Innovation.* London: Pinter, 1985.

Rickards, T.: *Creativity at Work.* Farnborough: Gower, 1988.

Rickards, T. and Freedman, B. L.: A reappraisal of creativity techniques. *Journal of European Training.* Vol. 3, no. 1, 1979.

Van der Ven, A. H.: Problem solving, planning and innovation. *Human Relations Journal.* November–December 1980.

26 Visual Confrontation: Ideas through Pictures

HORST GESCHKA
West Germany

Introduction

Today a number of creativity techniques are used in industrial companies to enhance creative thinking in problem solving processes. These techniques can roughly be classified as:

1   Brainstorming techniques, which are based on association chains among group members.
2   Analytical methods like the morphological tableau, or techniques which differentiate solution directions in a tree structure.
3   Confrontation techniques, which try to simulate elements of the creative process.

The different groups of techniques have specific application areas where they are most effective. Design tasks can be successfully approached with the confrontation techniques, especially by a method which uses pictures as a means of stimulation.

The Creative Process and Visual Confrontation

From studies of the creative process, we know that in most cases ideas are triggered by an objective or an observation totally unrelated to the problem field. This phenomenon, which we call intuitive confrontation, is known from many anecdotes about scientists and inventors. For example, seeing the rising water level when he climbed into the bathtub, Archimedes had the idea of how to measure the volume of the tyrant's crown. He cried out the famous word 'eureka' (I've got it!). Another vivid example is the invention of the ballpoint pen. The inventor went for a walk in the park where he saw children playing a ball game on the wet grass. The ball, rolling across a dry asphalt path, left a wet track; thus, in the ballpoint pen technique, a mounted ball is moistened by a liquid.

This phenomenon of intuitive confrontation can be used as a technique in creativity sessions. Instead of the individual problem solvers merely making observations of and receiving impressions from their environment, particularly during phases of relaxation and openness, they are provided with stimulating objects. Concrete words or pictures chosen mostly at random will serve this purpose.

The creativity techniques centred on this principle also try to make use of other characteristics of the creative process. The most important findings on this process will therefore be briefly presented here. Creative processes can be categorized into several phases as follows.

Processing of problems   In this phase, the person concerned with the problem fully identifies with the task, working on it intensively and thinking about it a lot. An inner urge to find a good solution builds up.

Incubation   While working on the problem intensively there will, of course, also be breaks and phases of diversion and relaxation. During the incubation phase, one continues working on the problem subconsciously. It is not yet known how this works exactly. Apparently, the relationships between subjects are broken up and fixed connections are loosened. Relaxing activities, phases of falling asleep and waking up as well as phases of light sleep, are the times for incubation. Brain research has shown that during these phases the current of the brain has a frequency of 7–14 hertz, the so-called alpha frequency. In this condition memories and elements of recollection are activated and associated to a particularly high degree.

Illumination   During the incubation an inner authority – the subconscious – checks all impressions, observations, thoughts and memories as to whether they contain signs of a possible solution. If this is the case then this idea suddenly breaks through into conscious knowledge. This is eureka, the enlightenment, the brainwave, the intuitive inspiration. As a rule, such ideas are very vague.

Implementation   The vague idea of a possible solution must be put into concrete terms and elaborated. After that follow all the other laborious tasks necessary to implement an idea. In the words of Edison, the inventor of the light bulb: 'Genius is 1 per cent inspiration and 99 per cent perspiration.'

The Confrontation Techniques

For the confrontation techniques, one undergoes the phases of the creative process in creativity sessions. In general the sequence of events of such techniques can be categorized as in table 26.1.

Established confrontation techniques are: excursion synectics, stimulating word analysis, visual group confrontation, and picture folder brainwriting. Such methods require more time than, for example, a brainstorming session. One and a half to two hours must be taken as typical. The application situation is also different from that of brainstorming; confrontation techniques are used when

Table 26.1

Phase of the creative process	Steps of the confrontation techniques
Intensive work on the problem	Presentation, clarifying discussion and reformulation of the problem
Incubation	Application of dissociation and relaxation techniques
Illumination	Working through a series of words or pictures in accordance with intuitive confrontation
Implementation	Evaluating and preselecting ideas; putting into concrete terms and building up the ideas

1   There is a serious problem, solutions for which are not at all obvious.
2   No satisfactory results have been achieved by other efforts (individual work, brainstorming or brainwriting).
3   The emphasis is on finding exceptionally original ideas.

Visual Confrontation

In visual confrontation, pictures are used for relaxation and dissociation as well as for the development of ideas. Two techniques make use of these principles: picture group confrontation and picture folder brainwriting.

Picture group confrontation   The steps in this technique are as follows:

1   Presentation of the problem by the moderator or the problem owner; problem discussion and analysis within the group: in this step, the group seeks a better common understanding of the problem, leading eventually to a more precise definition of the task.
2   Spontaneous idea generation in the group by means of a quick brainstorming session; ideas are recorded on a flip chart: this step frees the mind from existing or obvious ideas. Having purged oneself by this quick brainstorming, one is more open for newly developed ideas.
3   Checking of ideas and possible reformulation: the ideas generated are checked as to whether they are in line with the problem definition. Frequently, it is made clear by extraordinary ideas that the problem has not been fully understood. In this case the problem should be explained once more and, if necessary, the definition of the problem must be changed.
4   Relaxation and dissociation: by means of a series of approximately five pictures shown on a projector, the subject is switched off and concentration and peace are achieved. This process is assisted by suitable background music.
5   Development of ideas by means of pictures: a set of six to eight pictures is worked through. These are of a totally different character from the relaxation pictures. They show many clearly recognizable details. First

a participant describes the picture. This makes it very concrete for all group members. Individual participants then take up certain elements of the picture and from a principle contained therein derive an idea for the solution of the problem. The ideas are at the same time entered on a flip chart. Those participants who cannot at first derive starting points from the pictures are also stimulated by the group.

6   Further development of ideas: as the ideas developed by this method are often very vague and the concrete terms are on different levels, it is useful to make a preselection (e.g. by affixing dots) and to develop the suggestions given preference. This often results in the forming or combinations of ideas.

Picture folder brainwriting   This essentially follows the same procedure as picture group confrontation. Steps 1, 2 and 3 are identical. However, relaxation and idea development are done individually. Each participant receives a folder of carefully selected pictures and works through it, noting down emerging ideas on cards. After 20 minutes of individual work, the cards are passed to one's neighbours around the table for further stimulation.

The pictures used in visual confrontation sessions should be carefully selected to be rich for idea stimulation. They should fulfil a number of criteria, such as having clarity, positive appeal, many distinct objects and a harmony of structure. They should be provided in the form of slides or picture folders. During the session, the pictures are worked through at random or in the received order; they must not be selected or arranged by participants in relation to the given task. There is only one exception to this rule: if a picture is closely related to the topic of the session it is eliminated.

An Example

It is not necessary to give an account of an idea finding session in full length. This example is of the central idea generating step (i.e. step 5 in the preceding section).

Topic   Improved garden furniture.
Problem   Conventional garden furniture shows a number of disadvantages: it weathers quickly, the covers get wet, it takes up a lot of storage space, etc. A new range of furniture is needed to provide essential improvements.

Idea development from pictures
1   A picture shows a *balloon*. This generates the following idea: air cushions are integrated into the garden chairs which can be inflated when required and give additional stability. The cushions are emptied when the chairs are stacked away which reduces space requirements considerably.
2   Amongst other things, another picture shows a *desk with a drawer*. This results in the following idea: sliding elements are integrated into the

tables and the backrests of garden chairs which can be pulled out to form a tarpaulin. This covers the piece of furniture and can be fastened on the opposite side.

3 In one picture *stacked bales of straw* can be seen. This gives rise to the following idea: the furniture is to be designed in such a way that it can be joined as a compact whole; connecting pieces provide good support. The whole is covered by a tarpaulin and in this way the garden furniture can be left outdoors during periods of bad weather or even in winter.

4 A further picture shows a *shelter* in a mountainous region: this subject causes one participant to suggest offering a garden shed matching the garden furniture in style and design. Here the furniture can be stored quickly and compactly. The shed also provides further storage space.

Other problems to which this methodological approach has been applied include: fastening products for do-it-yourself purposes; applications for waste from blast furnaces; new bottle closures; foot care products; gas pipe systems; forms and means of identification of persons or objects; sales aids for cosmetic products; bench saw accessories.

Conclusion

The creativity techniques of visual confrontation can lead to original ideas and possible problem solutions through the use of different kinds of pictures. This approach of deriving ideas from pictures is particularly suitable for design tasks; the method has proved especially effective in this field.

Further Reading

Geschka, H.: Creativity techniques in product planning and development: a view from West Germany. *R&D Management*. July 1983.

Gordon, W. J. J.: *Synectics: the Development of Creative Capacity*. New York: Harper and Row, 1961.

Van Gundy, A.: *Techniques of Structured Problem Solving*. New York: Van Nostrand Reinhold, 1981.

27 The Role of Communication in Corporate Identity Projects

CHRIS LUDLOW
Henrion, Ludlow & Schmidt, UK

Introduction

The end result of any design project is, or should be, an implemented solution which meets the objectives and criteria laid down at the initial stage. The success of the design management techniques employed during the project can be judged by the end results. Effective implementation will only be achieved when the whole process is carefully and thoughtfully managed from start to finish. This will be true irrespective of how good is the original design concept.

Most design projects are run according to a defined process – from briefing through design concept and development to implementation. In cases where the time between briefing and implementation is short and the amount of work required relatively small, then the need for specific design management structures and inputs may be minimal. But when projects extend over a long period of time – months or even years – and the complexity of the project increases, frequently involving a series of subprojects combining to form a design programme, then planned and effective design management throughout the process becomes particularly crucial to its success. This is especially true in the case of corporate identity, which some would say is the most esoteric branch of design.

Importance of Communication

But what constitutes design management? Of course, views differ on the most effective structures and processes, and definitions vary just as widely. The common denominator in all cases is that successful design management can never be achieved in isolation: communication will always be a key factor. Skill lies in understanding why, when and with whom communication should take place, what results should be sought from it and how to apply the results to the solution of the core problem in order, eventually, to achieve successful implementation.

Such knowledge is fundamental to the success of those who practise in the area of corporate identity, where ways must be found for the client and consultant to communicate on both broad issues and detailed points throughout the whole process in order that understanding can grow between the two parties. The conventional communication process which takes place during a design project might be represented as an alternating two-way flow between client and consultant, as in figure 27.1.

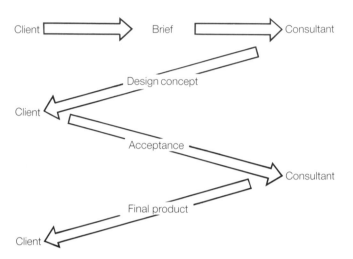

Figure 27.1 Design project communication

In the case of a simple project this sequence may be operable, but with larger ongoing projects, such as corporate identity programmes, frequent consultations and exchanges of information at a number of different levels are necessary. In the initial study phase of a corporate identity programme, the gathering of appropriate information is a vital foundation stone for the work which follows, as well as for the way in which identity should later be implemented and 'sold' throughout the organization.

The key to success, here again, is communication, but with the knowledge of which questions to ask and how to use the answers obtained. In asking those questions, or in obtaining information, various techniques can usefully be applied. Obviously, market research has its own role to play in providing a basis on which to work (although it is no substitute for truly intuitive innovation).

Information Gathering

Any amount of existing information – organizational, financial, strategic – can be obtained from the client in printed or written form. Each of these elements of information can be used throughout the investigative, conceptualization and implementation stages, being applied at specific points or influencing the manner in which the job is run or the nature of day-to-day communications with the client. But these remain relatively dry forms of information because they involve little interaction with the client.

One most effective example of how interactive communication of information can ensure success for a corporate identity project is the structured management interview. The manner of communication of the information in a one-to-one interview situation is as important to the long term success of the project as the information itself. Those whose views and concerns have been taken into account at the formative stages of a project are much more likely to cooperate in getting things put into action than is a manager who is suddenly landed with the task of implementation with no prior knowledge or background. In this respect, what is often thought of as a separate task – motivation – is seen to have roots in an early stage with good effects at the final stage.

The potential benefits of the interview process go right through the project, acting on the organization rather as holistic medicine acts on the body, identifying and treating the cause of the problem and thereby alleviating the symptoms. This slightly facile comparison should not be taken too far, but in one-to-one discussions between consultants and interviewee the same vital elements of trust and confidence are shared as between doctor and patient. Information obtained in this way can often be more relevant and useful than certain forms of market research, especially when future-oriented aspects are under discussion. In these situations, market research sometimes only confirms what is already known, whereas the potential of the organization can only be realized through full communication of individual views and aims.

For major projects, the appointment of a client representative at board level can further help to ensure good communications between client and consultant throughout the project. Such a sponsor or champion can help to bridge any gaps in trust or confidence which may exist, preparing the way for radical recommendations which may not be comfortable or easy for everyone to accept or act on. Presentations should also be viewed as opportunities for communication – not simply from the consultant to the client but also vice versa. Managers presented, at the end of the study phase, with the results of the views they themselves have expressed, may see long term implications which cause them to change those views. The consultant needs to be sufficiently receptive to appreciate the value of such a discussion and, if relevant, take it into account in the formulation of the design brief.

Effective communication within a design management structure can help to avoid some fundamental difficulties which can hamper the successful introduction and implementation of a large identity programme. These difficulties can include:

1   Lack of client understanding of the true nature of corporate identity and what they should expect from it.
2   The not unnatural reluctance of some managers to implement a

corporate identity which they do not understand and in the development of which they played no part.

3   Strains resulting from the cultural gap which exists between nations, in spite of superficial signs of mutual influence.

Understanding Corporate Identity

That clients are sometimes unaware of the true nature of corporate identity and the specific benefits which should be expected from undertaking such an exercise is clear from experience. Clients may not appreciate that to develop an appropriate identity will require considerable research on the part of the consultants and that the exercise will not be successful without considering also how the structure of the organization should be expressed visually, how internal communication of the identity should be carried out, and so on.

Communication of basic objectives and means thus becomes the key to helping some clients to define their expectations of a corporate identity exercise and, therefore, to get full value from it. Communication of this sort of information can, however, be a delicate matter because few clients want to be told that they are unaware of their real reasons for calling in a consultant; so a basis of trust must exist as a prerequisite for understanding and acceptance. Also, any assertions or proposals must be capable of being substantiated.

Beiersdorf, a vast West German chemicals and pharmaceuticals company, was not only little known in its domestic market but almost unknown elsewhere, in spite of the fact that it originated and owned world brands like Nivea. Managers considered that its corporate identity could be strengthened in order to increase corporate presence in relation to product brands. They also understood that the company name was difficult for people outside the German-speaking world to pronounce. They quite rightly wanted a solution, but an appreciation of what a good and effective corporate name and mark could potentially do for them, and what other measures would be necessary to research the problem and to establish a solution, were completely absent from their thinking.

In such circumstances, it would have been useless to embark on a simple redesign exercise because their expectations could not be articulated and, therefore, no criteria would exist for evaluating any proposals put forward. Therefore, a year-long programme of investigation and research was started with the goal of communicating to managers what their aims should be and what might be the objectives for the identity. On the conclusion of this study, a report was prepared for Beiersdorf which proposed a new identity structure in which the company name would be used as an umbrella identity over all the brands. It was demonstrated how this would be expressed in a whole range of applications and areas.

When this first report was presented to Beiersdorf, the eventual mark was represented simply and uncontentiously by an X. Extensive research had been done into all marketing and communications needs, strategies and structures and, in each individual area – such as the complicated matrix of links between market areas and distribution networks – the value of the presence of X as an umbrella was explained and quantified. As a result of this, both client and consultant were aware of what was being aimed for. The consultants could work towards a final concept within clearly defined bounds so that creativity was not expended in areas where it would have been wasted. The client could judge the consultant's proposals in an objective manner, guided not by whim or personal taste, or even by elusive consensus. Also based on the findings of the analysis, a set of corporate criteria, which sought to express the Beiersdorf corporate philosophy, were worked out and introduced.

The fact that the reasons for action and the criteria for judging the action were so clearly expressed meant that resistance to change through implementation of the proposals – something which is often feared by organizations – did not occur. Essential preparatory communications work had paid off, and Beiersdorf now has not only an effective and successful corporate identity but also criteria for image-related decisions in most areas, from new product development to advertising, through personal policy and architecture. Communication of these criteria to management assists in day-to-day decisions which are in any way related to maintenance of the corporate identity.

Motivation

Motivation as an issue has already been mentioned. Managers and staff to whom a new identity, its meaning and objectives have not been explained will often find it difficult to adopt it intellectually and emotionally. The communication of ideas for the purpose of motivation has its seed in the very first stage of a project, as explained earlier. However, motivation can demand particular effort in certain contexts. This is especially true for international organizations where the complexities of implementation make a positive attitude on the part of a very wide spectrum of staff a crucial element in successful implementation. Motivation, however, cannot just be tacked on as an extra. It is an integral stage with far-reaching effects and it is based on the communication of information.

An international firm of accountants, Coopers & Lybrand, had started the process of overhaul of their identity, but their size (400 offices), their geographical spread (100 countries) and their organization (an autonomous partnership) all presented potential obstacles to successful achievement of their corporate communications goals. For them, it was also a very unusual situation. They were usually the advisers, facing their clients: now their own advisers were facing them, asking questions which seemed unrelated to the central question of how they should look. Because of the potentially autonomous, decision-influencing nature of the partnership, and because of sheer numbers, it was proposed to undertake a large scale communications exercise in front of about 350 partners, timed to coincide with an annual international partners' meeting. At the same time, the consultants put together an implementation strategy which included training sessions at several of Coopers & Lybrand's worldwide control offices.

The management of design within the firm was to come about through the communication of the background to the project, the objectives and the rationale of the proposals to each of the different groups within the firm by whichever means was appropriate. The whole exercise was a resounding success and resulted in professional standards in identity being dovetailed with professional standards in accounting.

Communication was the key factor throughout the whole process. In a very significant way the organization spoke for itself, with the consultants helping to formulate and then express their aims and philosophy through their services, communication channels and corporate behaviour. The client communicated its philosophy and aims to the consultants who, in turn, formulated ways to communicate them to target audiences. To do this, the parties themselves had to communicate most effectively: communication through communication.

Communicating across National Boundaries

In overcoming the difficulties inherent in creating identities for organizations which operate internationally, and where the client and the design consultant come from very different national cultures, communication is again important to successful project management. An increasingly encountered interface today is that between East and West. The cultural gap between Europe or America and, say, Japan is still very wide and will remain so for generations. In the client/consultant relationship, where the consultant is Western and the client is Japanese, much groundwork of a rather delicate nature needs to be done before a trustful, fruitful working arrangement can be taken for granted. Only then can the client even begin to judge what an effective solution might be or might achieve in a market that, to the client's sensibilities, may be totally alien.

Mitsubishi Motors had tried, over a period of time, to introduce various Japanese originated ideas to solve the problem of low brand awareness in Europe. Periodically, executives from Tokyo would arrive to present ideas to a bemused occidental audience of hard-nosed, independent automobile distributors.

None of these ideas ever took root because none was appropriate to local conditions. Nor had any of those who were expected enthusiastically to accept and apply the schemes ever been consulted about their local market needs, about cultural conditions or about environmental considerations. In this uncommunicative, unsatisfactory but very sensitive situation, it became clear that there was no chance of a successful solution – and thoughts of implementation could not even be entertained – before the cultural ground was prepared and mutual understanding on certain very basic points reached. The consultants' first task was therefore to find common ground, to research needs on both sides and eventually to set in place structures to ensure practical support for what, by any standards, would be a mammoth implementation task.

Again the recognition, at the very first contact, that this was a situation in which communication would be both difficult and vital, was to have a profound influence on the whole project, right the way through from design to detailed ongoing implementation. In such cases, research is motivated not by quasi-academic motives but by the necessity to find out, and radically to deal with, certain preconditions which are pivotal to success or failure. The motivation is thus purely practical. The Mitsubishi project was a considerable success, especially when measured against the expectations of those who had prior experience of the problem. When understanding of the Western cultural background was allied to an appreciation of the Japanese desire to promote their own values in a way acceptable to international markets, then the way was clear to devise and present an effective and workable solution.

Because, in turn, the solution was acceptable to Mitsubishi and to the distributors and dealers, implementation was given a flying start. The trust engendered was such that the consultants were asked to advise on, set up and run the ongoing implementation of the programme worldwide, complementing the client's own activities in the area of dealer development. Thus the activity of design management has been largely in the hands of the consultants from pre-concept stage through to the logistics and politics of motivating individual distributors and dealers. It was the consultants' responsibility to take on the task in order to ensure the success of the project. In order for the client to give this task away, it was necessary for the consultants to communicate the need. It was also true to say that part of the previous lack of success in the Japanese attempts to provide dealer and distributor support was caused by problems in communicating ideas. The consultants therefore also took on this diplomatic role.

Conclusion

In all of these examples it is seen that, in order for the client to communicate to its target audiences, effective and full communication between client and consultancy is a prerequisite. Management of design processes is therefore seen to depend on both the nature and the quality of communication, with success occurring when the knowledge of what information to seek is accurate and experience of how to apply the knowledge is present.

PART V

THE LINKS BETWEEN MARKETING AND
DESIGN

28 The Relevance of Design Futures

JAMES WOUDHUYSEN

Exploratory Design Laboratory, Fitch RS, UK

Introduction: the Paralysis of Marketing Theory

It is remarkable how slowly things change in marketing theory. In 1957 the muckraking journalist Vance Packard shocked America with his best selling book *The Hidden Persuaders*. In it he exposed how manipulative psychologists, employed by US car manufacturers, would search out a whole category of 'upward strivers' – men who lusted after Buicks, Oldsmobiles, $10,000 Ford Continentals or $12,500 Cadillacs. Three decades later, many marketeers still feel they are doing something new when they treat consumers not in terms of income, but in terms of higher aspirations. Well-paid marketing executives give the upward strivers of Packard's day, now known as yuppies, a quite unrealistic prominence; meanwhile, we are left to reflect on the paralysis of international marketing theory.

For far too long, debates in marketing have centred around socio-economic variables (As, Bs, C1s, C2s, etc.), demographic and expenditure patterns, and psychographic techniques based on market segmentation by lifestyle. Most of these issues were around in Packard's day. There is nothing wrong with demographic or psychographic techniques, provided that they are applied judiciously and interpreted with scepticism. Nor, to take another province of marketing theory, is it wrong to engage in detailed analysis of direct market competition, as Michael Porter (1980) recommends. In the chaotic commercial conditions of the 1990s, it is certainly essential that, to retain or build market share, managers arrange for all employees continuously to measure and improve customer reactions to the design of their goods and services. This is what Tom Peters (1988) suggests. Yet, though marketing theory is often useful, we need to move on.

It is significant that Philip Kotler's classic textbook on marketing (1980) contains barely a single reference to design. Contemporary texts, too, still make the same error (McBurnie and Clutterbuck 1987). Yet if design is to be managed seriously – with all the seriousness that attaches, say, to operations or finance – then we cannot afford to rehearse any further the injunctions against 'marketing myopia' that Theodore Levitt (1965) made more than 20 years ago. We need to move on from the past of marketing theory to a perspective on design futures.

Market Research versus Research into Design Futures

Start with market research. It is relatively well known that conventional market research can only reveal what consumers already like. Market research offers little guide as to what consumers would like, in the future, once presented with a fresh alternative. By contrast, a perspective founded on design futures recognizes the four hypotheses offered under the following headings.

Trends in consumer behaviour

The successful designs of tomorrow should be based on an in-depth grasp of trends in consumer behaviour and consumer perceptions.

With consumer behaviour, a particularly important factor for design is not so much 'spend per household' as quantitative studies of individuals' use of time (Gershuny 1988) and of the sexual division of labour in the home. Who will be doing the washing up, who will be listening to the hi-fi – and for how long? For a domestic appliance to make the grade, these are the design futures questions that marketeers should ask.

With consumer perception, one must consider not only how a firm's goods and services will be positioned relative to those of direct rivals, but also the indirect impact of products and personalities which, often at the unwitting expense of the firm, will be likely to capture the hearts and minds of its audience. In the consumer's eye the sensual values of, say, a mobile phone may be no match for those that can be derived from four tickets to a rock concert. The job of the futures-oriented coordinator of marketing and design is to research these cognitive and emotional issues and, through a detailed assessment of the likely evolution of their dynamics, draw up an inspired design specification.

Public appreciation of design

Public appreciation of and concern for design will, in years to come, be an increasingly international phenomenon.

Design already makes the cover story in America's *Business Week* (April 1988). It is widely seen to have helped fuel the recent industrialization of Spain. It provokes Prince Charles and Margaret Thatcher to speak out. We are all design critics now.

Why is this circumstance likely to persist? To the extent that the 1990s turn out to be a non-inflationary decade, customers are unlikely to make the criterion 'cheapest' their main one when it comes to buying things. Moreover, in an atomized, individuated economic environment, where owing the 'right', distinctive possessions often makes the quest for identity more joyous, design promises to be the single most striking non-price factor affecting purchasing decisions (Woudhuysen 1988). It is time to drop defensive postures about the contribution that design – including its aesthetic aspects just as much as its cost saving ones – makes to commercial success. Design is and will remain an integral part of social life, which is why it must be integral, too, to the whole process of new product development and marketing. Forget the timeless definitions; the role of and for design grows out of the real competitive environment of the late twentieth century.

Design is multidisciplinary Different kinds of design will overlap more and more.

Just as architecture has begun to pay more attention to interior design, so it will become harder to say when the design of a shop ends and that of a leisure facility begins. Amid a retail context, product design, packaging and coordinated displays (visual merchandising) are all becoming as vital as lighting, signage and fixturing. In tomorrow's porous design universe, the executive office may more and more resemble the living room and the private dwelling be burdened with corporate-style 'home informatics' (Miles 1988). Nor will the overlap between different kinds of design be confined to environments. In what Shoshana Zuboff (1988) has termed 'the age of the smart machine', the graphic elements in product design promise to grow ever more critical.

For marketing and design directors, clear guidance through blurred environmental boundaries and inscrutable information technology emerges, through a design futures perspective, as a central customer requirement. In turn, the above hypothesis means that forming multidisciplinary project teams is the only sensible way to manage tomorrow's protean design project briefs.

Pedigree Design has an international and worthy pedigree.

Through surveying the international history of artefacts – how they looked and worked in the past, how they have been shaped by, and given shape to, the business and market forces of different national cultures – it is possible to engage in pattern recognition in design and so come to a more accurate estimation of its trajectory in years to come.

A business historian's approach to the world pedigree of design can be vital to the successful management of design projects. Moreover, being sensitive to the history of the visual – having at one's disposal images, not just words, about years gone by – is already part of the task of meeting customer requirements. Television commercials, movies, leisure facilities: all these have made consumers more aware of different eras, different national experiences. Without raiding the past for quick solutions, and without cultivating a fruitless nostalgia, marketeers need to remember the adage that those who do not learn from history are doomed to repeat it.

Two Criteria for a Design Specification

Two important criteria for design specification emerge from all this: intelligibility and the role of metaphor.

Intelligibility Research into design futures reveals that pressures on consumer time are not just a yuppie phenomenon. In Britain the working week for manual workers of both sexes in manufacturing has risen by more than an hour over the past seven years, and similar trends may be observed throughout the service sector (Department of Employment 1988). What follows from this is that intelligibility – the capacity of a product or environment to gain instant recognition and understanding by the consumer – must be pre-eminent in every design specification. Display, installation, emergency routines, safety, signage, the general level and quality of information provision: 'software' factors such as these deserve special attention if design and marketing managers are to satisfy the spirit of the times.

Metaphor   The fight to gain a share of the consumer's expenditure, because it is a fight to capture his or her imagination, means that there must be a way in which the finished design is not just functionally apposite and intelligible, but also a metaphor for wider social and aesthetic concerns.

Metaphors make sense because tomorrow's consumers will be sophisticated enough to demand of each design they encounter that it project an innovative and non-literal *Gestalt*. They will not be ignorant punters, but instead will expect that each design should surprise them and give them pleasure, often in an offbeat way. For marketeers, it is essential to specify design that symbolizes subjects and meanings beyond the ones that are obvious. This applies whether the project relates to a product, a company, a brand or a building.

Using a Design Futures Perspective

What has conventional marketing got to tell us about the contemporary significance of intelligibility or of metaphor? Not very much. But the experience of Fitch RS, a large multinational team of design consultants based in Britain and the USA, shows that companies such as Electrolux, Kodak, Metal Box, Philips, Polaroid, Reebok, Texas Instruments and Xerox, and store groups such as De Bijenkorf and La Rinascente, are prepared to pay for the results of a design futures perspective.

Fitch RS has a very commercial orientation. We employ, for example, 40 marketing specialists, both to improve our services and to give our clients marketing advice. Yet still we want our clients to grasp that, though well-designed products and services should always be informed by good market research, they should also be informed by a much more catholic – and, we believe, more dynamic – perspective on the future. That is why, uniquely among the world's design consultancies, we run an autonomous unit devoted to the task of understanding the future of design. In this taxing endeavour we use photographers, librarians and researchers to conduct a continuous and international enquiry into the economic, technological, social and visual trends of the next decade. This research, which is broader and deeper than market research and which encompasses both the world of images and the world of words, can open the eyes of clients to design and so encourage its proper management.

In terms of the four hypotheses previously outlined, we like to persuade managers about the benefits of design to the consumer, about its popular appeal, about its multidisciplinary character and about its fascinating track record. We think that, without a design futures perspective, the corporate vision and line management of design can all too often turn out to be superficial, subjective, and perhaps even cynical. The following case histories of our work show how we have used design futures concepts such as intelligibility and metaphor to come up with designs that are market winners.

Dillons bookstore

Situated in London's Bloomsbury, Dillons enjoyed an excellent reputation as a specialist, university bookstore, but had great difficulty in attracting the broader lay public. The Pentos Group acquired the store and briefed Fitch to help turn it into a popular bookshop with a strong identity among general readers, while still retaining the loyalty of the university customer.

The main task was to make intelligible the store planning of the building, which had previously hidden the wealth of books on offer and alienated the general customer. Fitch therefore reorganized the floor layout to create a large central axis and strong circulation routes. We sited the general book department in the key position on the ground and first floor; and we also introduced a shelving system to allow for a more attractive display of books. An assured and immediately memorable corporate identity, reflected in advertising, packaging and receipts, was developed. Today Dillons is, commercially, Europe's largest bookstore and ranks among the world's top places to buy books.

The Burton Group

To design the Burton Group's Top Shop chain of teenage womenswear stores, Fitch used a succession of metaphors – every three or four years – drawn from the realm of rock music. These metaphors have run from the Beatles, through Gary Glitter and John Travolta, to Boy George. They were not at all a matter of whim. They animated the environments which have helped make Burton a household name amongst British consumers.

Vacuum cleaners

Our case to a major client in domestic appliances was simple. They needed to shift the consumer perception of their wares from vacuum cleaners to home care products. 'Vacuum cleaners' described a technology (vacuum suction) in which few consumers were interested, and a process (cleaning) which had associations with menial tasks. Research by Johnson Wax in the USA revealed that women there set much less store on cleaning the home, and much more on cultivating a happy one. This research was certainly consonant with British findings, and ensured that the client's first goal in vacuum cleaners had to be to concentrate on the end result for the consumer, not the method by which that result was achieved.

We pointed out that what consumers wanted was a clean home in the fastest possible time. Thus the cleaning endeavour had to be intelligible in the widest sense. For example: as more men began to take on a proportion of domestic work, so the decals and literature associated with cleaners had to be as inviting and expressive for men as they would be for women. They had to be clear for naive, first-time or infrequent users as well as for frequent ones.

Intelligibility meant a distinctive range of accessories, a tidy cord and clear 'bag full' display. It meant that hassles surrounding bags had to be avoided at all costs – for, in a time-pressured environment, exasperation might be reported to family, friends and neighbours and so quickly undo the most expensive advertising campaign. Above all, intelligibility meant easy service: simple diagnostics, rapid maintenance and repair, unambiguous typography, graphics and layout, and exemplary writing in the instruction manual, service guarantee and warranty. Future consumer trends suggested other criteria for the design specification:

Weight became more critical to the extent that an ageing population makes us all 'disabled'; by the same token, handles had to be thought out with regard to the increased incidence of arthritis and inflammation of the tendons caused by excessive keyboard work.

Noise could never be allowed to block out that all-important telephone call, since the client's products would operate in the telecommunications dominated 'smart house' of the future.

Portability would be more and more demanded by a restless, mobile consumer market.

Multipurpose modularity would make sense. Popular obsessions with home ownership, DIY, better gardens, car maintenance and so on would put a premium on flexible cleaning systems of modular construction.

Professionalism would be among several appropriate design metaphors. With more and more people working from home or from the car, the boundaries between domestic and contract applications of cleaning would become more fluid. This meant that a 'professional' model had to be considered.

Midland Bank's Vector account

When we helped the Midland Bank design the name Vector for a new consumer account it launched in 1987, we wanted a metaphor for both scientificity and springiness. There was no market research that could have, by itself, come up with the highly evocative – and highly successful – result. The target market was financially aware but financially relatively unsophisticated As, Bs and Cs – 'opportunist' individuals prepared to spend money to save time. The age range was 25 to 44. These were people probably earning more than £10,000 a year, and going overdrawn a bit nearly every month. The account, which was to be backed by a card, was to:

1   Save the user time by organizing all financial affairs clearly and in one account.
2   Make money by interest on current account.

3 Give an automatic overdraft facility of up to £1000, the first £250 being free.

4 Send no offensive letters.

5 Ask a regular fee of £10 a month to replace all normal bank charges.

Midland's first ideas for a name were, frankly, rather primitive. The bank had begun to think in terms of brands, but toyed with initials (PDQ), numbers (AI) and attempts to express aspects of the brand – its universalism (All Konto), its product technology (Network) or its consumer benefits (Trust, Handy). Even a relatively meaningless name can, with enough advertising and public relations clout behind it, become a significant and popular brand. Ford's Cortina, which was named after an obscure Italian town, confirms this. However, Midland was looking for a name which would do much of its work for it – by being idiosyncratic enough to put the account several years ahead of its rivals in the financial services sector. On this criterion, none of the original candidate names reached the appropriate quality standard.

We therefore put together a team of Fitch and Midland senior managers and wordsmiths, plus external consultants, to arrive at a shortlist of names through a form of rigorous but open-minded brainstorming. This process was chaired by an experienced professional and ran over four half-day sessions. We spent a lot of time on nautical metaphors: Poseidon, Mistral, Dolphin. Gradually, we began to centre on twists to these which could give the account a precision feel, as well as a tone of movement: Helmsman, Skipper, Navigator, Voyager. In turn, that led to a stress on mathematical dynamism. But this was only half the battle.

As well as finding an exciting name, our metaphor needed to convey an element of luxury, good taste, of being up-market. To satisfy both requirements we arrived at a two-deck format. Our recommendation: *Indigo* to represent purple, silk, emotional richness; and *Vector*, a message of pointing and arrowhead accuracy. Midland did not go all the way. It dropped the word Indigo, but carried through the colour when it had the card designed. A year after its launch in May 1987, Vector had proved more profitable than first envisaged.

Conclusion

Compared with marketing or advertising, the business of design is relatively young. This is a handicap but also an opportunity. It implies that we do not have to obey all the textbooks that have, since the war, laid out familiar paths towards the consumer. Instead, through a sense of the present as history in design, we can engage consumers of the future in all their changing personality.

The American historians Stephen Kern (1985) and David Bolter (1984) have suggested that there come moments in human civilization when the ways in which people apprehend the physical world change dramatically.

Today, it seems probable that new social textures of time, space, energy and information are likely to make the concrete, commercial scrutiny of design futures much more relevant to business conduct than many of the more abstruse recommendations of marketing theory. There is still much to be learnt about marketing; but with a design futures perspective there is quite a bit about marketing that could, too, perhaps be forgotten.

References

Bolter, J. D.: *Turing's Man: Western Culture in the Computer Age*. London: Duckworth, 1984.

Department of Employment: *Employment Gazette*. August 1988.

Gershuny, J.: *Changing Times: the Social Economics of Post-Industrial Societies*. University of Bath, June 1988.

Kern, S.: *The Culture of Time and Space 1880–1918*. London: Weidenfeld, 1985.

Kotler, P.: *Marketing Management: Analysis, Planning and Control* (4th edn). Englewood Cliffs, NJ: Prentice-Hall, 1980.

Levitt, T.: Marketing myopia. *Harvard Business Review*. September–October 1965.

McBurnie, T. and Clutterbuck, D.: *The Marketing Edge: Key to Profit and Growth*. London: Wiedenfeld, 1987.

Miles, I.: *Home Informatics*. London: Pinter, 1988.

Packard, V.: *The Hidden Persuaders*. New York: McKay, 1957.

Peters, T. J.: *Thriving on Chaos: Handbook for a Management Revolution*. New York: Knopf, 1988.

Porter, M.: *Competitive Strategy*. New York: Free Press, 1980.

Smart design: quality is the new style. *Business Week*. 11 April 1988.

Woudhuysen, J.: The kind of design history managers ought to know. London Business School paper, 1988.

Zuboff, S.: *In the Age of the Smart Machine: the Future of Work and Power*. Oxford: Heinemann, 1988.

29 Links between Marketing and Design

LOWRY MACLEAN
Tomkinsons Carpets, UK

Introduction

Tomkinsons is a highly successful company manufacturing fashionable, value-for-money floor coverings. Rapid growth through the eighties has brought it a high profile within the carpet industry. Its Mr Tomkinson brand sells throughout the UK via a network of recommended dealers who are supported with exclusive products and a programme of national advertising. The company's policy is to develop products which avoid intensive price competition by being differentiated and yielding better margins. We aim to offer our customer a carpet that is better value than one that could be bought from our competitors – not necessarily cheaper or more hard wearing, but better by design and in more fashionable colours. Product value is the key to our profits. We are only as good as the products we generate; hence we place great emphasis on the efficient management of design.

New Consumer Demands

For the most part the carpet industry has only just woken up to the consumer demands of the eighties, with some companies still very much production led and only now beginning to employ marketing personnel. Until the early sixties, when new high speed machinery began to make its mark, carpet manufacturing was a long and labour-intensive process. Most carpets were of Axminster or Wilton construction and for most people they represented a once-in-a-lifetime purchase. Carpets were heavily patterned in deeper colours so they would be practical and soil hiding, and they contained many different colours so as to be versatile. Today, the majority of carpet sold is tufted and people are replacing them much more frequently. The accent is now on colour and fashion.

At Tomkinsons, the partnership between marketing and design begins at the outset, with the two departments sharing a joint approach to styling and promotion. Every product and point-of-sale item we generate should exhibit Tomkinsons' 'handwriting'. To achieve this, our marketing and design teams have agreed a general statement of how we see ourselves, who our customers are and what they want from us.

We try to see ourselves not simply as carpet manufacturers, but rather as the creators of an important element of interior furnishing. Carpet should not be a commodity item just to be walked on. Consciously or unconsciously, it is a token of unity in a home, establishing a colour relationship between furniture, fabrics and wallpapers, without which they would look entirely unrelated. We believe that when customers choose a carpet, they are choosing a colour, a mood, which will determine the total look of the interior. The primary concern is neither price nor fibre, but finding that particular shade of colour which matches the fabrics and wallpaper.

We always try to remember that our customer is not the retailer, but the end user – the woman in the high street. Traditionally the carpet industry has been male oriented and dominated, because of the need to shift heavy rolls and to cut carpet off the roll in store. With the advent of new machinery which can cut an order to size before it leaves the factory, retailers no longer need to stock as many rolls. There is a greater emphasis on presentation with more and smaller samples of carpet in the store. The customer now has a much wider choice of styles and colours, and in the majority of cases it is the woman who decides which carpet to buy. At Tomkinsons, we try to keep the focus on this woman consumer from the beginning right through to the end of the development process.

Team Approach

In 1985, we established a system for the planned management of ideas. We identified groups of people within the company who could generate and gather ideas, and then brought them together to form a product development team. The team includes members of design and marketing, as well as technical and production managers. The regular meeting of this team is the collection point for new ideas. These ideas may be generated by the marketplace, by new technology, or by members of the development team, who are allowed the space and freedom to work on their own pet products and concepts. We make sure that they get out and about and look at the market from their different viewpoints. One person may see possibilities where another may not recognize them, simply because of the nature of their work and interests.

It is the task of our marketing department to gather information which will be meaningful to our designers. This they obtain from market surveys and industry studies, as well as sales force feedback and regular trips down the high street. Together, the more creative members of both departments keep in touch with the latest styles and colours in fabrics, wallpapers and furniture. All members of the team scour the market for gaps, opportunities, innovations and trends – and meet every two weeks to exchange ideas. At these meetings, innovative ideas which do not fit into current plans are recorded and kept on file for the future. The market may not be right for a product just now, but the fashion end of the carpet market does change rapidly, and what was unpopular three or four years ago is often today's success.

The product idea is expressed in a brief from which the designers can work. The brief is the translation of marketing requirements into a working design direction, so although marketing staff may identify their customers by specialist classifications, the design team needs a clear picture of the person they are designing for. When an idea has been accepted for further development, work on the technical and creative aspects progresses simultaneously. Daily discussions, both formal and informal, ensure that members of the team are aware of any changes to the brief and the resultant production and marketing implications. Technical staff work closely with production managers, while a special creative team, made up of selected members of the design and marketing departments, starts to put together colours and images for the new range.

The visualization of an idea is always difficult when there is a team of people involved, all of whom read different meanings into words and descriptions. Our solution is for the two departments to work together from the start so that they both carry the same picture in their minds. The profiles of the target consumer groups are supplied by marketing, while tailored themes and looks are provided by the design side of the creative team. Together, these images form the starting point for design work.

Meeting the Brief

There are always alternative ways of meeting the brief. For example, we could take a winning design, repeat it in twelve colourways and meet the brief perfectly. Our design team, however, may decide to produce separate designs on the same theme for each of the twelve colourways. This is one of the ways that designers can add to the brief. A recent example of designers enhancing the brief is the development of a new Axminster range, 'Country Flowers'. The creative team, one designer and one marketeer, conducted in-depth research into what was available to today's more affluent Axminster consumer, and identified a gap in the market for a 'modern traditional' – a free-flowing design with an elegant floral theme in fashionable up-market colourings.

They visited the stores where our customers would shop to get an idea of the products and influences to which they were currently being exposed. All this information was developed into a revised brief which included sketches of the kind of look they were proposing. The design studio came up with a look that met the brief, but their enthusiasm was such that they worked all out to develop twelve individual designs on the same theme to make the range totally unique. When our marketing people started to develop the image of the collection and the point-of-sale material, they had a more exciting prospect to sell than they had originally envisaged. They had an anthology of the countryside which offered them greater scope for promoting and packaging the range.

It would, in fact, be restrictive for our marketing department to dictate the final content of a range. They direct the thinking but do not determine the product, and the design team then have the opportunity to enhance or improve the brief. Our designers may spot a winning look where marketeers saw only a commercial range.

Another success story, which was a direct product of the marketing/design liaison, is a range called 'Prime Movers'. As the name suggests, it is a range based around primary colours and is aimed at the younger, fashion-conscious market as well as teenage bedrooms. In scouring the market for opportunities, the creative team identified a market gap which may have existed for quite some time, but which was made more apparent by the growing popularity of primary colours for kitchens and children's playrooms, and an abundance of bold red, black and grey wallpapers. The designers put together a range of possible colours and designs for the marketing team to research.

When they started to look at what was available they soon found that while customers could find a red or a yellow carpet if they were prepared to shop around, nowhere was there a range of primaries that was well packaged, easily identifiable and widely available. The concept found plenty of support within the company. We had the right fibre, backing and construction to meet our target price. We set a launch date and decided on the distribution strategy. All we had to do now was to agree the content of the range. The design team were, however, championing an idea which went very much against what our experience told us would be successful. They were proposing a range of just four colours – red, blue, green and black – in a variety of different designs, namely pinstripe, polka dot, pawprint and plain. The traditional formula for success had always been to develop a winning design or style and reproduce it in a series of commercial colours appropriate to the market.

It could not be argued that the product the design team had come up with did not meet the brief. It met the brief exactly, and was innovative and eyecatching, but there was some concern about whether or not we could achieve the required volume out of just four base colours when our other ranges were operating with anything from 12 to 26 colourways. We researched the product through nationwide 'hall tests' as well as smaller consumer panels, rating the responses against known Mr Tomkinson and competitor products, and the response was overwhelmingly positive. Together, our marketing and design people devised a window display packed with bold images and bright colours to attract the identified target customers. Retailers wanted to keep the displays indefinitely because they were such a success. This response confirmed that we had hit our target market – achieved by involving the design team in communicating their narrowly targeted concept very forcefully at the point of sale.

Market Research

Another step in the development process which demands a close liaison is the formulation of questions for research. There are two types of research that marketing can commission to help the designers. The first of these is general research of the market and its trends. This information not only is valuable to the marketing department in planning the future product strategy, but can also give our designers an overview of the market and of the customers – who they are, what they want.

The second type of research is specific concept and colour testing. When we have developed either a range or a concept to meet the brief, our marketing and design departments will discuss the type of information and responses they need. Research results are only valid if we have asked the right questions, so we listen very carefully to our designers and our marketing team translates their requirements into a research brief. When research is completed the responses then have to be translated into useful suggestions and ideas for the design studio to work on when perfecting the range.

Different products require differing levels of research. We may only need to conduct one round of tests if we are asking respondents to select between possible new colours for an existing product range. A totally new product launch can take up to a year of research and testing, with an initial bank of some ten to twelve product concepts being refined stage by stage down to only a couple of winning looks.

Conclusions

In all its areas of operation, our product development team benefits from the support of senior management, who are committed to the new product development effort. One major advantage of an interdepartmental team is that the entire workforce is aware of the importance of new products to the company's future and will support the team in its endeavours. The team approach eliminates the kind of situation where one person has a pet project and sponsors it, because any new product must be supported by the majority of the team. This approach also ensures that those involved acquire the broad knowledge and diverse skills that will help them to be versatile and react quickly to the brief. Our designers have a clear understanding of the company's long term objectives, which is vitally important since we need to evolve new products that will not only be a success today, but will also fit into future product strategies.

A constant dialogue between the marketing and design teams also allows us to continually monitor progress against preset objectives. If the colour or fibre content is adjusted, we can pin-point the implications almost immediately. If a new competitor product is launched, or if certain styles of carpet show signs of losing their market share, the design team can react very quickly to this information. In the longer term, daily contact between the two departments avoids disappointments or negative reactions to the design team's suggestions at the regular formal meetings.

Design does not operate in a vacuum and cannot work without an understanding of both corporate strategies and market trends. Design cannot be, nor should it be, responsible for identifying market opportunities and threats, but it should be provided with such information by the marketing department. In the same way, marketing is not a separate function from design as responsibilities do overlap. Success is the result of the entire product development team working together, with strong leadership and a sense of challenge and enjoyment towards a common and clearly identified goal.

30 Robustness and Product Design Families

ROY ROTHWELL and PAUL GARDINER
Sussex University, UK

Introduction

A lean design and its frequently defective brief generally lead to an unprofitable lean product specification. Such products often immediately fail or enjoy only a temporary success until market requirements shift. On the other hand, a robust design leads to a design family of products. These families have uprated, rerated and derated versions which can meet upward, lateral and downward shifts as markets segment or develop. Robust product families in the past have tended to occur almost by accident or through a tradition of over-engineering.

Today, in many cases, the robustness is being deliberately designed in through the use of computer aided techniques which can explore and develop various possible alternatives. A family of product specifications can meet a broad range of changing market requirements. Robust designs permit economies of variety while at the same time profitably maintaining a central core of economies of scale for most of the basic subassemblies.

What is a Lean Design?

Sinclair's C5 electric car and GEC's early warning Nimrod aeroplane are two examples of lean designs which have been failures. This is somewhat surprising because they come from an individual and an organization that have previously proven records of successful innovative designs. The reason that they failed in the domestic and military markets respectively is that they were the outcomes of rather lean design briefs.

Nimrod The Nimrod airborne early warning (AEW) project was begun in the late seventies for several reasons. First, it was conceived as providing an advanced replacement for the very old Shackleton aircraft which, with their primitive radars, were being used for the detection of potential airborne attacks. Secondly, the European members of NATO could not agree on a common specification for an early warning aircraft. Thirdly, the Labour government of the day wanted to promote British high technology and the highly skilled jobs to go with it. In 1976 the RAF developed an Air Staff Requirement (ASR

400) describing its airborne early warning radar requirement which, it is believed, laid down that the system should be able to track up to 500 separate targets for 15 minutes, within a 360 degree radius of about 300 kilometres.

The Ministry of Defence rejected Boeing's offer of the already operational AWACs and in 1977 approved the development of 11 Nimrod AEWs at a cost of £817 million. It was then believed that the Nimrods would be cheaper and could be delivered earlier than the AWACs. After absorbing more than £1 billion and encountering numerous delays and technical problems, the whole project was cancelled in 1986. There has been some debate about whether the failure was due to the RAF or GEC, and in particular to the former's reinterpretation of the ASR 400 requirement to include overland operation in south-east England. This specification change created a massive data overload for the onboard computers in the Nimrods.

In fact, this massive and costly failure can be traced right back to the beginning of the project, and it appears that the RAF and GEC were equally to blame for creating what was a lean design solution to the ASR 400 requirement. The RAF opted for surplus Nimrods previously used for maritime reconnaissance, and GEC opted to use its existing 4000 series computers; from that point on the project was almost doomed to failure.

Compared with Boeing's 707 used for the AWACs, the Comet/Nimrod as an aircraft was itself a lean design (Gardiner 1984). The Nimrod is simply a smaller, more limited aircraft (see table 30.1). For drag and structural reasons a single rotating radome, as used on the 707 AWACs, could not be installed on a Nimrod airframe which instead had to employ bulbous front and rear radars. The two separated and smaller radars had inherently poorer resolution characteristics than the single larger rotating radar of the AWACs. In turn, this led to the question of a considerable volume of irrelevant information or 'clutter' that the onboard computers had to deal with.

The information burden was further increased by the reinterpretation of the ASR 400 to include both overwater and overland operation. During operation over the south-east of England, the Nimrod's computers now spent much of their time tracking fast cars and trucks, which were indiscernible from low flying helicopters. As a result of these problems the earlier GEC 4080 computer was replaced by a twice-as-fast GEC 4190 computer to better enable Nimrod to cope with the clutter and misinterpretation. Even as the programme was being cancelled, it was realized that the 4190 would itself have to be further upgraded to cope with the extensive data problems. Along with these hardware problems, there were many computer software difficulties that remained unsolved.

Finally, because the Nimrod is a physically small aircraft, there were a variety of cooling and cabling problems which, in service, would probably have considerably reduced the system's reliability. Nimrod simply could not be stretched in the same way as the 707. In short, the RAF's selection of a cheap but lean Nimrod, and GEC's selection of what was a lean radar

Table 30.1 Comparison of the Boeing 707 and Comet families

Boeing's robust 707 family

	Domestic service 100 and 200 series	Intercontinental service 300 and 400 series
Length	144 ft 6 in (44.04 m)	152 ft 11 in (46.61 m)
Wingspan	131 ft 10 in (40.18 m)	143 ft 5 in (43.61 m)
Engines (4)	12,500 lb (5670 kg) to 18,000 lb (8165 kg)	17,500 lb (7945 kg) to 19,000 lb (8450 kg)
Seats	179	219
Range	3015 miles (4949 km)	4865 miles (7830 km)
Production	around 1000	

de Havilland's lean Comet family[a]

	Empire service 1 and 2 series	Transatlantic service 3 and 4 series
Length (max.)	93 ft (28.35 m)	118 ft 0 in (35.97 m)
Wingspan	115 ft (35.05 m)	114 ft 10 in (35.00 m)
Engines (4)	4450 lb (2018 kg) to 7300 lb (3311 kg)	10,000 lb (4536 kg) to 10,500 lb (4763 kg)
Seats	36–44	81–101
Range	1750 miles (2816 km)	2590 miles (4168 km)
Production	around 100	

[a] Nimrods were modified Comet 4Cs.

and computing configuration, produced a Nimrod AEW that was almost inevitably bound to be a £1 billion failure.

Sinclair C5   Sir Clive Sinclair has had a long term interest in electric vehicles, but it became much more precisely focused when new legislation affecting electrically assisted cycles was introduced in the UK in August 1983. The main features were:

1   Pedal propulsion with electrical assistance.
2   Bicycle configuration not exceeding 40 kg and a motor power input up to 200 watts; or tricycle configuration not exceeding 60 kg and a motor power input up to 250 watts.
3   A maximum speed of 15 m.p.h.
4   Operators had to be at least 14 years old but no licence, insurance, road tax or helmet was required.

As a result, Sir Clive's C5 was conceived as an inexpensive, personal electric vehicle with a usable driving range of around 20 miles at speeds up to 15 m.p.h. Choosing a three-wheeled configuration and a sculptured plastic composite body unit produced a sort of rolling reclining lounge chair for one person. Without the battery, the vehicle weighed around 30 kg. Oldham Batteries produced a 400+ watt-hour lead-acid battery for Sinclair's C5 which, although a lightweight design, still weighed 15 kg. Nevertheless, at this weight a spare could be carried and the vehicle could still stay within the legislative total vehicle maximum of 60 kg. Carrying a second battery meant the C5's range could be doubled up to some 40 miles provided the batteries were swapped over during the trip. Although the C5 was promoted as a high technology electric vehicle, it very much remained an old technology design in relation to the conventional lead-acid battery and electric motor components used. Both the battery with its lead and the motor with its steel and copper were heavyweight components.

At a designed selling price of just under £400, Sinclair's C5 was in competition with lightweight mopeds. Because of the hugely different energy densities of lead-acid batteries and petrol, Sinclair's C5 was a technological 'energy cripple' compared with a moped. In actual operation, the consumer magazine *Which?* found the real situation to be even worse. The top speed was only around 13 m.p.h. and not the 15 m.p.h. claimed, and the realistic range was between 5 and 10 miles and not the 20 claimed (*Which?* June 1985). This report was equally critical of other aspects of Sinclair's C5 design including its inability to keep up with traffic flows, poor braking, potentially dangerous seating position and lack of reliability.

In the end the market niche for the C5 was just too small for viable commercial production. With a planned plant capacity of 200,000 C5s per year and sales of only 5000 units in the first ten months of the launch year of 1985, it is not surprising that Clive Sinclair's C5 production went into receivership with the loss of millions of pounds. The decision to stay with a conventional lead-acid battery technology which could not be improved, and the legislative framework established for electrically assisted pedal cycles, very much constrained the C5's design and its brief to the point that there was never any real hope for success. Essentially the C5 was a lean design for a non-existent market.

What is a Robust Design?

Lean designs are inflexible and unadaptable; robust designs are flexible and adaptable with respect to external changes – especially markets and user requirements. In a world where change is normal and constancy is rare, robust designs for products, processes and systems have a much greater likelihood for survivability and potential for commercial success. Except for certain military applications, this robustness cannot be goldplated against all possibilities, and a successful robust industrial design can adapt to most changes but not all. Among pre-war commercial flying boats and post-war ocean cruisers there were examples of robust design, but first large land-based multi-engined commercial aircraft and then long distance commercial jets were two major changes that even the earlier robustly designed planes and ships could not withstand.

Unlike lean designs, robust designs take more time, resources and effort, be it in terms of labour, computer aided modelling, financing or development work. In short, robustness does not come cheap: only leanness does. If to this story of robust designs we add the existence of 'tough customers' for the products, processes or systems, then this more frequently results in the production of 'good designs' (Gardiner and Rothwell 1985).

The key feature of robust designs is that they allow for change because essentially they contain the basis for not just a single product but rather a whole product family of uprated or derated variants. Further it should be noted that this remark also applies to process and system families and not just product families. At any one time not all the uprated, rerated or derated variants have to be in production or use. However, they potentially exist, and when tough customers demand new variants it is commercially possible to deliver them.

The world is constantly changing in terms of culture, science, technology and economics, which in turn throw up new ideas and new opportunities. Not all of these are always desirable but sometimes viable new combinations can be found. There have been Maglev and Hovertrain demonstrators, but their area of application is already hotly contested by long distance fast trains and short to medium range aircraft. Hovertrains and Maglev trains are probably best left as undeveloped new combinations of ideas and opportunities. On the other hand, deep offshore drilling for oil in the North Sea has proved to be a very successful combination of new ideas and opportunities. The debate about and the rejection or approval of these big new technologies have been carried out at senior corporate levels and in government departments. Robust designs also mature with open debate, whether it is for a whole new technology such as optronics or for a new consumer product such as the Sony Walkman.

Car Design

Sir Terence Beckett, while running product planning for Ford UK, was responsible for rejecting for ten years the idea that Ford make a car to rival Issigonis's radical Mini introduction at BMC. Based on precise reverse engineering of the Mini, Beckett's argument was that Ford could actually make one, but could not make money out of it. This product planning framework and open debate involved the same sort of discussion that eventually led to the development of a robustly designed Fiesta, which has been a top selling commercial success for Ford for the past ten years. Currently there are a dozen Fiesta variants ranging from a simple economy version to a special sporty performance car costing 50 per cent more than the cheapest.

The Fiesta product family caters for a whole range and variety of customer and market requirements.

Robust designs do not come off the shelf. They require substantial and different sorts of efforts, at different phases of their development (see figure

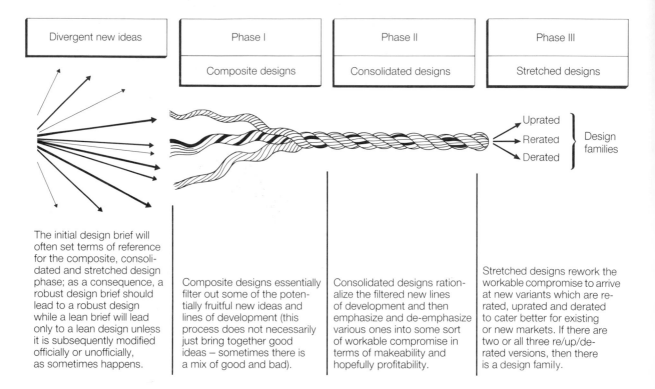

Divergent new ideas	Phase I	Phase II	Phase III
	Composite designs	Consolidated designs	Stretched designs

Uprated
Rerated   } Design families
Derated

The initial design brief will often set terms of reference for the composite, consolidated and stretched design phase; as a consequence, a robust design brief should lead to a robust design while a lean brief will lead only to a lean design unless it is subsequently modified officially or unofficially, as sometimes happens.

Composite designs essentially filter out some of the potentially fruitful new ideas and lines of development (this process does not necessarily just bring together good ideas – sometimes there is a mix of good and bad).

Consolidated designs rationalize the filtered new lines of development and then emphasize and de-emphasize various ones into some sort of workable compromise in terms of makeability and hopefully profitability.

Stretched designs rework the workable compromise to arrive at new variants which are re-rated, uprated and derated to cater better for existing or new markets. If there are two or all three re/up/de-rated versions, then there is a design family.

Figure 30.1 Robust designs

30.1). Much day-to-day design work is fairly conservative and almost routine because it involves using known techniques and solutions to produce something that works as expected. There can be a degree of hostility or scepticism with regard to new ideas and opportunities, which is often warranted and prudent because the new is often technically untried and unknown in the market. Nevertheless technical progress occurs and new ideas are taken up. Often, when there are many divergent new ideas around, the first phase in the development of a robust design is one of composition, whereby a reduced number of new, and often old, ideas are pulled together: attempting to continue too many new ideas at any one time is very risky.

The next phase is one of consolidation, whereby the degree of compatibility and the relative priorities of new and old ideas are established. Sometimes at this stage no matter how good a particular new idea appears by itself, if it does not fit well in relation to the overall design it should be rejected. For example, in the case of Ford's Fiesta, the prototype had a torsion bar type of front suspension without coil springs which left room in the engine compartment for the spare tyre. While this suspension worked well, later in the development programme it was discovered that the torsion bars might not have consistent characteristics when made in production volumes. The torsion bar idea was scrapped and a more conventional MacPherson strut and coil spring suspension was reintroduced. This left insufficient space for the spare tyre, which had to be moved to the rear of the car. This meant in turn sacrificing luggage space – a very precious commodity in design terms in a small car such as the Fiesta. In this design consolidation stage, a good but unworkable new idea was rejected and other design requirements were sacrificed; these were the right decisions to take, otherwise there could have been a very expensive recall programme.

Cars and Stretching

The Spanish car company SEAT began production of its Ibiza model during the mid 1980s. Initially five versions of the three-door model were available, followed by another four versions based on a five-door model. The stretching of the basic model with three doors to one with five doors is not as easy as it might appear. The Ibiza is in the compact hatchback car class. Its length is 143 in; other members of the same class, such as the Ford Fiesta, are longer at 144 in, and the Renault R5 is shorter at 141 in.

The Renault R5 comes in three- and five-door versions but some modifications are necessary. The three-door R5 has a 94.6 in wheelbase which has to be lengthened to 96.9 in for the five-door model. Having two different wheelbases does introduce some production diseconomies. A five-door model of the Fiesta cannot be made. Although the Fiesta is longer than either the Ibiza or the R5 it has the shortest wheelbase at 90.1 in. Compared with the Ibiza's 96.5 in and the five-door R5's 96.9 in, the Ford Fiesta wheelbase is just too short to squeeze in rear doors.

SEAT appears to have understood the criticality of the very long wheelbase within the limitations of the compact hatchback length which permits three- and five-door models:

Usually, after a three-door hatchback has established itself, its makers bring out the five-door version. And often, in gaining a couple of doors, it loses a little of its appeal. The rear doors look rather pinched. And so do the passengers. However, all is not lost. The SEAT Ibiza, a car that embodies German engineering as well as Italian style, was conceived from day one to accommodate five doors as well as three. (SEAT advertisement, *What Car?*, February 1987, p. 4)

Admittedly this quotation is taken from a SEAT advertisement, but it does illustrate well that the three- to five-door stretching idea needs to be brought into the design process at a very early stage.

Derated Aero Engines

In the stretching phase of a product family, derated variants do not necessarily lead to new products that are in any way inferior. By derating its RB211 aero engine, Rolls-Royce produced a lower thrust 535 engine series, and in the process actually ended up producing a superior engine. When Boeing, in the late seventies, was designing its new 757 aircraft, it initially called for a 32,000 lb thrust engine. Pratt and Whitney, General Electric and Rolls-Royce all responded with offers of engines. During the 757's design evolution period, Boeing upped the thrust requirement several times and eventually General Electric dropped out of the competition. Rolls-Royce became the lead engine supplier with Pratt and Whitney following later. The Rolls-Royce 535C engine entered service on 1 January 1983 and was succeeded by the 535E4 in October 1984. The 535s have been Rolls-Royce's most reliable and trouble-free engines. They are also the world's quietest engines on any 100+ seat civil aircraft. Boeing's 757 with Rolls-Royce engines is permitted night landings at some airports where all other commercial aircraft are banned because of noise regulations.

The original RB211 was built up of seven basic modules. By removing the large front low pressure fan module and replacing it with a scaled down fan, the derated lower thrust 535C engine was derived. The remaining intermediate and high pressure stages were then run at moderate air velocities, pressures and temperatures. This greatly contributed to the engine's reliability and its good long term fuel consumption figures. The 535C with a thrust of 37,400 lb was upped in the 535E4 to 40,100 lb. Some of the improvements came from changes in turbine blading and the incorporation of a new exhaust integrator nozzle. The biggest change was the redesign of the front low pressure compressor stage. A new wide chord fan blade design was used that reduced the number of blades and eliminated the use of snubbers linking each blade. Overall, the story is that the RB211 was derated to produce the superior wide chord 535E4 aero engine. This innovative new wide chord blade design is now being diffused back into the RB211 engine family to improve further the fuel and thrust ratings of the current upgraded variants (see figure 30.2). The designed-in robustness of the initial RB211 was crucial in enabling the subsequent modifications to be undertaken.

Open debate and testing during the consolidation phase is very important because solving design problems in a post-production phase can often cost orders of magnitude more. Computer aided design techniques now allow the testing of different alternative combinations of new and old ideas which, if done with older prototyping approaches, would have been too costly to explore. For instance, Rolls-Royce has found that it can reduce the overall cost of aero engine research and development and still produce innovative and improved designs:

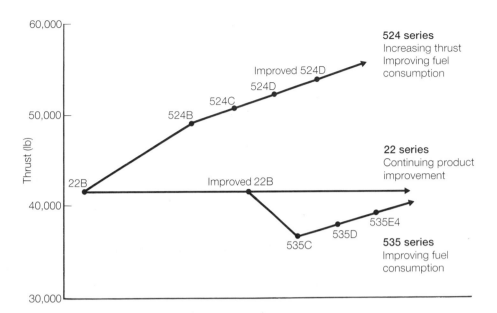

Figure 30.2 Rolls-Royce RB211 engine family
Source: Rolls-Royce Ltd, *RB211–22B Technology and Description*, TS2100, no. 20, July 1980; updated by authors to 1987

The traditional approach to combustion development was to carry out a large number of rig and engine tests to evaluate modifications whereas with this [computer] model it is possible to evaluate several geometries before committing to rig and engine testing to confirm the choice of final solution. Computer systems provide the basis for future reduction in R&D costs as well as significant reductions in the overall timescale for developing a new engine. (Ruffles 1986: 16)

Since the beginning of the seventies and particularly with the spread of CAD, it has become increasingly possible to develop a robust design which has the deliberately designed-in capability of being stretched. Rolls-Royce's new aero engine (the RTM322) will be produced in a number of versions and will be scaled to cater for different market requirements and power outputs (see figure 30.3). The core of the RTM322 is common to turboshaft, turboprop and turbofan versions. This RTM322 core, when scaled up × 1.8, becomes the core for the RB550 series which is expected to be produced in turboprop and turbofan versions.

With previous new engine introductions there has been a long tradition of continuing to refine a design over its production life. This 'growth' leads to improved specifications and a succession of mark numbers. Once this growth phase is undertaken by Rolls-Royce's advanced engineering group, the results will be reincorporated in the two core engine designs and the five possible variants. Rolls-Royce's director of technology believes that this approach makes it possible to get 'economy in technology' and to help keep down the cost of R&D (Ruffles 1986: 15).

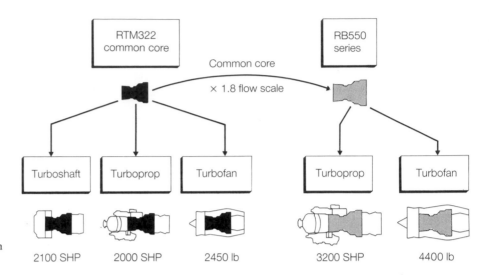

Figure 30.3 Rolls-Royce RTM322 engine: economy in technology

Photocopiers

Photocopying machines are for most companies a major capital expenditure, just like mini/micro computers and word processing work stations. Technologically, there is one big difference between conventional photocopiers and these other types of machine: word processors and mini/micro computers are electronic machines, whereas conventional photocopiers are essentially electromechanical machines. Having a working design lifetime ranging from tens of thousands of copies for the smallest machines to many millions for large machines, and having to handle dusty paper and toners, there are quite severe requirements on electromechanical designs if the machines are to operate reliably even with routine servicing. One of the best ways of handling the problem of heterogeneity in a market characterized by a very broad range of user requirements is to develop a basic robust design and then have variants of it by offering a limited number of add and drop option packs or modules. This has been successfully achieved by Canon.

In terms of total copier sales, Canon has been a market leader for a number of years. Canon also introduced the world's first digital copier in 1986, which is much more electronic in operation than the conventional electromechanical machines. We will turn to these new types of copier later. It is worth noting just how wide a range of functions and market uses the conventional machines have to serve:

Speed   8 to almost 200 copies per minute.
Capacity   500 to 70,000 copies per month.
Size   Portable to large desk type shapes.
Enlargement/reduction   35–400.
Exposure   Fixed to variable and automatic.

Colour   Black or up to six.
Paper handling   Single sheets to double-sided volume collating and stapling.
Price   Below £1000 to over £60,000.

The only area in which Canon does not compete is at the top end of the market for very large copier-duplicators costing many tens of thousands of pounds. This area remains the province of Kodak and Xerox. Below this very expensive end of the market, Canon has a number of different series and variants which cover most price and non-price specifications. Within most series, perhaps 80 per cent of all the components are standard and only the remaining 20 per cent can be altered to produce different variants in that product family. This ratio is by no means accidental and is firmly backed by extensive research and development. During recent years Canon has spent, as a percentage of sales, more on R&D than has been taken out in profits. Canon now has robust families of copiers to meet almost all copying requirements: the firm, through designing in robustness, effectively combines economies of scale with economies of scope.

It is likely that this situation will continue through Canon's development work on two new types of copiers which represent the next big step in the development of the reprographics industry: full-colour copiers and digitized laser copiers. Conventional copiers are electromechanical analogue machines which directly transfer an image from the optical system to the print mechanism. The new laser copiers are much more electronic and employ digital signals rather than analogue ones. In laser copiers the image from the optical system is converted into digital electronic signals for image processing. These signals can then be sent to a scanning laser modulator and finally to a printing mechanism (see figure 30.4). The importance of this new laser copier system is that the digital electronic signals for the image can be handled like all other electronic signals by various information technologies. Readers can be separate, and visual information can be digitally processed, stored or transmitted simultaneously to a number of distant printers. Laser copiers are becoming a lot more electronically intelligent. In the future, 'Canon will take color copying itself a generational step further, creating full-color laser copying combined with intelligent editing functions made possible through digital analysis, and control of color reproduction will result in images of remarkably sharper definition and vastly richer tone gradients' (Canon 1986).

Given Canon's previous performance, we can expect to see new robust families of laser colour copiers, but much of the robustness will stem from the utilization of proven electromechanical modules. This is again an example of combining something new with something old. Laser copiers still retain the well-developed optical systems and the print mechanisms of conventional analogue copiers (see figure 30.4). Essentially, the new laser copiers insert an additional digital electronic information processing step between the original optical and print systems. In this way the traditional photocopier

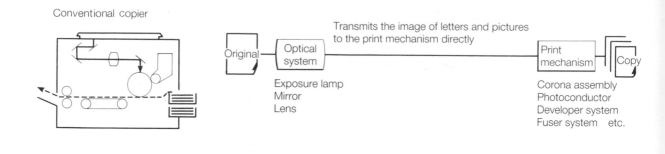

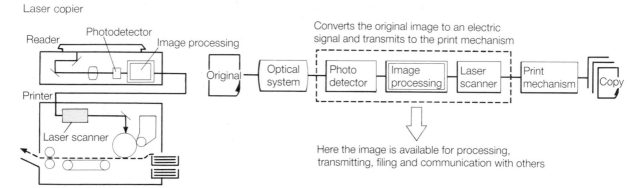

Figure 30.4 Conventional (analogue) and laser (digital) copiers
Source: Canon (UK), *Canon Laser Copier*, 1986

technology is taken one step forward and made more intelligent, while preserving much of the original robustness.

Aircraft and Flight Simulators

As we have shown, robustness need not be accidental; it should be designed into the original concept. For example, we can look at British Aerospace's BAe146 four-engine jetliner. A prototype BAe146 first flew in late 1981, and in the 146–100 and 146–200 versions it has subsequently been bought by airlines around the world for either passenger or freight use. In September 1986 it was announced that a 100-seat version, the BAe146–300, was under construction. What is significant to us is that British Aerospace knew that the basic design could be stretched for this large version and that it could be achieved in a way that responded to market demands:

Two years ago [1984] at Farnborough we announced that we were studying a 146–300 and suggested the form it might take. We knew that the basic design had plenty of stretch in it and we wanted to investigate how much of that stretch the market could use.

We have been receiving clear messages that the first requirement of our existing and potential customers is to capitalize on the very low block-hour costs of the 146. Service experience of the series 100 and 200 has given ample evidence of the remarkable economy of the aircraft and the brief has been clear – more capacity with the same economy. (BAe112/86 news release, 1 September 1986)

Along with the production of a new aircraft family such as the BAe146, there is a need for flight simulators. The most recent generation of flight simulators can reduce training costs by two-thirds when compared with using actual aircraft. Furthermore, simulators can be used for training exercises, which might be dangerous if performed incorrectly in an actual aeroplane. Rediffusion Simulation has produced two of the world's most advanced flight simulators for the BAe146. One is in use in Hatfield, England and the other in San Diego in the United States. We just commented on the built-in stretch of the BAe146; it is equally significant that Rediffusion Simulation has incorporated the same sort of stretch in its flight simulators. These simulators perform equally realistically for the series 100 or series 200 BAe146s, but in addition there is the built-in potential to do the same for the new series 300.

These are two clear examples of well-managed robust design configurations that allow for stretch as markets and users continue to evolve and develop. Robust designs offer considerable advantages to both manufacturers and users. The concept of robustness will, we believe, increasingly become a powerful tool in the competitive armoury of technically progressive, strategically managed companies.

Conclusions

While in the past design robustness has been achieved almost accidentally through over-engineering, today a number of major companies are deliberately designing in robustness at the product specification stage. Robust designs offer the manufacturer considerable flexibility in meeting evolving user requirements and the ability to cope with market segmentation as it develops. Through careful and imaginative design, resulting in design robustness, manufacturers can obtain scale economies in production (high commonality of parts) coupled to economies of scope (high product variety), which represents a powerful competitive combination. At the same time, these designs offer the customer greater choice in the range of product attributes immediately available, and greater confidence in the supplier's ability to satisfy future requirements as they emerge.

The basis for design robustness appears to vary across product types. In the case of aircraft, for example, it clearly depends on the degree of interrelatedness between the different critical operating parameters, i.e. the degree to which one performance parameter can be varied without detrimentally degrading other parameters. In the case of automobiles, robustness depends as much on the product/market strategy of the firm as it does on technical factors (Rothwell and Gardiner 1984). In some circumstances a lean design series can be transformed into a new generation of robust designs through a deliberate and comprehensive redesign process, as in the case of the API-88 hovercraft (Rothwell and Gardiner 1985). Finally, close and forward-looking interaction between leading edge suppliers and leading edge users during product specification can contribute significantly towards design robustness (Rothwell 1986).

The advent of CAD and flexible manufacturing systems has greatly facilitated firms' ability to produce robust designs, enabling them more effectively to combine economies of scale in manufacturing with economies of scope in product attributes. In a world in which international competition is intensifying and user needs are rapidly changing, the ability to produce robust designs will be a major factor determining a firm's profitability, flexibility and, in the longer term, survivability. Designers and managers ignore this message at their peril.

References

Canon: *The Canon Story 1986/87.* Japan: Canon, 1986.

Gardiner, J. P.: Robust and lean designs with state of the art automotive and aircraft examples. In Freeman, C. (ed.): *Design, Innovation and Long Cycles in Economic Development.* London: Design Research Publications, 1984.

Gardiner, J. P. and Rothwell, R.: Tough customers: good designs. *Design Studies.* Vol. 6, no. 1, pp. 7–17, January 1985.

Rothwell, R.: Innovation and re-innovation: a role for the user. *Journal of Marketing Management.* Vol. 2, no. 2, Winter 1986.

Rothwell, R. and Gardiner, J. P.: Design and competition in engineering. *Long Range Planning.* Vol. 17, no. 3, pp. 78–91, 1984.

Rothwell, R. and Gardiner, J. P.: Invention, innovation, re-innovation and the role of the user: a case study of British hovercraft development. *Technovation.* Vol. 3, pp. 167–86, 1985.

Ruffles, P. C.: Reducing the cost of aero engine research and development. *Aerospace.* Vol. 13, no. 9, pp. 10–19, 1986.

Consumers' Association: Sinclair C5. *Which?* pp. 274–5, June 1985.

31 Proposals, Briefs and Specifications

CRISPIN HALES
USA

Introduction

Proposals, briefs and specifications in design work are used to communicate the needs or intentions of one party to another. In this chapter the terms are set in context and some particular types are discussed. One procedure for compiling an engineering design specification is outlined, with reference to examples, and a short bibliography is provided for obtaining further information.

The terms 'proposal', 'brief' and 'specification' are used in a variety of ways depending on the circumstances. As they are all general terms it is recommended that descriptive qualifiers are always used ahead of each term to avoid confusion. Thus we have a project proposal, a project brief, a design brief, a design specification, a product specification, materials specifications and so on. Sometimes it may be useful to add secondary qualifiers. For example, one could be involved with a product design specification or a system design specification, depending on the nature of a particular project.

A proposal or brief is usually concerned with overall aspects of a project or problem while a specification is concerned more with detailed aspects of the project. There is a tendency for the term 'brief' to be used more by consultants and industrial designers and for the term 'specification' to be used more by engineering designers and managers. The term 'proposal' is more general than either of the others and is used here to refer to any document, other than a simple price quotation, submitted in response to a request for proposal (RFP), a call for bids or a call for tenders. Figure 31.1 shows the design process mapped in context within a project, within a company, within a market, within an external environment. Superimposed on this are various types of proposal, brief and specification to indicate how they relate to each other and to the design process. This map is a useful reference for the discussion which follows.

Project Proposals

Often a design project stems from an idea or an identified need in a market, and competitive bids are solicited to help develop the idea or solve the problem. Such a call for bids may vary from a simple verbal request up to a highly detailed request for proposal in several volumes, depending on the organization and the scope of the work involved. The response to this is termed here a project proposal. It is assumed that as the primary aim of a project proposal is to secure some kind of contract it will be a written document. There are many useful guidelines for the preparation and pres- entation of project proposals. Hajek (1977) covers those for large one-off projects in considerable depth, while Warby (1984) deals with smaller scale projects. The proposal provides a formal statement of what the offeror understands the project to involve and the approach proposed to satisfy the request. It is a document used for bid selection and evaluation in the first instance and for negotiating terms of contract with the successful offeror afterwards.

The bidder's list   The chances of submitting winning proposals depend, in the first instance, on obtaining the relevant requests for proposal as early in the bidding process as possible. This demands an intelligence system combining strategy, formal procedures and information collection tuned to particular market situations. The aim is to get in early, at the highest possible level of those shown in table 31.1. In an organization's main market area, the lowest level for survival would be the bidder's list or its equivalent.

Table 31.1   Levels of involvement at the RFP stage

Level	Qualifications required
'Sole-source' bids	The only, or by far the best supplier (may involve helping prepare the RFP)
Invited bids	Selected company based on reputation and/or past performance in the market
Bidder's list	Company of assessed capability with recognized qualifications and expertise in the particular market area
RFP distribution	Sufficient capability and enthusiasm to be considered a possible contender
Advertisements	General capability in the market area
Unsolicited proposal	An original idea put in the form of a project suggestion or funding request

The decision to bid   A conscious decision to bid or not to bid is a critical step. It involves complicated factors such as type of project, scope of project, what resources are available and what the return on investment is likely to be. An evaluation of the technical and financial risks involved is important and even the cost (in time and money) of preparing the proposal itself may be a factor in the overall equation. For guidelines on making such a decision see Hajek (1977).

Proposal preparation   A main guideline for proposal preparation is to 'be responsive'. Proposals detailing what the offeror would like to do as distinct from what has been requested are not likely to be successful; proposals are usually assessed

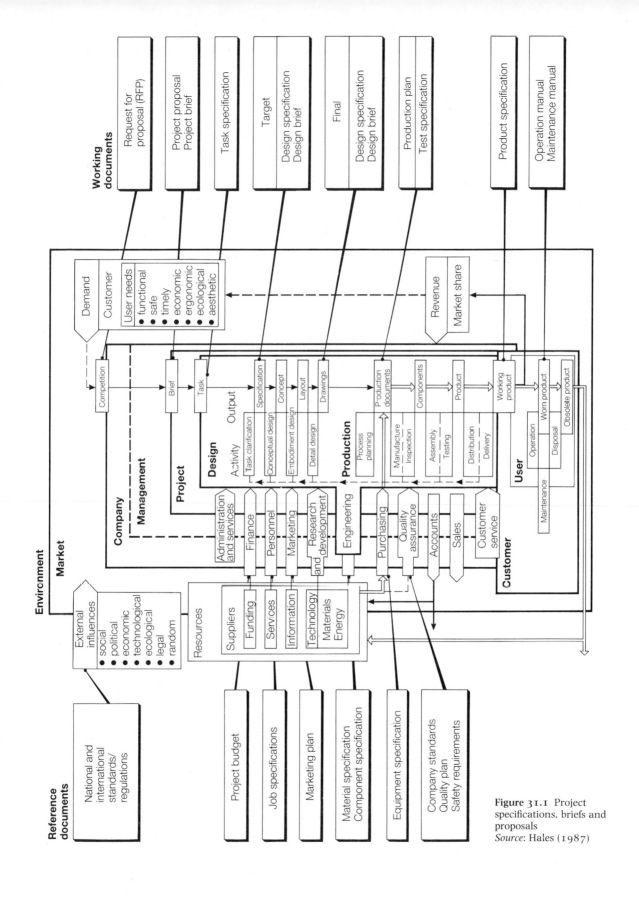

Figure 31.1 Project specifications, briefs and proposals
Source: Hales (1987)

according to criteria drawn from the RFP document and irrelevant material gains no credit. Requests for proposals usually contain specific instructions as to the content and presentation but, if not, the following is a useful guide as to structure and contents:

1   Executive summary of proposal.
2   Background and qualifications of organization.
3   Statement of the problem as understood.
4   Technical discussion to show an understanding of the problem and how it is proposed to find solutions.
5   Statement of work to be carried out.
6   Management of project including resources, key personnel and organization.
7   Project plan.
8   Estimated costs, suitably itemized.
9   Concluding statement.
10   Supporting appendices.

Project Briefs

The initiation of a design project may also come directly in the form of a request to carry out specified work, commonly termed a project brief. Again such a request can vary from a simple verbal communication up to a detailed document closely defining the work to be carried out. Topalian (1980) discusses project and design briefs in some depth, and most of the general points mentioned regarding project proposals also apply to project briefs. However, whereas the main aim of a project proposal is to secure a contract, the main aim of a project brief is to define the work to be carried out and the results expected. It is thus more the equivalent of the work statement section of a project proposal than the overall document. It is also something which may be developed and used as a working document during the course of a project, doubling as a design specification.

Much of the confusion surrounding use of the term 'brief' arises from the fact that the same document, with only a few revisions, may be used as the 'request for proposal', the 'proposal' itself and the 'design specification' during the early stages of a project. Unless it is made very clear which version of the brief refers to which aspect of the negotiations, complicated communication problems are likely to arise.

Task Clarification

In order to carry out a design project, particularly one with a high engineering content, two things need to be established at the outset, each being a complementary result of the 'task clarification' phase of the design process.

The first is a clear statement of the problem to be solved, for which solutions will be sought. The second is a set of requirements and constraints against which to evaluate the proposed solutions.

The former is termed a 'problem statement' or 'definition of the problem' while the latter is termed a 'specification', a 'target specification' or more correctly a 'design specification'. Both of these are essential if a solution to the problem is to be found which satisfies all parties. Considerable effort, and possibly some preliminary design work, may be needed to help establish what the real problem is before compiling the design specification, but it must be done. Finding solutions to the wrong problem is unacceptable design practice.

The nature of design problems

Design problems are not the same as analytical problems, although analytical approaches are often used during the course of design work. The characteristic differences between design problems and analytical problems are

Table 31.2 Comparison of analytical problems with design problems

	Analytical problems	Design problems
Problem area	Well defined	Poorly defined
Problem statement	Precise	Vague
Information state	Sufficient	Insufficient
Solution	Correct solution obtainable	No single solution

shown in table 31.2. In general a design problem summarizes what is undesirable in a particular situation, and the problem is considered solved when an improvement in that situation is achieved which is acceptable to all parties. This will be a compromise solution as distinct from a correct solution.

Defining the design problem

On many projects insufficient time is spent in clarifying the task. It is good practice to define the problem in writing as a first step. This is not as easy as it sounds, for design problems are rarely what they appear to be on first sight and they may change with time. It is often helpful to try and formulate the problem at a higher level of abstraction (more generally) than first stated, and the following are some essential questions to be asked (Pahl and Beitz 1984):

What is the task really about?
What implicit wishes and expectations are involved?
Do the suggested constraints actually exist?
What paths are open for development?
What properties must the solution have?
What properties must the solution not have?

Project planning

Once the design problem has been appropriately formulated it is important

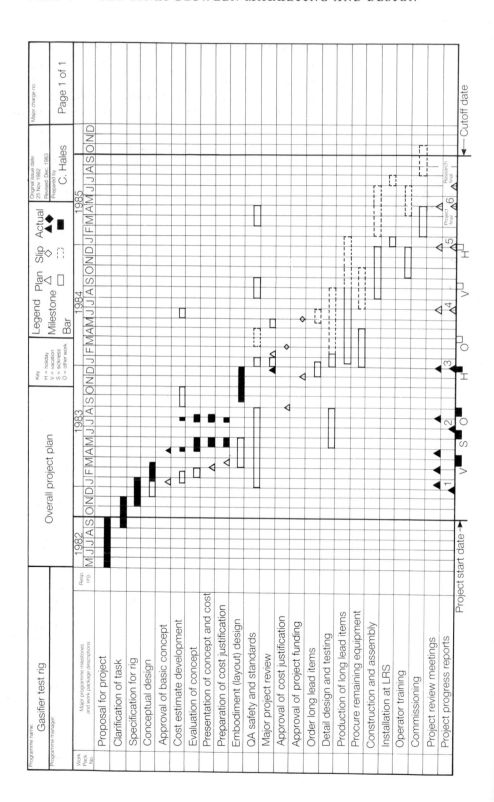

Figure 31.2 Project plan
(Gantt chart) for a small
design project

that the vague initial ideas on how the project should be structured, scheduled, and managed are set down more formally (task specification). The earlier this is done the better, even if there is not enough information, as only then do the real constraints of time, human resources and financial resources become apparent. At this stage, reporting procedures can be set, preliminary tasks allocated and approximate time schedules estimated together with design team requirements. Figure 31.2 shows a typical preliminary project plan for a small project. For large projects, techniques such as PERT (Hajek 1977) are recommended to establish initial priorities (see also Leech and Turner 1985; Turner and Williams 1983).

Design specification When the design problem has been appropriately formulated a list of requirements in the form of 'demands' and 'wishes' can be drawn up regarding any potential solution to the problem. Demands must be met under all circumstances. If minimum demands are not met, the proposed solution is not acceptable. Wishes should be taken into account whenever possible butwarrant only limited increase in cost. Demands and wishes must be quantified whenever possible and it is sometimes helpful to rank their importance.

Table 31.3   Design specification checklist

Main headings	Examples
Geometry	Size, height, breadth, length, diameter, space requirement, number, arrangement, connection, extension
Kinematics	Type of motion, direction of motion, velocity, acceleration
Forces	Direction of force, magnitude of force, frequency, weight, load, deformation, stiffness, elasticity, inertia forces, resonance
Energy	Output, efficiency, loss, friction, ventilation, state, pressure, temperature, heating, cooling, supply, storage, capacity, conversion
Material	Flow and transport of materials, physical and chemical properties of the initial and final product, auxiliary materials, prescribed materials (food regulations etc.)
Signals	Inputs and outputs, form, display, control equipment
Safety	Direct protection systems, operational and environmental safety
Ergonomics	Man–machine relationship, type of operation, operating height, clearness of layout, sitting comfort, lighting, shape compatibility
Production	Factory limitations, maximum possible dimensions, preferred production methods, means of production, achievable quality and tolerance, wastage
Quality control	Possibilities of testing and measuring, application of special regulations and standards
Assembly	Special regulations, installation, siting, foundations
Transport	Limitations due to lifting gear, clearance, means of transport (height and weight), nature and conditions of despatch
Operation	Quietness, wear, special uses, marketing area, destination (for example, sulphurous atmosphere, tropical conditions)
Maintenance	Servicing intervals (if any), inspection, exchange and repair, painting, cleaning
Costs	Maximum permissible manufacturing costs, cost of tools, investment and depreciation
Schedules	End date of development, project planning and control, delivery date

Source: Pahl and Beitz (1984)

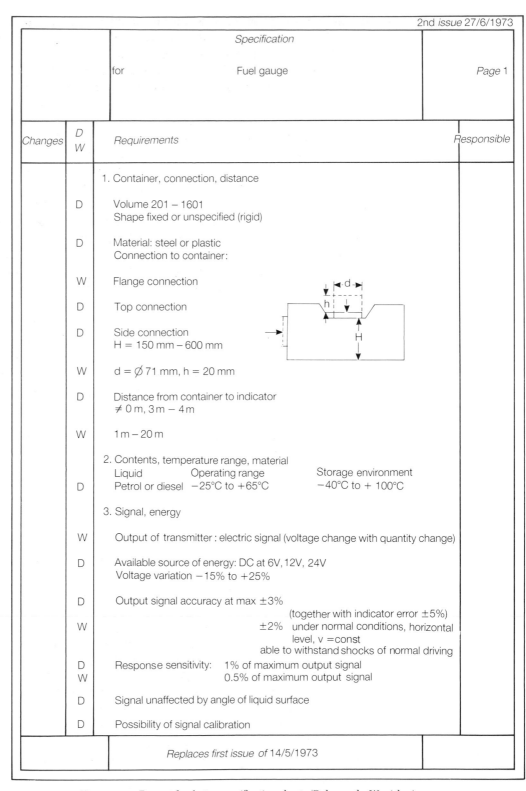

	D W		
			2nd *issue* 27/6/1973

Specification

for   Fuel gauge   *Page* 1

Changes	D W	Requirements	Responsible

1. Container, connection, distance

D   Volume 201 – 1601
     Shape fixed or unspecified (rigid)

D   Material: steel or plastic
     Connection to container:

W   Flange connection

D   Top connection

D   Side connection
     H = 150 mm – 600 mm

W   d = ∅ 71 mm, h = 20 mm

D   Distance from container to indicator
     ≠ 0 m, 3 m – 4 m

W   1 m – 20 m

2. Contents, temperature range, material
   Liquid   Operating range   Storage environment
D Petrol or diesel −25°C to +65°C   −40°C to + 100°C

3. Signal, energy

W   Output of transmitter : electric signal (voltage change with quantity change)

D   Available source of energy: DC at 6V, 12V, 24V
     Voltage variation −15% to +25%

D   Output signal accuracy at max ±3%
                                    (together with indicator error ±5%)
W   ±2% under normal conditions, horizontal
                              level, v =const
                              able to withstand shocks of normal driving
D   Response sensitivity: 1% of maximum output signal
W   0.5% of maximum output signal

D   Signal unaffected by angle of liquid surface

D   Possibility of signal calibration

Replaces first issue of 14/5/1973

Figure 31.3 Format for design specification sheets (D demands, W wishes)
Source: Pahl and Beitz (1984)

The design specification The design specification is a formal document compiled from the list of demands and wishes. If circulated to all involved it ensures that everyone is formally consulted and has the opportunity to contribute. It identifies the source of particular requirements (person's name against each item) and, if updated during the course of the project, it provides an accurate working record.

By combining the list of demands and wishes with the results of project planning, a comprehensive design specification can be compiled providing the maximum design freedom within the given constraints. This is best structured according to a checklist such as the general one offered by Pahl and Beitz (1984), reproduced in table 31.3. It is suggested that design specifications should be compiled on a set of standardized specification sheets, for example using the format shown in figure 31.3. (Other formats and checklists are given in the following publications: BSI Handbook 22, *Quality Assurance*, 1987; BSI PD 6112, *Specifications*, 1967; BSI PD 6470, *Management of Design*, 1981; Defence Standard 05-67/1, 1980; Macdonald 1985; Oakley 1984; Institution of Production Engineers, 1984; Verein Deutscher Ingenieure 1987.)

Once the design problem has been defined and the requirements have been listed in the form of a design specification, a firm base has been established for the project to proceed through the conceptual, embodiment and detail design phases. Solutions to the defined problem may be sought, and the resulting concepts may be evaluated against the design specification.

Example: Gasifier Test Rig

The task, as defined by the client company and formally set down in a project proposal, was to design a materials test rig to operate under particular high pressure, high temperature conditions. Although the main needs for the rig had been identified, it was seen as having several likely uses and the requirements were thus 'ill-defined'. No design specification existed. A series of rigs had been constructed and operated by the same project team, so this project was seen as another in a progressing sequence; however, as this rig would involve the difficult problem of handling flowing coal at high temperature and pressure, the design task was considered to be both novel and complex.

A problem statement was prepared, and by questioning all project participants against the checklist shown in table 31.3 a comprehensive list of demands (essential requirements) and wishes (desirable features) was generated. The detailed design specification compiled from this was a 20-page document listing 308 requirements and constraints, of which 217 were demands and 91 were wishes. It was formatted as shown in figure 31.3 except that 'contributor' was used instead of 'responsible' as the column heading for the names of people who contributed. Thirteen people con-

tributed directly, and 34 requirements came from the 400 ideas generated
by a brainstorming session involving 15 people.

The requirements fell into four categories: function, production, operation
and general information. Each requirement was labelled with the name of the
contributor and the document was circulated to all the project participants for
review and modification by a set date. A total of 92 corrections, clarifications
and additions were made, involving 72 demands and 20 wishes. Once the
specification had been agreed by all parties, only two items were changed
during the rest of the project and these were caused by specific external
influences. Details are given in Hales (1987).

The procedure used was regarded as most effective by the project team
and was adopted for use on other projects. Previously, researchers needing
a test rig would sketch out the requirements in the form of a concept, and
submit this either to the senior design engineer in the company or to
an outside supplier. Design work would begin and there would often be
misunderstandings and problems leading to disagreements and wasted effort.
A major reason for this was the lack of involvement of groups such as safety
specialists at the task clarification stage. Important requirements would be
omitted from the initial list and continual changes would be made during
the rest of the project.

Table 31.4 shows that for the gasifier test rig over 40 per cent of the
design requirements came from sources other than research staff and, in
particular, 19 per cent came from the services staff responsible for manufacture. The procedure used for this particular design specification almost
doubled the list of requirements which might have been expected had normal
company practice prevailed, and it ensured that a comprehensive set of
criteria was prepared for the selection and evaluation of conceptual solutions
to the design problem. It also forestalled a number of later difficulties in the
project. Out of the 533 hours spent on task clarification and conceptual
design, preparation of the specification took 170 hours or 32 per cent.

Table 31.4 Breakdown of design specification for gasifier test rig by source and type of requirement
listed

Source	Function of rig	Production of rig	Operation of rig	Information for design	Totals by source
Research staff	77	48	36	21	182
Management staff	7	4	7	7	25
Services staff	4	18	9	27	58
Other sources	18	10	3	12	43
Totals	106	80	55	67	308

Source: Hales (1987)

In theory, the output from the conceptual design phase, following the
task clarification phase, should be the concept which most fully satisfies the
requirements of the design specification. Only those candidate concepts

which satisfy every demand in the specification should pass through selection to evaluation, and then the most suitable concept should be determined by evaluating the remaining candidates against the wishes.

For the gasifier test rig this meant that any candidate concept would have to satisfy 217 demands to be selected and those selected would have to be evaluated against 91 wishes. This presented the problem of how to deal with such a full list of requirements; in practice, the selection and evaluation procedure was based only on those requirements judged to be the most important. A full evaluation procedure, involving the detailed weighting of criteria, was found to be unnecessarily complex for this particular project and highlighted the need for flexibility when applying a systematic approach. The systematic preparation of a design specification where the source of every requirement was recorded was of particular value as it avoided many potential later disagreements.

Conclusion

In general, the use of a clear structure and a systematic approach helps to coordinate and control complex design projects. However – and it cannot be stressed strongly enough – the approach must be applied flexibly. A systematic approach will not replace inventiveness, creativity, intuition and experience but it complements them and allows them to be used to maximum effect.

References and Further Reading

British Standards Institution: *Quality Assurance*. BSI Handbook 22 (3rd edn). London, 1987.

British Standards Institution: *Guide to the Preparation of Specifications* (under revision). PD 6112. London, 1967.

British Standards Institution: *The Management of Design for Economic Production*. PD 6470. London, 1981.

Defence Standard 05-67/1: *Guide to Quality Assurance in Design*. London: HMSO, 1980.

Engineering Council: *Managing Design for Competitive Advantage*. London, 1986.

Hajek, V. G.: *Management of Engineering Design Projects*. New York: McGraw-Hill, 1977.

Hales, C.: *Analysis of the Engineering Design Process in an Industrial Context*. Eastleigh: Gants Hill, 1987.

Institution of Production Engineers: *A Guide to Design for Production*. London, 1984.

Leech, D. J. and Turner, B. T.: *Engineering Design for Profit*. Chichester: Wiley (Ellis Horwood), 1985.

Macdonald, R. M.: Drawing up the purchasing specification. *Proceedings of the Institution of Mechanical Engineers*. Vol. 199, no. B1, 1985.

Oakley, M. H.: *Managing Product Design*. London: Weidenfeld and Nicolson, 1984.

Pahl, G. and Beitz, W.: *Engineering Design* (ed. Wallace, K. M.). London: Design Council, 1984.

Topalian, A.: *The Management of Design Projects*. London: Associated Press, 1980.

Turner, B. T. and Williams, M. R.: *Management Handbook*. London: Business Books, 1983.

Verein Deutscher Ingenieure: *Systematic Approach to the Design of Technical Systems and Products*. VDI 2221. Beuth: Berlin, 1987.

Warby, D. J.: Preparing the offer. *Proceedings of the Institution of Mechanical Engineers*. Vol. 198B, no. 10, 1984.

32 Prospective Design

JEAN-PIERRE VITRAC
J.-P. Vitrac Design, France

And just as water has no stable form, no permanent conditions exist in war.

Sun Tzu: *The Art of War* (500 BC)

Introduction

Prospective design is a new approach, allowing firms to satisfy consumer expectations which may not yet have been formulated. If design, the creation of products or systems, is to develop, business firms need a suitable business tool.

The Problem

A firm's strategy is generally founded on analysis of what has happened in the past, together with observations of various kinds. It will try to make use of tangible, concrete information, preferring to base its actions on certainties. When research or chance does bring a genuine innovation within the firm's range, it will be exploited only if it is cut down to acceptable proportions and made to fit into more or less normal market conditions. In short, it will not be used unless it can be made to conform.

The products of new technologies have been marketed in this way, so that their over-familiar images partly hide their novelty: examples are the compact disc, the video recorder and the microwave oven. Appearing much less interesting than their original technical concepts, they lose in impact and 'magic' power.

Two major factors are now undermining these attitudes, leaving most of the actors concerned (i.e. industrialists) in a state of anticipation, unsure of themselves and wondering what methods, procedures and practices they should adopt. There have not been many really useful surveys and analyses of these factors, which interact with each other, for the very good reason that they are phenomena of society which are difficult to quantify or assess.

The first is the shift, on a worldwide scale, of the centres of creation and production. New regions have come to the fore, regions capable of innovation, even if they once used to copy the products of Western markets and Western customs.

The second, and the more important, is the discovery that consumers exist who, when confronted with something new and original, can be tempted by it, contradicting all the criteria previously applied to them. We have here a new identity. Hitherto, consumers have been credited with a mainly traditional personality (except perhaps in the youth-oriented market): unadventurous, usually of moderate intelligence, reactionary, with classic tastes, disliking change and innovation, settling into habits of consumption which have given certain products a virtual monopoly.

It hardly needs to be emphasized that though such an image may remain alive in the minds of some managers, it is now wholly out of date. On the one hand, then, we have unforeseen new products which question a number of market assumptions and are reviving consumers' interest; on the other, we have industrialists taken aback by the situation and employing controlling methods where creative methods are required. On the one hand we have more alert consumers who see consumption not as a constraint on their activities, but as an optimum pleasure which they would like to enjoy within their personal means; on the other, we have unwieldy industrial and commercial structures and static channels of distribution encumbered by rules. In a large number of instances a wide gap can be seen opening up between a firm's identity, its proposed products, and the deep-seated, rarely revealed motivations of consumers.

For example, a study of domestic lighting equipment revealed a manufacturer whose communications strategy was concerned with technology, profitability, the nature of the lighting source, etc. These were all factors which had turned out, at the end of the study, to be of minor importance compared with genuinely novel concepts of the product, where it quickly emerged that affective reactions and a sense of relating to the object had priority.

Consumers accept a new idea more easily than is generally believed, as long as there is a good reason for it (i.e. as long as the concept of a product's image or function can be seen at once). It is only a short step from the customer as king, with much time spent researching in order to give just what he or she wants, to the customer as 'groupie', the fan of a certain product or brand, who plays the firm's game and likes its idea and its image. One need only mention Sony, Benetton and Swatch as examples of brands successfully based on image and identity.

Every firm therefore needs to establish new rules for the game. The large number of products on offer, coming from all quarters, the fragmentation of markets, the increasingly unpredictable diversity and development of cultural behaviour – all have a very disturbing effect on classic methods of analysis and investigation and tend to invalidate them.

It is therefore necessary to display creativity, using design as a means of formulating new hypotheses. The designer's eye, a new way of looking at things which is both instinctive and rational, observing the existing trends which it promotes, can make use of a highly developed and operational industrial culture. Finding new openings and creativeness are becoming the categorical imperatives of business development.

Accordingly, design needs to be brought in at the very start of a project and used as a way of thinking. Prospective design can thus claim to be an investigatory method using concrete proposals, and a method of producing analyses and syntheses. It is indubitably a very fruitful and vigorous method, bound up as it is with today's major currents of change and apparently based, at least to start with, on the irrational element of creativity.

In using the word 'irrational', it is in order to emphasize and, at the same time, defend the unpredictable nature of creation, an instinctive and intuitive process for which not everyone has the aptitude. Strengthened by being put into practice, gaining purpose and direction from experience and knowledge of markets and the public, the creative impulse will then be rationalized and made concrete by professional expertise and understanding of the way in which products are formulated.

Many manufacturers have great difficulty in formulating a new concept of their product, drawing up specifications, defining a product policy and working on development in the medium term. Analysing their existing potential and making better use of it, giving their image direction, organizing creativity and building bridges with new operational sectors all constitute major problems, and it is a fact that everyday administration hardly encourages the development of creativity in industry. Accordingly, many firms find it difficult to embark upon what may be called intangible investment, of which design is a typical manifestation.

Intangible Investment: Prospective Design

Prospective design, as a generator *par excellence* of concrete creative hypotheses, operates by grafting a creative, effective, stimulating and organizational entity on an industry either at certain crucial moments or in the long term, its essential function being to initiate fresh thinking.

The first priority is to set no limits to hypotheses, thus providing the best possible conditions for decision making in product development. The idea of investing energies in this method of prospective design proceeds from a desire to free creativity, getting the creative process to contribute all it can, and breaking away from the over-traditional structure of the usual patterns of thinking in industry.

Consumers do not create products; ideas are the result not of chance, but of systematic research and investigation. Specifications cannot properly be drawn up for products unless the maximum possible amount of data and information is available. Prospective design, then, should be employed as a concrete strategic method, brought into play ahead of marketing considerations; it can be described as a course of action which we might call the 360 degree strategy of creative investigation.

Developing a Firm Creatively

Firm Y manufactures plumbing fittings. The market stagnates; it is hard to predict developments other than technological ones (and not many of those). The question is, what should the development be for this market, and what should be the firm's own development?

Firm X specializes in deodorant products. This is a typical market with that kind of economic battle where there is room for only one brand leader if the firm's place on the market and in distribution is to be assured. But what about tomorrow? How long will it last in this particular market? What will future products be like? What will consumers want? How should the firm diversify?

Firm M is in the clock and watch making industry. There are phenomena of fashion to be taken into account, a technology that has been mastered, a rapidly changing market, with products whose primary function is often hidden behind their image; they are utilized in a more lucid, almost a wayward manner. What attitude should the firm adopt? What will its new strengths be in the future? How is it to operate within its own field in an original and profitable way?

In the face of a great many diverse and complex questions, calling on knowledge and expertise of all kinds, a firm may have difficulty in responding. It is unlikely to have at its disposal any method or system enabling it to approach and resolve such problems in a total rather than a sporadic manner. It may make use of statistical data on markets, investigate distribution and consumption, have surveys of a more or less specialized nature carried out in the hope of seeing some new idea or concept emerge. But its thinking will still be based on analyses of existing factors, and it is extremely likely that the product or products resulting from such a course of action will merely follow existing trends, with little novelty or media appeal of their own, yet requiring a great deal in the way of communications strategy if they are to be successfully launched.

Most specifications drawn up by such methods are restrictive. They tend to be mainly drafted in terms of constraints and not with an eye to the objectives. Industry wishes to make use of creativity to open up new opportunities, yet still blocks its way, hampering it with objections even before the first step has been taken. Consequently, we have to go further, liberating industrial thinking – and the only way to do that is to provide industries with new and concrete hypotheses.

Methodology

The first step is to set up a structure within the firm concerned which will be responsible for the project as a whole. The composition of this structure may be a permanent one, or it may be set up only on occasion around one or more persons with responsibility for the project, people instrumental in marketing, technology, research, trade, etc. In this case, the operational figures will mostly come from inside the firm, but they may sometimes be outside advisers.

This team, working together with the consultant designer, plays an essential part: since its task is to allow new ideas to surface, it is the requisite 'intelligent' interface between the outside world (this is where the consultant designer comes in) and the whole mechanism of the firm itself. It represents not only the communication factor in general, but the first link in a chain which, keeping the objectives always in mind, will be best able to communicate them to the people in charge of their final application. Indeed, the consultant designer cannot be seen as separate from the firm itself, and is not just an ordinary supplier of services but the motive force in a group with its eyes on the future.

The first question the firm will put is both complex and simplistic. Complex, because it reflects a situation in which problems of market development, image, communication and internal organization are all mingled. Simplistic, because it can often be put in a word or a couple or words followed by a question mark. Plumbing fittings? Cutlery? Television? Jogging shoes? And so forth.

Before any answer is attempted, it is advisable to steep oneself in the culture of the firm, using any analyses that have been carried out and may be relevant to the perception of its image, the definition of its products, the typology of the consumers it has in mind, and any other factors. This is a study in depth and also a study of the resources, both human and material.

It is essential to know what objectives have not been realized, what hypotheses have already been advanced, what moves were made towards certain operations if they came to nothing or were not finalized. Though all this contact making may seem rather laborious work, it will give rise to a parallel intuitive perception which should favour prospective creation. That creative impulse will then organize itself into several phases, three covering the prospective design itself and the fourth the putting into practice of their findings.

First comes the phase when the field is clear for the creation of concepts: this is very much the designer's work. A whole set of creative hypotheses is presented in the form of sketches, covering the entire area to be explored. The idea is not just to suggest those concepts we ourselves may have selected in advance, but to offer future consumers a creative panoramic vision of the world in which we are about to work, and all the possibilities that may one day confront them.

There are two objectives. The first is to place the new concepts permanently in relation to those that already exist, or at least in relation to concepts which have arisen from, or are very close to, those in force in present markets. Concrete information cannot be obtained until we have a system of references and standards available. The second objective is to let consumers judge the quality of our proposals for themselves, which means that we must keep our minds open.

In accordance with these principles, the group responsible for the project will not be able to intervene in this phase except in a creative and synthesizing manner. Experience suggests, and indeed it is all too clear, though the matter is seldom mentioned, that a vast number of projects have been aborted even before their proposals had been thoroughly examined, simply because of what may be described as those industrial filters which act as brakes: the sales department which registers an instant negative reaction, the technologist who pronounces the manufacturing operation impossible or too expensive before studying it, the marketing person who cannot identify target consumers.

During this phase, concepts are always presented in concrete form: the idea of the product is visualized schematically, with each item in it dedicated to a single concept. It is therefore very important to know if the presentation as a whole fills the desired gap, leaving the minimum of room around it. The main questions to be asked are: what is consumer perception in this sphere of activity, and what is the extent of creative development it will absorb? We can thus assess the risk factor and get an idea of what the future consumer will be like.

Once this preparatory work is over, we enter into the reactive phase, which is much more than simple ratification of the hypothesis proposed. It is a genuine approach in depth and gives an idea of how far we can go. To do this, we confront consumers with our various proposals. From the moment when there are proposals in existence, a number of expectations can be revealed. Go into a shop and you can buy only what it has in stock; our task is to create what is not in stock yet, something that must respond not to some fantasy of its creators but to the subliminal needs which this reactive phase allows us to uncover.

In the course of this presentation, consumers are first placed in relation to their present surroundings. Then, they will be immersed in new worlds; however, we are not making Utopian suggestions, but proposing something that could be realized in the future. The strength of this method is that it provides consumers with a perfectly plausible and instantly applicable projection of their own world, even if it looks quite different from what they are used to seeing every day, and they will end up wondering why these products are not yet on the market.

However, if we look in detail at the development of most products we shall find that it is a gradual process depending more on the present state of knowledge and technology than the market and a genuine development of needs. How can consumers be expected to be more inventive and open to new ideas than manufacturers, when all they can do is react in a routine manner to proposals put to them? What we are trying to do, first and foremost, is bring out people's deep-seated reactions of approval, disapproval or indifference in relation to our proposals. Most of all, we need to understand how they can fit into the lives of those concerned.

The importance, at this stage, of the whole notion of the concept has been recognized; it goes far beyond the simple fashionable image. Above all, and even before they assume an analytical attitude, people have a strong affective position which becomes all the clearer when the concepts put to them are genuinely relevant. They will tend, naturally enough, to show no interest in anything that seems to be purely formal aestheticism.

A very common attitude is the accepted belief that a firm whose business it is to manufacture a certain product will supply quality items without any problems. However, that quality is only the technical aspect of the item concerned and its making, and arguments depending entirely on it will not be totally persuasive. Persuasion in essence arises from a concept which naturally responds to some obvious need, even if it did not exist the day before.

To return to methods, it is a skilled business to stimulate this kind of consumer group. The people who do it are not just operating objectively, but need to make provocative suggestions which will help to show what may lie behind various reactions, to get past opinions and reach certainty (or uncertainty), and to adapt the concept.

Throughout this phase, the group of responsible persons will be with the consumer group, for a very simple reason: though the consultant may always be slightly distrusted, distrust generally evaporates before consumers. Consumers represent the final link in the industrial chain, and it is what they say that is really instructive. Future concepts will thus be partly ratified even before they have been selected and put into production, and future risks in decision making are much reduced.

The third phase is a phase of synthesis, but not just simple synthesis: in essence it also provides guidance for the organization of future developments. As decision taking is a very complex operation, we lean on convictions at this point: the plan of action is suggested, taking into account the firm's initial analysis of means and opportunities, as well as consumer reaction to the proposals in the preceding phase.

Decisions will be relevant to both the short and the medium term. The aim is to conserve the firm's active strength and its image while developing that image along certain simple lines and, as a parallel process, being able to introduce, now and in the future, one or more ideas which must not break totally with the firm's image but prepare the way for clearly differentiated but now totally identified images or activities. The strategy is not one of rupture but of consolidation and voluntary development.

At this stage the group as a whole hardly needs persuasion, since each individual member will be on the receiving end of the experience. All that now remains is to enter into the active phase, an area which is usually the exclusive province of the firm itself.

First, specifications are drawn up on the basis of the preceding observations, and the normal, classic course of product design is initiated. This process is simplified because the specifications are bound to be very full and because, instead of setting out in a state of uncertainty or from hypothetical suppositions, we can set out at once from certainties, their effect having already been verified. This is an excellent way of introducing innovations, and eliminates a great many of the risks attached to older methods – risks which have hitherto been major factors in limiting creativity.

Finally, the impact of this method on the internal planning and structure of the industrial firm should be emphasized. It brings the firm back into the front line of development, opening up long term opportunities for it to take a series of effective actions.

Example: Mecanorma

Although research usually concentrates on a range of products or a whole sector of activity, the Mecanorma study was based on a single product – a tubular-tipped pen for graphic and technical designers.

In the face of considerable competition from very well-known brands, undisputed market leaders, the Mecanorma product was gradually losing sections of its market. The firm's marketing department was not sure how to deal with the situation, since the results of a consumer survey were rather negative and gave no indication of how to react except by letting the product die a natural death. The survey showed lack of consumer interest in the product's image, technology and use. It was seen as an ordinary, standard implement which could not become anything else.

A phase of prospective creation was initiated with a view to stimulating the product again. It comprised the following steps:

1 Reformulating the product. Considering all possible ideas, from the disposable pen to the modular pen. Giving the pen new kinds of action, automatic release, manufacturing it from new materials, etc. The objectives were:
   (a) To make users react to new proposals, on the basis of sketches
   (b) To extend their present view of the basic product
   (c) To make it a live, modern, interesting and even futuristic product.
2 The consumer reaction phase: the new ideas were stimulating, interest was lively, new ideas and lines of development for the product came up. For instance:
   (a) Playing up the idea of transparency
   (b) Enhancing the image of the product's technology
   (c) Improving automatic release of the cap, manipulation of the pen, ease of use, the importance of leak-proofing (making the principle obvious)
   (d) Making it an object of active communication, etc.

3 Synthesis and realization of specifications now that data on the real factors involved were available.
4 Design of the product.

A new product was developed from factors contained in the synthesis. Not only did sales of the product rise within two years from 300,000 to over a million items, but its strong image also allowed the firm to penetrate export markets, particularly in Japan.

Conclusions

The question of what to make and sell in the future is something every industrial firm must consider, though some, given their positions as market leaders in their specific area of operations, may be more or less certain already.

It follows that the range of applications for prospective design is extremely wide. One firm may need to find uses for a material which is too technically complex for the manufacturing of products for the general public. Another firm may find it requires total reformulation of a product because the data provided by market analysis do not allow specifications to be drawn up. In a sector where most firms are in free fall, or where there is increasingly strong competition, new products, means of distribution and targets may all have to be identified.

Whether we are dealing with a system (the fast food industry, for instance), with the exploitation of a material, a single product, or a series of products allowing a firm to diversify, the question is always the same, and the problem is posed in terms of creativity. What all these industries have in common is an obvious deficiency in terms of research structure. How would anyone decide which way to go in order to reach Rome, Tokyo or New York, without any idea of the geography of the world? Perception of their own sector by industrialists is often far too narrow. Lacking the ability to step back and consider it, they do not exploit their own potential and image. Everything is capable of development, and can be the subject of creative action; everything can change and evolve, and a great many industries need to do so.

The key, then, is to be found first in an open minded attitude towards investment in creativity, and then in relating to partners who share this vision of a whole. We are concerned no longer with service but with creation, with cooperative movement – with the future of products and products of the future as an essential condition of development.

33 Innovation by Design

COLIN CLIPSON
University of Michigan, USA

Introduction

Throughout the 1970s the big three automobile companies in Detroit and their counterparts in the United Kingdom foundered in the face of competition from Japan, Germany and Scandinavia. Poorly organized to meet rapidly changing markets, energy crises and customer performance shifts, they lost sight of their customers, and lost a grip on emerging technologies necessary to competitiveness. Products of poor quality, poor performance and short life cycle had little chance against high quality, dependable Japanese and European products.

It is taking a long time to redesign the business processes of these industries. In some sectors of manufacturing there has been almost a reluctance to change. Poorly trained managers who found themselves faced with new, rapidly changing technologies developed technophobia out of fear of things they did not understand.

New and improved products are essential to the well-being of any business. Unless a firm can upgrade its products and services to remain ahead of the competition, the margin of profitability will be significantly reduced with attendant loss of market share.

However, historical reviews of business failure rates show that the incidence of new product failure is very high and costly. In a classic study of product development, Booz, Allen and Hamilton reported that only two out of ten new products were a commercial success and that seven out of eight hours of technical effort failed to result in commercial success (Jain 1984). As Bacon and Butler (1981) point out, the seriousness of this problem is not generally recognized; most firms are poorly equipped to deal with technical change and continue to operate with organizations that themselves are poorly designed. Why have these conditions become more critical in recent years?

Technological Change

If we look at the development of computers over the past 30 years we see accelerated change of a phenomenal nature. From its first operation in the 1920s, the computer only reached commercial feasibility and success in

1964. Similarly, the transistor took decades to become incorporated into commercial products such as hearing aids, navigational instruments and computers. The silicon microchip and integrated circuit followed similar if shorter processes from invention to commercial realization. In the last two decades, rates of change have become much shorter with more rapid translation of technical advances into commercialized products. Since 1953 the United States has been subjected to rapidly reduced lead time on computer innovation, from several years lead time in the 1950s and 1960s to zero and lag time in the 1980s.

The implications of this accelerated diffusion of technology and rapid translation into innovative products are that companies have very little time to rest on the strength of any one currently successful product. The cost of developing new innovative products is extremely high; companies must be able to defray the high front-end cost associated with translating inventions into commercially successful products by marketing them as widely as possible around the world.

Economies of scale in product development may be achieved by companies who manage to work together, or find some cooperative ways to share the costs of developing new products. Even when adding new products to the product line, a company can avoid going through product development by joining with other firms, acquiring other firms or getting products from them (Terpstra 1984).

'The bottom line is that companies that choose to develop domestic markets may find themselves totally blocked out by competitors that are well entrenched and ready to launch offensives on others' home markets' (Ohmae 1985). Significant differences can be detected in the technological edge of the high technology industries over the medium technology industries:

High technologies   Electronics, computers, communication equipment, fine chemicals, pharmaceuticals, office equipment.
Medium technologies   Steel, light electrical industry, automobiles, petroleum refining, textiles, non-ferrous metals, paper and paper products, fabricated metal products, ceramics, earth and stone.

In a comparison of Organization for Economic Cooperation and Development (OECD) nations, the high technology group had 1.49 times the sales growth, 2.8 times the productivity growth and 2.75 times the profit growth of the medium technology industries. It is becoming very difficult to make a profit in old line industries (figure 33.1). For example, VCR products are growing at three times the rate of other audiovisual products (Ohmae 1985).

Redesigning Organizations

Given the problems implied by technological change and the blurring of the traditional divisions between various sections of the manufacturing and

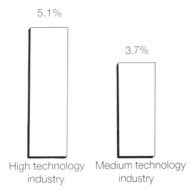

Figure 33.1 Technological change: net profit/sales ratio for the world's leading companies, weighted average 1980 and 1981
Source: Japanese Ministry of International Trade and Industry, *Economic Analysis of World Enterprise: International Comparison*, 1982

marketing systems, corporations themselves have to be redesigned in order to withstand the rigours of change and adaptation to new market conditions. Some general conditions of sound organization design are as follows (Clipson 1985):

1. The chief executive officer and corporate management understand the critical link between design engineering and manufacturing engineering in their business and believe that effective design for quality is an essential part of the corporate business strategy.
2. The corporation understands that design is not merely styling and market hype.
3. The corporation values long-term profits and does not heavily rely on discounted cash flow analysis. Investment in research is a key business strategy.
4. Organizations are managed by people who know the business and know how the various aspects of the business are related and how they operate. This implies an orientation obtained through practical experience in an industry rather than reliance on the belief that a good manager can manage anything.
5. Organizations that have horizontal and adaptive structures are able to meet the changes in market need more readily than those with cumbersome hierarchical structures, and create multifunctional development groups to plan new product developments.
6. Organizations that stay close to their customers' needs have a competitive edge.

The road to profitable innovation is littered with well-intentioned wreckage. Bacon and Butler (1981) summarize a few of the short cuts and legends to be avoided:

The better mousetrap   All you have to do is build a better version of brand X and the world will beat a path to your door.
Another Xerox   All you have to do is find another superproduct with a powerful technology to exploit.

The gift of genius   Find a brilliant inventor to set loose in the laboratory and your problems are solved; the new products will start to roll off the line immediately.

More and more companies are taking R&D abroad but many still prefer to centralize this critical activity at home. The numerous advantages of centralization include realization of economies of scale, easier communication, better protection of knowledge, more leverage with domestic government, and ease of coordination between R&D, marketing and production. On the other hand, undertaking R&D abroad is encouraged in response to subsidiary pressure and foreign government incentives. It serves in the adaptation of home products abroad, makes use of local talent and is considered an effective public relations tool. Carrying out R&D abroad also broadens the base for seeking new product ideas, is cost saving, and is closer to markets. Finally, it could be carried out as a continuation of the R&D activities of a firm acquired abroad. In general, it is expected that the future will see more companies enlarging their R&D activity abroad, using the sharing or transferring of technology as an effective way to enter these markets (Jain 1984).

In well-designed organizations, the ability to change and remain dynamic is a key ingredient of continued competitiveness. Design implies change and improvement, solving technical problems and meeting new needs. Translating inventions into marketable products and services requires a delicate balance between uniformity and diversity; innovation flourishes in this milieu and needs the pressure of both. *Uniformity* implies standardization and common procedures; the pressure for uniformity includes the economies of standardization, the need for interchangeability in product systems, the need for control of processes, the need for a standard quality product, specialization of markets and the need for common management controls. On the other hand, there will be pressures for *diversity* in a dynamic environment. This is expressed in a number of ways such as regional and marketing diversifications, the attraction and risk of moving into new and untested product markets, broadening and changing technological resources and, more radically, changing the fundamental directions and goals of the organization.

The twin pressures of diversity and uniformity are tangible phenomena in the design process. The relationship between innovation and design at a company such as Hewlett-Packard is not a simple linear action, but a holistic process of interrelated actions – technical, social, economic, strategic and aesthetic – all subject to these twin pressures. Innovation is acknowledged to contribute to economic growth, social benefit and survival in the competitive world, yet many organizations are not innovative and mimic the innovators to stay in the race. Consumer product manufacturers such as Philips have remained successful in their international markets by redesigning both their approach to product development and their product manufacture. According to Robert Blaich, Director of Design, design for manufacture and assembly is at the centre of Philips's compacted development time.

Most organizations have difficulty in being innovative and designing well. There is evidence to suggest that many corporations do not have an organizational structure and process that allow the lateral integration of marketing, research, design engineering, financing and manufacturing for the successful conclusion of their business enterprises; in these organizations, ideas get thrown over the wall rather than worked on collaboratively. Many corporations have top heavy, vertical organization structures and new ideas just take too long to move through the business process. Such organizations can be said to be poorly designed. If a product or service is to be successful in the marketplace, then the organization itself must be designed to meet its conditions. The interrelationship of organizational attributes is the key to any successful business/design activity. Neither invention nor innovation and translation by design can take place in an inflexible or chaotic environment; nor can they take place if flexible design procedures for capitalizing on new market conditions are lacking.

One of the most critical problems for organizations in designing well and competitively is their response or lack of response to change and, more specifically, to rates of change in markets' technical and economic conditions.

It must be remembered that there is nothing more difficult to plan, more uncertain of success, or more dangerous to manage than the creation of a new order of things. For the initiator has the enmity of all who would profit by the preservation of the old institutions, and merely lukewarm defenders of all who would gain by new ones. (Machiavelli 1948)

New Product Strategies

Basically, new product development takes place in two ways. In the first a company or industry discerns the need for a new product or product range, develops markets and sells to the widest possible market. To do this, it has to not only come up with single products but be able to place them in product systems and even combine them in innovative ways. Using existing unconnected product breakthroughs, Sony took a small tape player, married it to lightweight headphones and made the Walkman an entirely new product that, after a slight hesitation, became a world beater. The company originated and broke through with a new idea, using high quality elements from its flexible audio product system.

In the second the end user develops an idea for a new product that is beyond the present vision of the industry and brings it to the attention of a manufacturer or industry. Many medium to high tech industries have 'users as innovators'. For example, at IBM in the Installed User Programs (IUP) Department, 30 per cent of IBM leased software for large and medium computers is developed by users. In other cases, such as medical electronics, commercial applications very often lag behind the clinical front line until user specialists take a hand.

More and more companies are taking account of these two approaches to commercial innovation. Companies like 3M, Honda, United Technologies and Apple are providing the means by which new ideas can be recognized and commercialized whether they are insider or outsider generated.

Successful companies use integrated design management as a means of ensuring that all products, communications and services of the organization are serving the overall business. Organizations that are successful in integrated designing are often started and run by outstanding individuals who know the business, like Morita, Hewlett and Packard. They are people dedicated to seeing that corporate goals and cultures do not deviate from the chosen path. Accelerated rates of change demand improved organization and procedures to increase the ability to adapt to new conditions and to conserve money by well-designed, economical manufacturing; keeping this in focus often requires a single vision.

Three business strategies have to interface with designing:

1 Develop market/user understanding thoroughly before design development.
2 Focus on commercially viable translation of ideas, i.e. innovation for successful commercial products.
3 Design a well-integrated organization to support product development in the milieu of ever changing conditions, with product systems or cascading development from one product to the next.

Invention is the first stage in the process of technological innovation. A tidy distinction between invention and innovation does not exist, even though there is a qualitative difference in the activities. They are frequently inextricably linked ... Invention is best treated as the subset of patentable technical innovations. Inventions typically involve minor improvements in technology. Three general theories of invention exist: one attributes invention to the individual genius; another considers it to be an inevitable historical process, proceeding under stress of necessity, where need dictates and technology complies; and the third and most realistic approach sees invention arising from a cumulative synthesis of what has preceded. (Parker 1974)

According to this view, the occurrence of invention is not certain; an act of insight is required. It is likely to occur to an individual directly concerned with the problem. This individual, however, is not suddenly struck by a brilliant idea. When and if the act of insight takes place, it is conditioned by the specific problems encountered, and occurs through a synthesis of previous knowledge. By this synthesis and the act of insight, the inventor may overcome a discontinuity. This theory conforms with the concept of technology building on technology, where progress is not a random process but a synthesis of what has gone before. It is true, however, that necessity hustles invention forward and that great inventors do exist, but these are not typical occurrences. Theories of invention must describe the usual. Economic factors are predominant in the motivation to invent. The primacy of economic forces, however, does not imply how research should be organized. Keeping the sources of invention as wide and diffuse as possible is considered to be the best approach.

Innovation may occur after a considerable time interval from invention. Successful innovation can greatly improve the economic performance of companies, enhancing growth and profit rates. The term 'innovation' covers all the activities of bringing a new product or process to the market. It tends to be a time-consuming transformation process which is both management and resource intensive, and is more expensive than invention. Development risks are divided into technical and market risks. The amount of technical risk involved depends upon the size and complexity of the desired advance. The market risk is, to a much larger extent, beyond the control of the company, being dependent on the achievement of an adequate market. When selecting projects, the criteria applied by management indicate a fear of the high risk of the market. Research allocations are typically modest and payback periods required tend to be short.

R&D is expensive but is a necessary cost. Commitment to research differs among industries, depending upon technological push and customer pull. The relationship between a company's size and its technological capabilities is unclear. Large companies do not appear to be unchallenged. In terms of relative expenditure, very big companies may carry out proportionately less research than smaller ones. In terms of major inventions, lone inventors and small companies may have a comparative advantage, while the large company's forte may lie in development and follow-up improvements. The same ambiguity seems to exist regarding the influence of specific market structures. The most crucial factor in the organization of R&D probably relates to the number of independent centres of initiative. By having numerous potentially creative units, an enterprise may greatly improve its chances (Parker 1974).

Conclusion

An 'invention is the solution to a problem, often a technical one, whereas innovation is the commercially successful use of the solution'; so say Bacon and Butler (1981). From the start, new product development activity must focus on innovation and continuous product improvement. By examining all aspects of innovation prior to major expenditures on development – such as market need, technical, production and marketing requirements, and a basis for protecting the product from competition – the product solution can be found. Designing in its various forms (product, graphic, packaging, advertising, etc.) is the key to translating the invention into a commercially

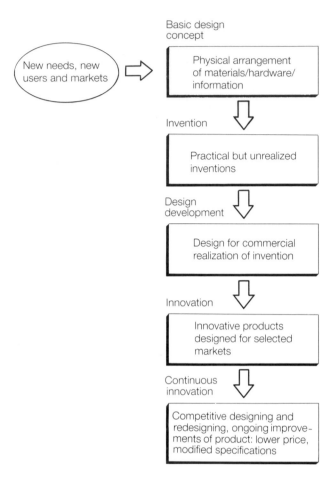

Figure 33.2 The process of invention, design and innovation

viable innovation (figure 33.2). Effective designing is the *only* activity that can make this tangible.

Note

Issues are drawn from the findings of 'The Competitive Edge' project; see note at the end of chapter 11.

References and Further Reading

Bacon, F. and Butler, T.: *Planned Innovation*. Ann Arbor: University of Michigan, 1981.

Clipson, C.: *Business/Design Issues*. Ann Arbor: University of Michigan and the National Endowment for the Arts, The Competitive Edge Project, 1985.

Clipson, C.: *Design for World Markets*. Ann Arbor: University of Michigan, 1987.

Jain, S. C.: *International Marketing Management*. Boston: Kent, 1984.

Machiavelli, N.: *The Prince*. Chicago: Great Books Foundation, 1948.

Ohmae, K.: *Triad Power: the Coming Shape of Global Competition*. New York: Free Press, 1985.

Parker, J. E. S.: *Economics of Innovation*. London: Longman, 1974.

Terpstra, V.: The role of economies of scale in international marketing. In Hampton, G. M. and van Gent, A. P. (eds): *Marketing Aspects of International Business*. Boston: Kluwer-Nijhoff, 1984.

PART VI

MANAGING DESIGN PROJECTS

34 Assembling and Managing a Design Team

MARK OAKLEY

Aston Business School, UK

Introduction

With a commitment made to embark on a design project (or perhaps a series of projects), steps must be taken to select one or more designers who can be organized, where appropriate, as a team. One prerequisite to building a design team is a sound knowledge of the firm's track record in earlier design work; for this reason, careful attention needs to be given to design auditing.

Design Audits

Design audits serve much the same purpose as financial audits – basically, to review the return (or potential return) being achieved on the resources employed, to check whether the level of resources is adequate for the tasks involved and to highlight relative successes and failures.

In addition to financial audits, many other kinds of checks are now routinely carried out in most companies. For example, reviews of manufacturing performance are commonplace, as are regular analyses of marketing achievements. Senior managers use the information provided to assist them in making the correct decisions about new investments and initiatives. In view of this, it may seem surprising that it is still rare to find design activities being regularly appraised in a similar manner. The explanation probably relates to the insecurity which many managers feel about design in general and a misguided notion that design can only flourish if left to itself.

In reality, design activities can and should be reviewed on a frequent basis so that the returns on investments can be measured and, if necessary, adjustments made to improve performance. Two kinds of design audits should be conducted. First, regular across-the-board audits of all design projects and design results are needed, typically summarized in reports every six or twelve months. Secondly, pre-project audits should be undertaken before the start of all significant design projects, particularly to compare the resources needed with those available and to assess the chances of a successful outcome.

Regular Design Audits

The outcome of a financial audit is often reduced to a stark bottom line and, much as design staff might resist it, something similar is possible with design audits. Generally, it ought to be possible to examine design activities, calculate the time and cost of the work done and relate this to the effects achieved by the design results. For example, if six months and half a million pounds have been spent on designing a new version of a product which gains an extra 5 per cent share of the market and has a larger profit margin than the old model, then clearly the basis is there for the development of some kind of design performance index against which other projects can be compared.

Of course, there is the danger that such an approach becomes too simplistic and ignores the many variables which may be influencing the total picture. However, providing margins of error are acknowledged, information in this form can give a feel for the value for money or return on investment which design efforts are achieving. In particular, the relative returns from expenditure on different types of design work may become more apparent – enabling managers to redirect resources towards those areas of design most likely to have the greatest success.

If conducted thoroughly and regularly, design audits can provide managers with much essential information in addition to the basic indicators just discussed. The most important issues to be examined are as follows.

Success/failure rates and causes

In any company it is inevitable that some design projects will be more successful than others. Indeed, some projects may have to be abandoned before they are completed and others may reach a conclusion but produce only a result which is judged to be a failure. There is a tendency to want to quietly forget about the failures, but this should be resisted because much can usually be learnt which may help avoid similar problems in future. Also, it is important to have an idea of the relative numbers of failures and successes so that a benchmark for future improvement is established – and, possibly, also for comparison with other firms in the industry.

The design audit should attempt to identify and evaluate the factors which have influenced successes and failures. Typically these might include:

1   Presence or absence of a competent project brief.
2   Correct prediction of resources required.
3   Competent management of project (including ability to work within time and cost restraints).
4   Quality of working relationships between designers and others both inside and outside the company.
5   Availability of skills and effective deployment.
6   Ability to respond to any changes in specification.

7   Whether progress reviews are held at appropriate times and correct decisions taken about further work, new directions or abandonment.

8   General quality of project management.

9   Performance of outside design expertise, if used.

10   Support, interest and influence of top management.

Design skills weaknesses

If the audit reveals a shortage of certain design skills, the implications must be addressed by managers. One of the dangers in design work is the temptation 'to do what we are equipped to do', meaning that only certain types of projects are tackled (possibly ones not relevant to the company's needs). Worse, the right kinds of projects may be attempted but with quite the wrong kinds of skills.

In order to maintain the correct levels and balances of skills, managers must provide adequate training and other opportunities for designers to keep up to date with advances in their areas of specialism, or to move into new ones. Where new design directions need to be followed, the answer may lie in the recruitment of additional staff. Such a decision will depend on whether the new design direction is considered to represent a long term trend; if not, it may be more economical to subcontract some design projects outside the firm. Only by regularly auditing design activities is it possible to estimate with confidence the skills provision required.

Project management competence

The best designers in the world will be unable to produce good results if they are working on poorly administered projects. Where projects have a habit of failing because budgets are exceeded or because vital time deadlines are missed, the fault probably lies in the supervision of the project.

Similarly, successful results may fail to materialize because the wrong designers have been used, or because the wrong questions have been asked; in short, because of incompetent project management. As in the case of design skills weaknesses, nothing can be done unless the problems are identified in the first place, further underlining the importance of regular design audits.

Pre-Project Audits

In addition to regular audits across the whole range of design activities, managers should also make a point of carrying out an audit each time a new project of any substance is proposed. The purpose of this audit is to confirm that the new project is within the capabilities of the design department or group which will be undertaking it. The more complete the knowledge from the regular reviews, the less difficult it should be to do a pre-project audit.

It is just as important (sometimes more important) to carry out a pre-project audit when preparing to use an outside consultancy as it is to do one before using an internal group. If a consultancy is reluctant to provide information, it is better not to use it but to seek alternative help. After all, the request is basically for evidence of a track record plus confirmation that adequate skills and capacity are currently available for the proposed project – not an unreasonable request to make of any subcontractor.

Who should Conduct Design Audits?

It is tempting to say that it depends on the firm. For regular audits, the main requirement is that the auditor should be disinterested, that is not someone directly involved with the design work. However, some knowledge of design activities and problems may be necessary. In large companies, the task may be carried out by a design manager from another division or perhaps by a design-aware manager from another functional area such as marketing – or even finance! Another possibility could be to hire someone from a reputable design or management consultancy.

In smaller companies, the need to minimize expenses might rule out using anyone other than a member of staff – and in any case, the scale of issues to be investigated is likely to be small and already reasonably well understood. However, it is still important to do design audits; even in small companies, false impressions and delusions can lead to poor design decision making.

Pre-project audits are less vulnerable to possible bias and it is usually quite adequate if they are undertaken by the leader of the proposed project or another manager from within the design area. The main task is to check that the new project can be undertaken with confidence and, therefore, there should be little reason for such a person to recommend an unwise course of action for which he or she will then be subsequently responsible. However, where there is any doubt, a disinterested person should do the audit or run a double check if the scale of the project warrants it.

Assembling the Design Team

After the design audit, the next stage will be to review the suitability and availability of any designers already working for the company either as full-time employees on the payroll or as outside consultants who may have been engaged on a fee or retainer basis. The word 'designer' is used advisedly and in a broad sense; the staffing of a design project may involve many who do not have the word 'designer' in their job titles. Hence, what is really important is the correct identification of the design, management and other skills needed for the project.

If the skills already available to the company are insufficient or inappropriate for the planned project, then the next stage will be to recruit extra staff or identify sources of additional help. Managers faced for the first time with the task of hiring design skills may experience some difficulty identifying the best sources of supply, judging the competence of different candidates and deciding on fair fee levels. The world of design consultancies is something of a minefield; there are many good ones who charge reasonable fees, but also many of doubtful competence who provide poor value for money.

As a glance through any design magazine shows, it is a profession subject to much 'hype', exaggeration and self-proclamation; hiring is very much a case of 'buyer beware'. Fee rates vary widely and, despite the efforts of some professional associations and agencies, there is no reliable system of certification or regulation to guide customers. Many designers lack the sense of urgency and timing often so vital if a project is to be a commercial success.

Confirming Design Skill Requirements

The deliberations involved in drawing up the brief and carrying out the design audit should have resulted in a fairly comprehensive understanding of the work that the design project will entail. In turn, this will have led to an impression of the skills that must be available – an impression that must now be brought into clear focus.

Starting with design skills, a list should be drawn up of all those which the project will need to draw upon. In the first instance, no attempt need be made to quantify the skills or fit them into any time schedule; what is necessary is to see the *range* of design expertise which is going to be required. Managers should not worry too much about the terminology used, as long as all aspects are included. As an example, for a project to design and develop a new, plastic-bodied electric kettle the following list of design skills might be drawn up:

Electrical design skills
    Heating element performance
    Heating element construction and manufacture
    Switch design
    Circuit design
    Prototype construction and testing skills.
Plastics design skills
    Body/lid/other parts design – choice of material and so on
    Fabrication/moulding of parts
    Model building skills.
Mechanical design skills
    Design of metal components
    Design of steam operated cut-out.
Styling/graphic design skills
    Shapes, colours, decoration, and so on
    Packaging, instruction leaflets.

Production design skills
   Designing new/modified production system or method
   Transferring/installing new design in production.

The list could be longer; the writer is not an expert on electric kettles! However, it does illustrate that even for a relatively simple product the range of design skills may be quite wide. The next question is, of course, the extent to which these skills overlap and can be rationalized. Fourteen separate skill requirements are listed in this example, but it may be that just five people can provide all of them, corresponding to the five groupings used in the list. Perhaps even fewer may be needed; if there is much experience in the firm of this type of product, one person may be used to dealing with, say, both the plastics design and the mechanical design aspects.

Beware of taking this process too far, though. If it was decided that just one or two designers could deal with all the tasks listed, it is very likely that the consequence would be inferior results in at least some of the areas. There is often a temptation to do this either because of an unrealistic confidence in the breadth of skills of certain individuals or simply to avoid spending any more money than seems to be absolutely necessary. Typically in a project like this one it will be the appearance of the product which will be compromised, and perhaps also the packaging – precisely the aspects most important at the point of sale.

With design skills identified and assessed in this manner, the way will be clearer with respect to any additional hiring or recruitment that may be necessary. Whether it is best to appoint more people to the full-time staff, or wiser to engage consultants on a fee basis, will depend on factors such as the anticipated work load during the project and the likelihood that it will be followed by other projects with similar staffing requirements. Short term needs may be best catered for by using consultants, and long term ones by using permanently appointed staff – although these are by no means the only considerations.

Appointing Designers

Recruiting designers to the permanent staff is a course of action which may involve major expense and long term commitment. Unless the project is large scale or part of a continuing programme of design work, it is doubtful whether such recruitment is a wise step. The key questions concern the type of design skills which are missing and the urgency surrounding the project.

The type of design skills which are most suited to provision by means of permanent staff appointments are those relating to technology, especially where this is of a complex and gradually evolving kind. In general, the more specialized the technology, the less likely it is that short term help will be of value – and probably it will not be available anyway. By contrast, the design skills most suited to temporary engagement from a consultancy or by means of short contracts are those concerned with styling and appearance; indeed, this is often an ideal way of bringing fresh ideas into a project.

If the project and its staffing are matters of great urgency then there may be no alternative to the substantial use of consultancies and other outside contractors. If the base of design skills in the company is just too small then the only realistic course may be to place the *whole* project outside. When this has to happen despite the previously declared intention of a company to do its own design work, managers should ask why this situation has arisen. It may not be possible to do anything in respect of the current project, but at least future performance may be improved.

More than likely the reason will be a failure to maintain a correct balance and progression of design skills in the company. In the absence of regular audits of design skills and achievements, design activities may have been neglected, only coming under scrutiny at times of concern such as when the need for a new product arises. In consequence, no planned recruitment may have taken place, leaving serious knowledge and skills gaps. More than likely, these will have been made worse because little training and development of design staff have been undertaken.

Using Design Consultants

Design consultancies range from one-person businesses to large, multi-disciplinary groups which may employ hundreds of people. As in other fields, small consultancies may be preferred because they can provide more attentive service and may well charge lower fees. The main disadvantages may be a relatively slow speed of working and a limited range of services; very small consultancies tend to offer services only in their chosen specialisms.

If necessary, several small consultancies may be engaged, each attending to aspects falling within their particular areas of expertise. However, the disadvantages of difficult control and coordination of a number of small groups may soon outweigh any advantages of lower cost or more personal service. Especially for larger projects, the use of a more substantial, broad-based consultancy may be the inevitable choice. Once decisions have been made about how much of the design work to subcontract and the preferred size of consultancy, one can be selected by reference to a design directory, by personal recommendation or by a search of local or wider sources of information.

Managing Design Teams

Setting up design teams for new products and services (or systems such as corporate identity programmes) will all be in vain if the organization of the company is ill-suited to the special demands of design work. The nature of

Table 34.1 Comparison of organizational features

Features of industrial operations	Features of design projects
Rational, standardized, predictable	Irrational, novel, unpredictable
Operations accurately timed	Accurate timing of activities usually impossible
Long runs of identical products/services	Activities frequently changing
Creativity and initiative not developed in workforce	Highly creative personnel essential
Work closely controlled, essential for profitability; risk eliminated	Profitability related to skill, chance, judgement, intuition, etc.

these demands can be seen if features of design projects are compared with those of more typical industrial operations, as in table 34.1. In the light of these different characteristics, managers may need to consider the implications both for the organization of the project itself and for its relationship to the rest of the company.

Some years ago, researchers Burns and Stalker (1966) addressed these issues and analysed the management styles of a sample of firms involved with new products and systems; this work remains important today. It was found that within these companies there were styles ranging from what was described as 'mechanistic' (very formal, hierarchical, bureaucratic and inflexible) to what was termed 'organic' (informal, based on teams and highly adaptable).

It was concluded that mechanistic systems work satisfactorily only where conditions are intrinsically stable – flow line production departments, for example – or in situations where close control of highly specialized work is essential. Mechanistic forms are not likely to prove satisfactory when applied to innovative design projects, which always demand some degree of flexibility and freedom of approach.

Organic styles are much more appropriate and hold out better prospects for success when products, services or systems are being designed. The following are some of the features that may be found in organic systems:

1 Unifying theme is the common task: each individual contributes special knowledge and skills, and tasks are constantly redefined as the total situation changes.

2 Hierarchy does not predominate: problems are tackled on a team basis rather than being referred up or down the line.

3 Flexibility: jobs are not rigidly or precisely defined.

4 Control is through the common goal rather than by institutions, rules and regulations.

5 Expertise and knowledge are located throughout the organization, not just at the top.

6 Communications consist of information and advice rather than instructions and decisions.

Managers responsible for design and related activities need to consider how they can promote such organic features. This may be a delicate matter, especially within those firms which are otherwise organized along precise and inflexible lines. Rather than imposing any particular form of work methods, managers must take actions to assist creative work so that organizationally desirable features predominate in place of undesirable ones.

Apart from these considerations of management style and attitude, there are a number of more mundane points that firms have to consider, for example:

Accommodation   Are workshop or laboratory facilities available? Is the environment satisfactory for creative work? (A corner of a noisy factory may not be the best place.)

Technical back-up   Is there a competent drawing office? (Does it have up-to-date computer equipment with trained operators?) Is there a company library of technical specifications? Can normal plant be borrowed for test purposes?

Customer information   Is there a way of checking requirements directly with customers? Is the marketing department able to provide information?

Links with the manufacturing or operating system   What will happen when new ideas are ready to go into normal, commercial operation? Will existing systems take up the products or services – or will extra plant be needed?

Relating Design Work to the Rest of the Company

These questions lead directly to consideration of the organizational problems which arise out of the relationship between a design activity and the rest of the firm. It is not always easy to decide who should be responsible for design work within the firm. Design is often considered to be a mainly technical activity, but giving control of it to a technical department – such as manufacturing or production engineering – may result in failure.

This may happen either because management styles inappropriate to design are applied or simply because these departments have resistance to anything new which may cause disruption. Resistance may take the form of constant rejections of new designs, refusal to supply information and help, or simply general obstruction – all while paying lip service to the need for new products or services. These attitudes may be particularly acute in long established firms where previous design work has been limited to minor improvements and modifications.

Such problems have been widely discussed and various solutions proposed. If the design activity is made an independent part of the company away from other departments, problems may still exist. Its manager may have to bargain with other departmental heads for cash and resources. Because design is a long term activity and its wealth creating effects are not always immediately obvious, it may seem to senior managers in other functions that it is just a wasteful consumer of the income which they work hard to create. Some companies try to overcome this by directing design operations through a steering committee which represents all major interests.

Others appoint a design manager or product champion whose job is to push through all barriers and solve whatever organizational difficulties arise.

Appointing a Design Manager

The role of the design manager may be crucial in achieving successful results. However, in many respects the qualities necessary in order to be effective in this job are substantially different from those traditionally expected of managers. The main requirement must be the design manager's ability to deal with change and ambiguity, as table 34.2 summarizes.

Table 34.2   Differences in managerial roles

The traditional manager	The design manager
Technical and analytical skills highly developed	Additional skills needed to deal with ambiguity and conflict
Knowledge based on experience	Knowledge based on structured updating
Expects continuity of tasks	Able to adapt to unpredicted events
Guided by standard procedures	Many one-time decisions
Seeks stable relationships	Tolerates temporary groupings
Expects rational behaviour	Accepts diversity of approach
Task-driven	Goal-driven
Stresses physical activity	Combines action with reflection
Individual approach to problem solving	Encourages team approach to problem solving

The implication of these special characteristics associated with design management is that great care is needed in selecting the right person to fill a post as design manager. Also, in addition to organizational skills, companies often demand a high level of technical knowledge and/or the ability to work successfully with customers or clients. It may be unrealistic to expect all of this from one person and it may be better to appoint two or more people whose skills are synergistic.

Conclusions

Selecting or recruiting designers of the right kind for a particular project is fundamentally important to the success of that project. An important preliminary to any design project is an effective system of design auditing so that managers have an accurate understanding of the causes of success and failure in design work in their companies. Great care needs to be taken to optimize the organizational environment in which a design team is to operate – and to ensure that the appropriate style of design management is practised.

Reference

Burns, T. and Stalker, G. M.: *The Management of Innovation*. London: Tavistock, 1966.

35 Product Evaluation

BERNARD GRENIER

Institut de Recherche sur les Formations Industrielles Supérieures, France

Introduction

The question of whether or not a product is a 'good' one is often asked nowadays. It confronts the consumer contemplating a purchase and faced with a choice between the products of two or more rival brands. It confronts the manufacturer wondering whether an old model should be continued in its present form, taken out of production, or remodelled.

It confronts the designer who has to opt for one out of several possible solutions. It confronts the members of a panel awarding a prize or conferring a label denoting quality when asked to give one product preference over another. Generally speaking, it confronts all professional advisers whose task is to give guidance to purchasers or potential users.

Difficulties of Product Evaluation

There is no getting around the fact that evaluation can give rise to extremely divergent or indeed contradictory opinions. The difficulties are almost as great for experts as for the general public. One person may think, for instance, that the best cameras still come from Germany, someone else that all Japanese models are now well in the lead, someone else again that Swiss cameras are the best, while a fourth person may cite the fact that photographic equipment made in his or her own country was taken on the space shuttle as proof of its superiority.

In point of fact, the difficulty arises from the synthetic character of evaluation formulation, which may involve a great many heterogeneous criteria.

A product may be aesthetically either more or less in line with contemporary taste. It may be easier or more difficult to manufacture. Its purchase and use may be more expensive, or cheaper. It may be sold in greater or smaller quantities. Its technical performance may be on a higher or a lower level, and involve more sophisticated or less sophisticated technical devices. It may be easier or more difficult to use, manipulate and maintain. It may deteriorate at a faster or a slower rate. The service it gives may be of higher or lower quality. It may be safer or not so safe.

It will be seen that these various criteria arise from various different viewpoints: cultural, utilitarian, technical, commercial or economic. Product evaluation, accordingly, seems to depend largely on the evaluator, first because of the particular point of view to which he or she gives precedence, and second because of the different degrees of objectivity that evaluators bring to their task.

It is clear that a correlation exists between the points of view just mentioned, and the objectivity of the evaluation. There is no denying the fact that an appreciation of a cultural nature is far more subjective than an appreciation of a technical or commercial nature, and that appreciations of a utilitarian or economic nature depend very much on the circumstances in which the product is observed. Though its technical performance will be identical, a snow-plough is not as useful in the tropical jungle as it is in the Arctic. And when we come to the matter of expense, the costs borne by the manufacturer and by the user are not the same.

Obviously these are extreme, even exaggerated examples. Reality is usually less extreme – with the result that the problem of evaluation is heightened. It is far harder to distinguish between apparently quite similar circumstances than circumstances which are radically different.

In the face of these difficulties, many people resort to employing only those two criteria which are the easiest to apply: the commercial criterion and the technical criterion. Some see the ultimate test of a product's quality as whether it will sell well; others are less concerned with the product's commercial success – which they consider a secondary issue – than with technical performance which is outstanding or at least based on outstanding technology.

These criteria, of course, can have the merit of objectivity, since it is relatively easy to apply them: in the first case, by looking at sales figures, balance sheets and profit margins; and in the second case, by conducting technical trials in standardized and thus incontestable conditions, and using standardized methods and units of measurements.

However, evaluation of the same product by means of these two criteria can lead to very different conclusions. On the one hand, products which are technically advanced from the objective viewpoint may turn out to produce poor commercial results. On the other hand, some will claim that technically mediocre products can be successfully marketed by efficient commercial methods (advertising, distribution, a dominant market position). So while most technical qualities may be evaluated almost immediately, most commercial qualities do not show their definitive value until the product is no longer on the market, and therefore its evaluation is of only academic interest.

Looking at the many reasons that may exist for product evaluation, we can see that the easy solution offered by taking only technical and commercial criteria into account is not very satisfactory. For one thing it ignores many aspects of cultural, utilitarian and economic factors, but most important of all it is retrospective rather than prospective, oriented towards the past rather than towards the future.

Evaluations are useful at an early stage – when the product is being developed, at the beginning and not the end of its life. In fact, once a product has been pronounced 'good' at the technical or commercial level, the main concern is to confer approval on the people who devised, manufactured and sold it. But if one wants to adopt a more strategic attitude and induce people to buy the product, or devise superior products, it is probably a better idea to emphasize the product's usefulness, or the relationship between the technical, functional, commercial, economic and cultural features of outstanding products. Thus it is useless to economize by making a complete analysis of the product in all its dimensions and showing all the ways in which those dimensions overlap.

As it is obvious that the ideal product does not actually exist, it is clear that a product may be judged good by some criteria and bad by others, and thus it will seem better or worse to different evaluators, depending on the way in which they are concerned with the product and their role in the evaluation process.

It would be ridiculous if this fact led to all methods of evaluation being considered invalid. On the contrary, it ought to lead to the search for an objective and thus incontestable basis for assessment, one that could be taken into account thereafter by anyone and related to their own requirements and preoccupations.

Instead of arousing dispute, such a method would allow people in general to take decisions affecting themselves in full knowledge of the facts: it would help them in selecting a purchase, deciding to launch a new product, choosing technology with an effect on some particular kind of performance, picking a style compatible with a product's convenience in use, or devising more competitive products.

Bases of Product Evaluation

In the light of the complexity of the problem, we need to recognize the factors it comprises. Three basic categories may be identified and are outlined in the following sections.

The material components of the product

A product is devised in order to provide human beings in its proximity with a service. It seldom acts autonomously. Product and human beings exist in the same environment, which is employed as an intermediary in the relationship between them (figure 35.1).

The environment itself has three components: the natural environment (natural elements, other living creatures) and the two aspects of the artificial environment, namely material factors (other products, equipment) and methodological factors (the way it is used, regulations for its use). It has

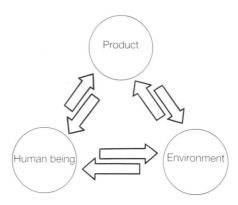

Figure 35.1

frequently been found necessary to devise a part of the product's environment specifically, parallel to the product itself, and this then becomes its own environmental system (figure 35.2).

Figure 35.2

If we look more closely, we shall see that there is a hard technological core within the product, and an interface zone between this core and the environmental system. This interface links the product to its environmental system (figure 35.3).

Core   This is the most technical part of the product, a network of components relating only to each other; these are purely material and technical relationships (in a car, for instance, the engine or the gearbox).

Environmental system   This comprises two parts, one which is devised specifically for contact with the product (with a car, for instance, road systems, other cars, the Highway Code), and another which is completely independent of the product's existence (still using the car as an example, bad weather, cold, wind, sun, sharp bends, steep hills and slopes, pedestrians) but of which one must take account in devising the product.

Interface   This is the peripheral part of the product, a combination of factors,

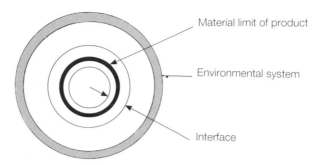

Figure 35.3

internal or external, ensuring the relationship between the technical core of the product and the human beings who use it on the one hand, and the environment on the other. The internal part is both material (packaging) and organizational (mode of use, after-sales service) and, taking the car as an example again, includes the steering wheel, seats, dashboard, maintenance manual, etc.

The circumstances of the product's life A product has its own lifespan: it is devised and developed, it comes into existence, it functions, it deteriorates, it is reconditioned; it dies, it is recycled. Definition of the product is the business of its devisers in development offices. Development of the project is the business of designers in development offices and testing laboratories. Manufacture of the product is the business of workers in factories. Commercialization of the product is the business of sales staff using distribution channels and their opposite numbers, purchasers. Utilization of the product is the business of users of various kinds (those who operate or repair it, benefit by it or are adversely affected by it) in a number of different places. Recycling of the product is the business of recyclers operating in various areas.

All these agents and environments are radically different. They are not, therefore, concerned with the product in the same way. To simplify enormously, it may be said that a good product is:

A well-devised product to its deviser
A well-developed product to its designer
A well-made product to its manufacturer
A well-sold product to its seller
A well-utilized product to its user
A well-recycled product to its recycler.

But 'well' cannot mean the same thing to all these people: does it denote ease, low cost, speed, profitability, intelligent use, or what? It may be mentally satisfying to suppose that good conception, good development, good manufacture, good commercialization, good value in use and good recycling potential ought all to converge, but we are well aware that the facts are often quite different. It is not very likely that a badly-made product will end up selling well, but a well-made product can never be sure of commercial success in the long term.

The criteria of evaluation where devising a product is concerned are extremely complex, since logically they ought to include all other aspects: how can it be so devised that it is well developed, well made, well sold, well used and well recycled? Success in all other phases of the product's life depends on its original conception, and this is what makes evaluation of the product, in the final resort, almost equivalent to evaluation of its conception. Indeed, it is hard to see how a poorly conceived product could be good at a number of other levels.

The types of performance characteristic of the product

Remembering to take their radically different natures into account, it is useful to distinguish between three main types of product performance:

Technical performance   This concerns the relation of material factors to each other (figure 35.4). It is of a physical nature and can thus be accurately gauged with the aid of a well-defined and accepted system of measurement (for instance, speed of movement, power, frequency bands).

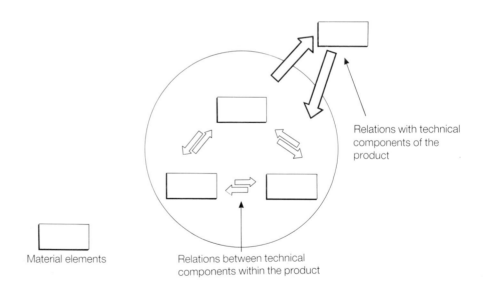

Relations with technical components of the product

Material elements

Relations between technical components within the product

Figure 35.4

Performance in use   This concerns relations between the product and the human beings in its vicinity, either directly or indirectly (figure 35.5). They are of a sensory, physical or mental nature, and are conducted by means of the interface (efficiency, convenience, safety). It is difficult to define the way in which they are gauged and there is no standardized system of evaluation.

Economic performance   This concerns factors resulting from the preceding relations which can be translated into financial terms and are the subject of

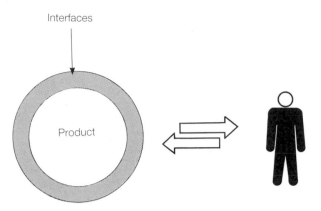

Figure 35.5

monetary transactions (costs, receipts, profits, investment). Their common unit of measurement is money.

A system of relationships exists between these main types of performance; its general appearance is as in figure 35.6. However, this system is far from being deterministic. For one thing, mutually unequivocal relationships

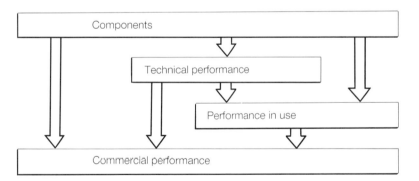

Figure 35.6

between elementary performances of different types do not really exist. And for another, there is considerable functional variation between different states and circumstances of product life. As mutually unequivocal relationships do not exist:

1   Every elementary performance depends on several components and on elementary performances at an earlier point in time.
2   Every component or performance at an early point in time influences several subsequent performances.
3   No component or elementary performance of one type automatically involves an elementary performance of another type.
4   Technical performance is not an end in itself, but an intermediary in the choice of components appropriate to performance in use and economic

performance, as concluded from research. (Thus high level technical performance means not necessarily satisfactory performance in use but desirable performance in use, and economic performance may necessitate a certain degree of technical performance.)

Variations of performance at different stages in a product's life are summarized in table 35.1. It is obvious that the groups A, G, M and B, H, N are very different from each other, since the product is changing its state, its environment, its evaluators and its criteria of evaluation.

Table 35.1

Circumstances	State of product	Environment	State of performance		
			Technical	In use	Economic
Definition	Idea, concept, specifications	Development office	A	G	M
Development	Prototypes	Development laboratories	B	H	N
Manufacture	Separate parts, subsystems	Factory	C	I	O
Commercialization	Inactive completed product	Distribution channels	D	J	P
Utilization	Active completed product	Many different environments	E	K	Q
Recycling	Product deteriorated	Recycling channels	F	L	R

Causes of Difficulty in Product Evaluation

The analysis above allows us to identify causes of difficulty in analysing the product. For the sake of simplification, they may be put into three categories.

Insufficient distinction between the nature of performances

Performance in use must not *a priori* be assessed by standards of technical performance. For example, a vacuum cleaner will not necessarily be any better at removing breadcrumbs from a shaggy carpet because it has a more powerful engine.

Insufficient distinction between circumstances in the product's life

For instance, performance in use does not concern use alone, it also concerns manufacture – but they are not the same. A product which is easy to make is not necessarily easy to use. Similarly, economic performance does not concern production alone, it also concerns the user – but they are not the same. For example, costs interest only those who are paying them. The costs of manufacture concern the manufacturer, the costs of utilization concern the user who is exploiting the product: a product which is cheap to make will not necessarily be a cheap product to use.

The utility of one and the same product is not the same when it is used in different circumstances. We must not confuse a product's evaluation by appropriate criteria with consideration of those criteria in the light of the requirements and preferences of the product's users. The performance of a product varies at different stages of its life: its performance at the concept stage is only hypothetical, its performance on leaving the factory is nominal (and will not be absolutely identical in all models in a series), and its performance deteriorates as it ages (and does so in different ways according to conditions of use).

Inadequate functional analysis in defining the terms for assessing performance

For example, reliability may be evaluated in terms of technical functions but also by its consequences – technical, in use and economic. Likewise, convenience in use and aesthetics have a reality and an appearance which are equally important but distinct; they can be gauged by professionals on the basis of very precise experimental procedures, or by interviewing users on their general opinions.

Obviously, depending on what exactly is being assessed and the manner of its assessment, the rules will not be the same. Nor will they be equally useful to the evaluator: for instance, appearance has a considerable influence on sales when the product is first launched, while the reality of actual performance has considerably more bearing on sales in the long term.

Conclusions

There is a need for stricter standards in evaluation. If it is to be really useful, product evaluation must include an explanation of the criteria and circumstances of evaluation, and the means of assessment employed. Consequently, all evaluation should be preceded by a wide ranging functional analysis of the product's technology and utilization.

Functional analysis of technology involves a precise understanding of all the ways in which material factors relate to each other. Functional analysis of utilization involves a precise understanding of all the ways in which the product and its environment relate to each other, with particular attention paid to the human beings concerned.

This functional analysis should take place early in the process of developing a new product in the general context of creativity, specifications, market research, standardization, regulation, tests, industrialization, commercialization, the after-sales period, computer aided planning and so on. The development of a methodology of functional analysis is beyond the scope of this chapter, but table 35.2 indicates the main points which may need to be evaluated, either in retrospect to establish the facts or provide a verification, or in advance of a design project to test a hypothesis or set an objective.

Table 35.2   Product analysis (main points to be evaluated)

Performance in production

In use
Convenience of production
Safety of production
Harmful effects of production
Economic
Quantities sold
Price
Balance sheet
Direct costs of production
Profit margin
Investment
Return on investment
Socioecological costs incurred
Time
Performance linked to influence of time on the above[a]

Performance in utilization

In use
Instrumental:
    Efficiency of utilization
    Convenience of utilization
    Safety of utilization
    Harmful effects of utilization
Perceptual:
    Aesthetic
    Symbolic
    Appearance of instrumental performance in use
Economic
Purchase price
Exploitation costs:
    Energy
    Consumables
    Spare parts
    Repairs
    Insurance
    Destruction
Any profits from exploitation
Time
Performance linked to influence of time on the above (e.g. durability, reliability in use)[a]

Technical performance

Measurements
All physical measurements used to assess relations of internal material factors to each other or to material factors outside the product, e.g. dimensions, weight, volume, hardness, viscosity, discharge, speeds, frequencies
Time
Performance linked to influence of time on the above (e.g. technical reliability)[a]

Component parts of product

Nomenclature
Structural and functional layout

---

[a] Obviously all performance changes with time: the characteristics of its variation also constitute performance.

36 Design Evaluation

BARRY TURNER

B. T. Turner Associates, UK

Introduction

To evaluate is to find out or state the value of something. In the context of design work this means a systematic review throughout the design process to ensure that the requirements are being met. A design review is a planned and systematic study of a design by technical experts who are not necessarily directly involved with the design under consideration. The main purpose of evaluating a design as it is produced is to ensure that new or improved products or projects can be introduced with minimum risk. Those who take part in any type of design review examine the design and the methods used from all points of view. Their basic objective is to judge if the product has a high probability of satisfying the specified requirements of performance, cost, reliability, maintainability and safety.

To achieve this end it is necessary to have several different types of review, of which some are formal and use a structured approach, while others may be more informal and focus on some particular aspects. Broadly speaking these divide into five basic types: requirement reviews, preliminary reviews, intermediate reviews, final reviews and verification or appraisal reviews. The requirement and verification reviews are special types as they seek to achieve accuracy and completeness at the detail level, whereas the preliminary, intermediate and final reviews are often directed at the system or major units of hardware.

The Design Sequence

From the large project which can take several years to complete, to the small project which may only take months to execute, many people are required and all play different roles. In a complex system, the design often proceeds through several stages which may be identified as requirements, feasibility, design study and delineation, development and production. The process starts and ends with the customer and, while the design department occupies the core activity in turning ideas into the hardware required by customers, there are many other participants. Figure 36.1 illustrates this point.

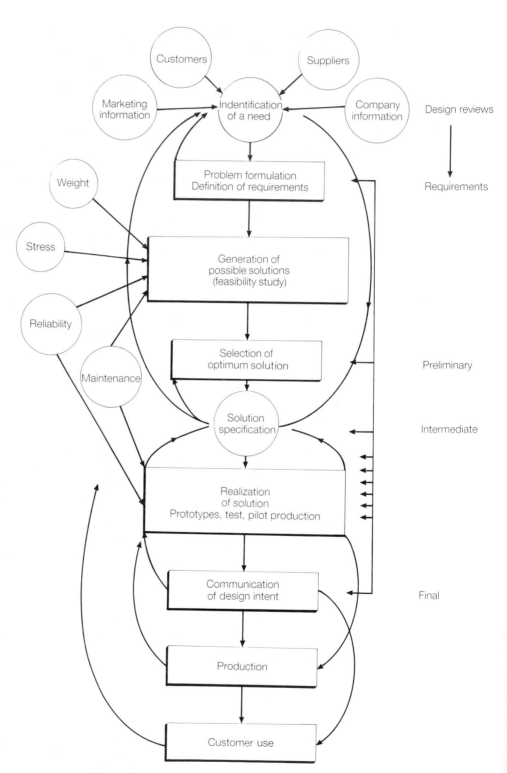

Figure 36.1 The design sequence

Definition of   This process starts when customers recognize that by some future date they
requirements   will be ill-equipped to survive and cannot solve the problem by ordering
more of what they already have. The new situation must be met with new
means. The formulation of the requirements is a task calling for considerable
judgement.

Feasibility study   With the general problem stated in an acceptable way, the feasibility of
various solutions can be examined. The conflict here is between the mind
which is untrammelled enough to dream up fresh ideas and the mind
disciplined in a hard school which subjects all ideas to searching criticism.

Design study and   Out of a number, one scheme appears more feasible than the rest. That is to
delineation   say it appears likely that the customers' several stated requirements can all
be met. In the design phase proper, the various demands are reconciled and
a compromise reached.

Development   Development may be defined as the controlled modification of drawings.
Modifications are required not only because the drawings are in error but
because our imaginations are limited and our knowledge of material properties incomplete.

Prototype or pre-   Models or prototypes may be made so that performance, all drawings and
production   possible means of manufacture can be proven. As a result of this work further
development may be required.

Production   If all that precedes this stage has been done properly, raw materials can be
manipulated and bought-out parts can be assembled to give customers what
they require. What customers get is in the message finally delivered to the
factory. To preserve in its long journey at least the spirit of this message,
critics must be appointed who will frequently compare the current intention
with the original aim. As responsibilities are subdivided and diffused it is
very easy to forget the original aim and undertake design for its own sake.
Figure 36.2 indicates possible points for formal design reviews.

The situation would be difficult enough if only the customer and supplier
faced one another, but it is made worse when suppliers are pitched in two
camps, the designers and the manufacturers. Each of these three groups
contributes its own special difficulties.

Requirements Reviews

Customers cannot be trusted when they say what they want – even less
when they say they know what they want. They are very apt to confuse
needs and wants. It is very important that this confusion is resolved as early

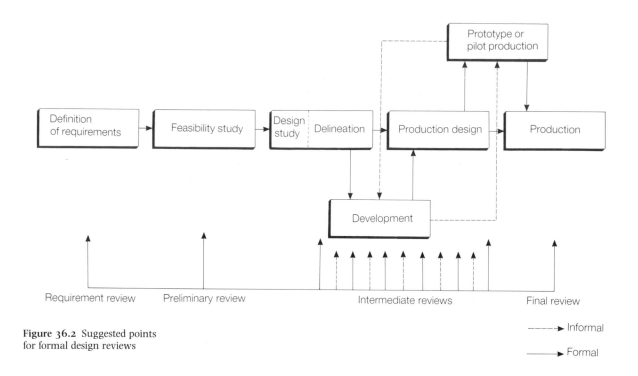

Figure 36.2 Suggested points
for formal design reviews

as possible when a very new kind of entity is to be designed because as it comes into existence it modifies considerably the situation within which the need was expressed. Customers may also ask for more than is needed, thus jeopardizing their chances of getting what is adequate. The several features which are needed may not be ranked correctly and too much emphasis may be placed on secondary features.

There is a very human tendency to play a role other than the natural one. Acting as amateur designers, customers may describe a problem in terms of a plausible solution. This is usually meant to be helpful but more often conceals the essential nature of the problem. An unfortunate consequence is that it may not be recognized which decisions should be taken by customers and which by suppliers. The latter can help customers to make up their minds by quoting costs, although these are seldom available at this time.

It is often thought that an agreement in numerical terms is more effective than one in words but this can be illusory. Both have their place but the mixture is often wrong. It can be difficult to place a new requirement correctly between extremes which either ignore the limitations of industrial arts or fail to recognize the essentially novel element in the future situation. Because the interface between customer and supplier is not rigid but yields to pressure from both sides, the demand–response reaction is not simple but instead each party is implicated in the other's problems. It is for this reason that requirements reviews are essential. The requirements review is aimed at examining the appropriateness and completeness of the design requirements by the customer and supplier for a particular system, project, product or component (see figure 36.2).

While a requirements review is often neglected, it is a vital part of the formal design review process. The design team usually prepares its view of the requirements which is then circulated and commented upon by the reviewers. Following formal discussion, the necessary reconciliations are made as required. It is absolutely essential to establish at the outset of the design process what the factor priority is for the particular design being considered. This is achieved at the requirements review through interactions amongst the specialists and the design team.

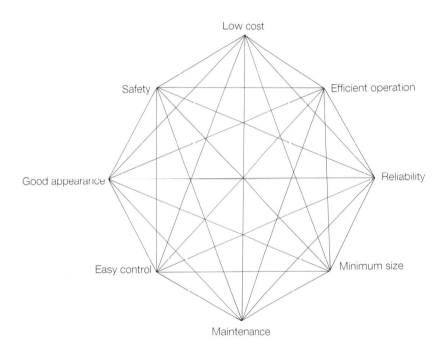

Figure 36.3 Factors in design work

Figure 36.3 suggests various factors which may have to be taken into account in design work. For an early warning radar system, the main facets of the diamond which will form essential requirements are likely to be reliability, efficient operation, ease of maintenance, etc. At the other end of the spectrum for a domestic appliance, it is likely that the main factors will be good appearance, ease of control, safety and low cost. The output of the requirements review is generally a design specification which becomes the foundation on which the design is developed.

Preliminary Reviews

Designers have three principal tasks: to understand what is needed; to choose the optimum combination of elements; and to formulate the proposals so that they can be detailed as manufacturing instructions. Design is not accomplished by one person even at the scheme stage, so it is important that someone is in a position to examine all contributions to the complete entity. Judgement exercised at this stage can have the most pervasive influence on the whole job. There is a lot to be said for this person being not too much of an expert in any of the subjects dealt with.

The fragmentation of design responsibility introduces its own problems. Ambiguity of intention can be avoided by the original designer producing much detailed information. But there are several objections to this: it takes too long: the designer should be concentrating on broad issues; the drawing office is reduced to a poor quality detail office. It is important that designers learn the difficult art of communicating generalities. They must also learn how to criticize and to stimulate fresh endeavour without having to become personally involved too often. They must be able to harness additional creative effort (so as to save time) early enough even though there is a risk that the design will turn out to be a characterless compromise.

The criticisms made at this stage benefit by being delivered personally. Senior designers who regularly visit the drawing office and make decisions which they are prepared to explain can do wonders for morale. They are also training their successors. A further important feature of the design process is the recording of decisions. Formal design reviews are a good management discipline for ensuring accurate recording of such decisions.

A preliminary design review can take place when a proposed solution to the design problem has been formulated. At such reviews it may be necessary to look at possible alternative solutions to ensure that the correct choice has been made. At this stage of the design process there may be several attractive solutions, although full comparison may not be possible until the sub-problems generated by these solutions are solved.

Intermediate Design Reviews

These are reviews which are taken at intervals throughout the design process after the layout design has been completed. They are carried out before production drawings are prepared and their purpose is to evaluate the design against the detailed requirements laid down in the design specification. Such reviews need to be carefully planned and controlled and may be supplemented by informal reviews where deemed necessary. Such informal reviews can be conducted on a side-by-side basis around the drawing board or a computer graphics screen. Individual specialists from manufacturing, stress analysis, quality assurance, materials and other functional groups may carry out a dialogue with the individual designers of the team on their particular design problems to ensure proper solutions are achieved.

The manufacturing process best suited for the bulk order may not be ready at the time when the early models are to be made. Prototypes must then be made in a different way. Which way is best is worth subjecting to independent scrutiny. It is all too easy to spend too much time sorting out the problems of how to make prototypes when this time should be spent on the object itself. Both in this phase of manufacture and in the early production run it is helpful to remember that we have set up a learning process. It is during this time that the designer realizes that people do not always read drawings but make assumptions about what was intended. Worse still, if the drawing is wrong, someone with a misplaced sense of charity may correct the hardware and leave the drawing untouched.

At this time the part played by inspection is vital, but an inspector can only safeguard the designer's interests if furnished with clear, helpful criteria by which to judge the product. These take the form of either local standards or specific inspection instructions. Manufacturers also have a critical role to play. When asked to make something, they are entitled to comment on its reasonableness when compared with the resources available. Some work-shops regard every drawing, no matter how ridiculous, as a challenge to their craftsmanship and ingenuity rather than their intelligence. It is for these reasons that the final design reviews and verification reviews are carried out.

Final Design Review

This is a review which is carried out after a prototype or pre-production unit has been made and tested. The main purpose of such a review is to ensure that all aspects of the design detail have been considered from a manu-facturability and reliability point of view. The main focus is therefore on the accomplishment of performance against the cost, producibility and reliability objectives.

Design Verification or Appraisal Reviews

The main purpose of these reviews is to provide assurance that the design detail is correct. They are performed by specialists and practitioners as detailed studies, often in isolation from each other. Typical examples would be: failure mode and effect analysis (FMEA); safety studies; reliability analysis; and electromagnetic compatibility (EMC).

Generally, the participants of design reviews are only mildly interested in computational accuracy and its effects. However, in a design verification review, day-to-day design effort is scrutinized to ensure that the correct design method has been used and the results obtained are accurate. Such reviews do not use the usual group meeting. To obtain maximum effect, the verifier should be another designer who is capable of and experienced in executing design work of a similar type but who has no direct responsibility for producing the design under consideration. This form of technical inspection of designers' detailed work may often require 'signing off' to signify the verifier's concurrence and approval. For some government contracts such as the design of nuclear power plants this may be mandatory.

Administration of Design Evaluation

Clearly, for large and multiple projects or big product changes, the administration of the various design reviews can become onerous. Careful planning is required if the necessary documentation, drawings, test results, etc. are to be made available. Responsibility rests with management who must establish a policy with respect to all formal reviews. Management needs to address itself to such questions as: when should a review be conducted? How many will be needed? At what point in the design cycle should they be held? Having made decisions on these points, management should select a design review date at least six weeks ahead of a proposed review and advise all the necessary parties of the objectives and scope to be considered.

Planning for design reviews Any type of design review has effectively three phases. Phase 1 is an introduction to the project – the stage of the design process at the time of the review. Phase 2 is a discussion with all those concerned on the design, and phase 3 is the production of final conclusions and actions proposed. Table 36.1 sets out a possible planning schedule for a design review meeting.

Table 36.1   Planning schedule for design review meetings

Activity	Days before and after design review
Design review date circulated	$D-35$
Agenda published. Personnel assigned to specific aspects of design. Invitations to review meeting circulated	$D-30$
Initial drawings, diagrams, test results sent out with other relevant data on particular work packages	$D-15$
Preliminary meeting by design department with up-to-date drawings and other data	$D-5$
Final run through by design department	$D-1$
Design review	D
Issue design review summary report	$D+10$

In the introduction phase, the review coordinator must state clearly the objective of the design review. It is desirable, but not essential, to have a coordinator who is not directly concerned with the project or product, so that complete objectivity can be obtained. Normally, this would be a project manager or product leader. Having set the scene, the coordinator should hand over to the designer to introduce the design. Table 36.2 sets out a possible guide for this introduction phase.

Table 36.2 Possible topics for the introductory phase of a design review

Background data and information
Reason for project, product or system
Summary of requirements with factory priority list
General design philosophy; any major changes from existing type of projects or product lines
New features; reasons for these
Assumptions used in the design work
Alternative designs considered
Reasons for selected design approach
Problems and risks to be considered
Possible design guides and rules to be followed

Next, the designer should proceed to the second phase of the review and discuss the main details of the design. Here, the reviewers should ask probing questions to try and unearth any hidden weaknesses in what has been proposed. In particular, reviewers will want to question the design assumptions that have been used and the areas of uncertainty. Here, the experience and judgement of the individual reviewers becomes of paramount importance. It is vital that all participants realize that the purpose is not to criticize but to provide constructive suggestions. It must not become an attacking meeting where the design team goes on to the defensive. For this reason, design calculations, developmental model tests, trial results and test reports must be provided as evidence for supporting the design.

A proven tool for reviewers is a design interrogation checklist which allows them to comb through a design systematically. General checklists can be developed but where possible they should be tailored to particular types of projects or products. Part of a typical example for a guided weapon is indicated in table 36.3 and one for a heating and ventilating plant in table 36.4. It is imperative that areas of 'software' as well as hardware are covered, such as operating instructions, spares lists, packaging/handling, labels as well as any computer programs.

In particular, the reviewers should address themselves to questions of the type set out in table 36.5. Answers to questions like these provide the output for a formal design review and the review coordinator must ensure that they are asked and answered.

The final phase concerns the follow-up action to be taken. The responsibility here rests with management – specifically, if there is one, the project

Table 36.3   Example checklist: guided weapon

Question	Answer		Chairman's comments (answer 'no' with your comment)
	Yes	No	
2 *Function*			
2.1 Have the following effects on function been evaluated: changes within specified limits of			
Power supply parameter values			
Environmental conditions			
Ageing			
Variation in component tolerances?			
2.2 Have all circuits in which special attention is required to quality of insulation been shown by theoretical and practical investigation to be satisfactory under all intended conditions of use?			
2.3 Have test acceptance tolerances been adequately checked?			
2.4 Has the provision of special test methods been considered in			
Factory			
Field?			

manager. It is imperative that each action item is investigated objectively and its solution documented as part of the follow-up. The solutions given may entail further tests and studies to evaluate whether the reviewers' concerns were unfounded or whether some additional calculations or an actual change in the design itself are needed.

For this purpose, it is desirable to have a formal paperwork system which demands a written answer so that the appropriate action is signed off and agreed by all parties concerned. A typical example of one such form is given in table 36.6.

Information and data for review meetings   As previously mentioned, it is imperative that for all review meetings the information and data should be distributed to all participants well before the meeting is convened (see table 36.1, D−15). Such information, which depends on the milestone at which the review is being undertaken, describes any interfaces with other items. The information might include layouts, drawings, specifications, anticipated customer needs, cost data, test reports, malfunction reports, relevant quality control analyses and any other pertinent data. A typical component review for a company-made part might include the following:

Detail drawings   Pictorial representation, description of materials, finishes, dimensions, tolerances, fabrication and assembly instructions.

Table 36.4   Example checklist: heating and ventilating plant

Question	Answer		Chairman's comments (answer 'no' with your comment)
	Yes	No	
4 *Environment*			
4.1 Are the following supported adequately, and stiff enough to stand the specified vibration requirements: Chassis plates Circuit boards Cantilevered structures etc.?			
4.2 Are all screws and other attachment devices locked, and by approved methods?			
4.3 Are all external connectors provided with protective caps fitted with approved retainers? (design rule 80.02)			
4.4 Have anti-vibration mounts been employed where appropriate to minimize stresses?			
4.5 Are all power dissipating devices mounted so that air circulation is adequate for cooling requirements?			
4.6 Are items subjected to forces during assembly and dismantling adequately strong?			
4.7 Are air filters and desiccators provided?			

Table 36.5   Typical types of questions that can be asked at formal design reviews

Do the test results support the conclusions drawn?
Are the design assumptions reasonable?
Are there areas where the risks appear to be higher than normal?
Are there items of significance that have not been subjected to analysis, such as failure mode and effect analysis (FMEA)?
Does the design satisfy all or only some of the stated requirements?
Are there requirements which were intentionally not satisfied?
Is that acceptable to the customer?
What problems remain to be solved?
Is there adequate assurance that these problems can be solved in a reasonable manner and in an appropriate timescale?
Are the design methods used appropriate?
Has a maintainability, reliability and accessibility appraisal been undertaken?
Does the design appear to be satisfactory to proceed to the next phase of development or production?

Table 36.6   Typical review form

Arising from design review			
Subject			Arising no.
Chairman		Stage	Sheet 1 of
Feature		Requirement	Heading
Background			
Question		Reply by project group	
Reviewer	Date	Signature	Date
Group recommendations		Comment of project group	
Chairman	Date	Manager	Date
Management decision		Action	Authority
The review group's recommendation is accepted. rejected. to be studied and reported on.			

Installation drawings   General configuration, any attaching hardware, spatial location of mountings, etc.

Component specification   Functional characteristics, test requirements.

Parts and materials lists   Setting out specification and class of items.

Reliability analyses   Failure mode and effect analyses.

Stress analyses   Any relevant calculations.

Attendance at design review meetings   Participants will include personnel from engineering, manufacturing, construction/site engineering, purchasing, reliability, quality control and cost

Table 36.7 Review group responsibilities

Function of group manager	Responsibilities	Type of design review		
		Preliminary	Intermediate	Final
Chairman (often project manager)	Convene and conduct meetings and issue reports	✓	✓	✓
Design engineer(s)	Prepare and present design and substantiate decisions	✓	✓	✓
Reliability	Evaluate design for the specified goals, etc.	✓	✓	✓
Quality control	Ensure that functions of inspection and test are adequately conducted		✓	✓
Manufacturing engineers	Ensure that design is producible with minimum cost and time		✓	
Operational engineer(s) or user/customer representative	Ensure that installation, commissioning and maintenance considerations are taken into account in the design	✓	✓	✓
Product safety	Ensure that design is able to meet all known regulations	✓	✓	
Packaging and shipping	Ensure that the product is capable of being handled without damage		✓	✓
Materials engineers	Ensure that materials selected will perform satisfactorily and review possible alternatives		✓	
Purchasing engineers	Ensure that acceptable bought-out items and materials to meet cost and time targets are obtained		✓	
Design engineers not associated with the design and consultant specialists (industrial designers)	Evaluate overall design to ensure that performance, cost, time, aesthetics, ergonomics, etc. are satisfactory	✓	✓	✓
Marketing	Ensure that customer requirements are fully stated and understood	✓	✓	✓

reduction. In addition it may be desirable to invite specialists from other areas to collaborate, but no design review should involve more than a dozen people. The actual constitution of any design review depends upon when it is held in the design process. In the preliminary design review, for example, marketing should be represented. Other participants will be needed for intermediate and final design reviews. Sometimes it will be desirable to have consultants and customer representatives present, and they must have their responsibilities agreed upon prior to participation. Each member should have at least the technical competence in his or her own area that would have been essential had the participant been the designer. Table 36.7 sets out possible design review group responsibilities and membership for the different types of review meeting.

Conclusions

When design evaluation is carried out in accordance with the procedures outlined, it has been found that very often considerable cost improvements are achieved together with earlier delivery dates. The combing through of the design during the design process leads to fewer design changes at the end of the work by ensuring that the detailed prescription and selected concept are correct. Where such reviews have become the normal working procedure in realizing a new design, there is no doubt that better design staff capabilities are obtained which in turn leads to accelerated design maturity.

However, it must be appreciated throughout the whole organization that the design review approach leaves responsibility for essential design work in the hands of the designers. The reviews are to help them to recognize and solve problems and to ensure that any appropriate corrective action is taken.

37 Managing Designers

DOMINIQUE BAUHAIN
France

Introduction

This chapter results from a study carried out in 11 design departments in French companies. It examines why a firm may or may not choose to permanently integrate designers within the system, how the firm can manage designers, their range of activities, their objectives, motivation, etc. and, finally, what conclusions can be drawn.

Today's designers can offer a wide range of abilities to a firm. Design, in its long history, has never stopped redefining its identity, evolving within the constraints of the environment. There is a constant tension between what design would like to be and what it can be within the economic context. There is a contradiction between the sensitive and artistic nature and the multifaceted demands of a firm.

The appearance of design in industry can be explained like other innovations: design is introduced as an aid to the resolution of a problem at a given moment. Rarely is a design department established at the creation of the firm. There seem to be two fundamental reasons why a firm resorts to design:

1 Design is seen as a tool to help increase competitiveness when sales are lagging or the image of the firm is deteriorating. Design only appears because there is a problem to solve.
2 Design is considered by the firm's directors as a cultural attribute. Thus design is seen not as a tool but rather as a value, and the firm, through its products, services and premises, shows a commitment to quality.

The second reason is less often found than the first. Yet even if the first reason is essentially commercial, a firm's commitment to design may evolve with exposure as the designer progressively adapts to the demands of the firm. By acting as an integrator of language, knowledge and requirements, the designer is increasingly offering more complete services to the manufacturer than just talents as a plastics technician, decorating artist or stylist. The designer becomes animator in the conception of the product and may well participate in strategic policy decisions.

Reasons for Permanently Employing Designers in the Firm

Research amongst a sample of 25 French firms has revealed some of the reasons for operating permanent, in-house design departments. There are as many companies in the sample producing consumer goods as producing industrial and professional goods. Moreover, most of the companies are big: six firms are the largest employers in France (Renault, PSA, Thomson, Philips, Matra and Bull); nine earn the highest revenue in both French and European markets (Renault, PSA, Philips, Bull, Radiola, Matra, Poclain, Allibert, Essilor).

From observations of the sample of companies, four reasons can be put forward to explain a preference for permanent integration, as outlined in the following sections.

The amount of work

A design department can only be justified if the amount of work makes it cheaper to integrate than to subcontract. If the firm produces only a few new products a year, which do not sufficiently occupy a full-time designer, the firm would be better served resorting when needed to an outside design company than having its own under-employed designer.

However, there are firms who do only introduce a few products every year, but because these have a lengthy period of development they justify the full-time presence of designers. Automobile manufacturers, for example, can totally justify the permanent presence of many designers.

The complexity of the product

As the complexity of product conception increases, the more important it becomes for different departments, people or services to communicate and the more necessary and the better the coordination within the firm must be. Integrated designers can reduce information delays, as they are well informed about the products and structures of the firm. A much longer period of time might be required to properly brief a consulting designer. It is for these reasons that a high correlation is observed between the firm's size, the complexity of the product and a designer's integration.

The complexity of the product, or of the conception process, can lead to integration of designers in order to reduce uncertainty or alleviate the problems associated with subcontracting, such as long delays, high costs, and insufficient attention to detail caused by lack of understanding by outside designers of the organization's characteristics. Indeed, two common criticisms made of subcontractors are as follows:

1   The difficulty of 'industrializing' the work of outside designers, as it may not be suited to existing technology or production systems, or may be just too expensive. Project results are often not feasible and do not correspond to requirements because of the consulting designer's inability to work within constraints, or because of difficulty in getting access to the information.

2   Coordination may not be sufficient. Very often, subcontractors are used only from time to time. Without it being the fault of the outside designer, results may not be coherent with other simultaneous work done on other products or in other fields (such as development of a communication plan or changes in the distribution policy, for example).

Management preferences

Management preferences might act directly on design integration, if such a department is viewed as strategic or prestigious. Senior managers may desire to possess a skill leading to a certain superiority, giving them prestige in the environment. Other firms may want to be seen to be associated with famous designers, as a boost to apparent technological competence. Resorting to a specific designer, to have products designed by someone in particular, is a strong communication point for certain firms.

However, the integration of designers is only possible if it is interesting for the designers themselves. The most often adopted solution in this type of negotiation reflects a compromise: the firm secures with a designer a permanent but non-exclusive collaboration. The designer may even become a permanent employee while retaining the possibility of working for others.

Pressure exerted by an external partner

One final reason may explain the design department's integration within the firm: the pressure exerted by either a parent company or a foreign partner on its subsidiary. Such integration, implemented under an external influence, may be intended to standardize structures, thus leading to homogeneous and efficient management.

To conclude on this point, using a subcontracted designer often precedes the designer's integration into the firm. After either positive or unsatisfactory experiences, the permanent integration of a designer can seem to be more profitable or more efficient in permanently coordinating new product work in the firm. In other words, for many firms subcontracting is a step in their development – it is a way to learn about design.

Table 37.1 sums up the reasons for and against the integration of designers within a firm, as opposed to subcontracting. The major drawback of integration – suppression of creativity because of the routine of repetitive working conditions – can always be alleviated by entering into competition with outside consultancies.

Characteristics and Development of Integrated Design Departments

Position in the organizational structure

The location of a design department may be related to the history of the firm, particularly in the manner the firm is structured with respect to the environment. The reason design enters the firm may influence the function

Table 37.1   Subcontracting or integrating design

	Subcontracting	Integration
From the firm's viewpoint Advantages	Lower cost if few products to be designed Possibility of changing designer employed	Lower cost if many products to be designed Easier coordination with other departments in the firm Designer's greater knowledge of the firm Possibility of subcontracting to induce sense of competition
Disadvantages	Lack of continuity in work if designer employed is changed Higher cost if much work involved Greater difficulty of coordination with other departments in the firm	Lack of creativity
From the designer's viewpoint Advantages	Variety of work Preservation of creativity High degree of autonomy in relation to client	Security of employment Continuity of work More means at designer's disposal
Disadvantages	Greater commercial risk	Status often ill-defined Lack of power in relation to other departments in the firm Routine, restricted sphere of operations

it will be assigned, the role it will play and the position it will occupy. Generally, the person who brings design into the firm is the one who needs it, for whom design will solve problems and by whom this new function will be managed. Beyond this, three cases may be distinguished.

Design as the solution to commercial aesthetic problems

Design is thus attached to the marketing department. In certain firms which produce rather bland goods, where the simple and standardized technology calls for a strong differentiation to enable recognition by the consumer, commercial and marketing functions tend to dominate. Design answers their requests by creating products suited to the target market. The designer holds a specific talent which is lacking in the marketing team; the latter would like to be able to translate what they believe to be the demands of the market with a drawing or a model – but they do not have the skill to do this.

Design solves a problem of task coordination

When the complex technology for producing a product demands the interaction of a number of different specialists, the designer can play the role of coordinator in the differentiated conceptual process. He or she can synthesize the constraints linked to the conception. In firms where the products are the result of a complex and evolving technology, the designer is thus most naturally integrated within a technical department.

Design needs to be autonomous

Design must not be dependent on any functional department. Certain firms choose not to attach their design department to a technical or a marketing department, in order to guarantee its freedom and preserve its independence and creativity. Designers who are attached to either a technical or a marketing department may well be subjected to their respective pressures, resulting in a restriction of choices and actions. However, direct linkage of design departments to top management is rarely observed, and so it may be difficult to achieve an appropriate design culture and proper motivation.

Range of activity of the integrated design department

The range of design's activity often relates to its location in the organizational hierarchy, which itself may highlight the kind of problems it is expected to help solve.

When attached to a *marketing department*, the designer generally works with the product's shape, as priority is given to the product's seductiveness. Design may also address packaging, graphics, etc.

When attached to a *technical department*, the designer is more likely to be responsible for coordinating the conception of the product – responding to the constraints imposed by different departments. The attachment to a technical department essentially implies work on new products, whose development may be long and difficult, but designers may also work on the product surroundings – packaging, sales booths, graphics and so on – to help achieve coherency.

When design reports directly to *general management*, designers may still work on the conception of products, although their responsibilities are usually enlarged. When top management's motivation for design is strong, design may be seen as a qualitative factor relevant not only to products but also to all events or communication processes within the firm's environment. Design, whose management is carefully planned within such firms, serves neither as just aesthetic added value meant to increase sales nor as a simple tool to coordinate the different tasks in the conception of a product. It is the cultural vector of the firm, the coordinating mechanism for all the elements of the firm's image.

Role of an integrated design department in product development

In the sample of firms studied, it was noted that designers tended not to be systematically consulted during the research leading to the conception of new products. And while designers are creative individuals able to generate new ideas, they are often required to contain their creative abilities within the bounds of a specification. Indeed they are not often the originators of new products, as they do not have access to information concerning the market, nor are they associated with strategic policy decision making.

They are often found executing and sometimes rebelling against decisions taken by the marketing or research and development departments. As for their place in the conception process, it depends on three factors:

1   The distribution of authority in the firm.
2   The nature of the decision (complex or simple) and of the environment (uncertain or foreseeable).
3   The power or influence the integrated design department has within the firm.

For the first two factors, which are highly correlated, it was observed that the more stable the environment, the less complex and removed from existing know-how are the products; and the less committed is the firm to its products, the more the decision makers are in control. The decision making process returns to algorithms which have already been tested through experience. In this case, the designers' opinions tend not to be sought and they may be involved only very late in the conception process, after all the parameters have been decided.

In the opposite case, where a high level of uncertainty and complexity in the decision making process exists, the number of decision makers increases. One of these is the designer, who can help control the uncertainties linked to the environment, changing preferences, perception of future trends, etc.

The acquired authority of the integrated design department

Three factors can influence the development of the integrated design department.

Changes in the firm's activity

These can modify the position and role of the design department, and are often reflected in changes in the size of the firm. Its growth or contraction may permit the development or cause the disappearance of the integrated design department. Similarly, strategic choices such as the specialization of products or diversification of activity, the integration or subcontracting of certain activities, and production standardization can all modify the range of activity, the status and the place designers have in the firm.

Results obtained through design

These are rarely measured as they can be difficult to estimate. The designer serves as a value adder but this may be difficult to isolate from all the work done to conceive, to finalize and to launch a new product. Design results are often accompanied by other modifications. At the end of the day, what may matter most is whether design has effectively modified the image of the firm in the outside world.

The strategy of the design department

Different departments seek different objectives which do not necessarily coincide with the objectives set out for the firm. They may not want to cooperate or follow the rules of the game without profiting themselves.

The integrated design department is usually a numerical and cultural minority within the firm. As a result, it tends to fall into conflict with the majority, who put pressure on it to conform. To exert influence, this minority of designers must be able to improve in three areas. The first is their own competence. Designers have a particular talent: the ability to give form to concepts. Their perception of trends allows them to play the role of intermediary between the firm and its environment. With these abilities, they can create a dependency for information, which becomes stronger still as uncertainty increases. Thus the designer may exert influence in the 'subjective domain'. The second area for improvement is their low status in the firm – perhaps defined by a contract which curtails their rights and obligations. It is quite worrying that designers do not attach enough importance to this essential element, often because of lack of training and knowledge. The third is their capacity to mobilize themselves as a collective group; to seize opportunities a group must be united.

In practice, design departments may adopt other practices:

Apathetic behaviour Designers manifest apathetic behaviour towards requests, using the time factor and hoping that their resistance will bring consciousness, awareness and change.

Exchange behaviour Designers cooperate only with an eye to immediate profit (such as being congratulated for effort in increasing the firm's sales).

Investment behaviour Designers follow a long term policy to train and inform other members of the firm of their activities. They patiently try at each opportunity to develop their awareness of their abilities.

Eventually, the position of the integrated design department will evolve: it largely depends on the activity of the designers, their adaptation to the firm, and the firm's experience with design.

Conclusions

The quality of a designer's contribution, whether integrated within the firm or not, depends primarily on the existence of a strategy and clear objectives. It is not the designer's role to take the place of a strategic decision maker.

Designers must be given a framework and provided with motivation. One must watch for the division of tasks, the specialization of activity which impoverishes their range of activity. They should be encouraged to participate in all design-related activities either taking place or being planned; for example, they should be involved in promotion campaigns for products they have designed. Information may be the determining factor for their successful integration within the firm, but this must not be achieved at the expense of contact with the outside world. Loss of contact with the surroundings tends to diminish creativity and lowers the chances of renewing it. Although difficult, it is beneficial to define for designers precise objectives which encourage participation by them in the firm's development and daily life.

The integrated design department is still often regarded as an unprofitable investment. The recommendations in this chapter will certainly enable the department to contribute to the wealth of the firm and to the progress of the designers themselves.

38 Design Protection in Practice

DAN JOHNSTON
UK

Introduction

The starting point for company policy about design protection is company policy about design generally. The policy should be a progressive one to encourage innovation, to develop designs that involve new thinking. Hopefully, it will not be a policy to copy others.

Of course, there will always be those who find it easier to be copyists. They try to profit by exploiting other people's successes, by copying designs towards which they have contributed nothing. It is natural therefore – and quite proper – that the real entrepreneurs, the risk takers, should seek protection against unscrupulous competitors. Innovators need to ensure that their contribution by way of new ideas, painstakingly developed into successful products, will lead to good commercial benefit.

Law in the UK (and industrial countries generally) has grown up to protect inventors, designers and innovative manufacturers against plagiarism. The motives have been justice for the creative individual, advantage to the particular industry and benefit in the end for everyone.

Forms of Design Protection

Patent law   Patents are for inventions and the law makes a sharp distinction between patents and designs. However, in a general way designers are called upon to be inventive and many designers achieve inventions. Certainly many product designs are best protected by patent(s). To be patentable an invention must be 'new', it must involve an 'inventive step' and it must be workable in industry. Undeveloped ideas, scientific discoveries and mathematical methods (e.g. computer programs) are not patentable. The maximum term of patent protection in the UK is 20 years. When formal Patent Office fees and the fees to patent agents are totalled, patents may be thought to be expensive. However, much of the cost is spread over the years.

Design registration   This scheme protects the appearance of products. There must be some 'aesthetic' element in the design that influences the purchaser in choosing to buy it. Purely functional designs and methods of construction are not eligible for registration in the UK. A registered design, like a patent, has monopoly protection. Protection against copying is for the design or something very similar, and the first person to apply has priority over others who come up later with designs which are too close. The maximum term of protection is 25 years. The cost of registration is less than for patents but if many designs are involved the cost can be considerable.

Copyright   Under the 1988 Copyright, Designs and Patents Act, copyright protection for designs is retained in certain areas. Copyright does not give monopoly protection. The protection given is specifically against copying; if two people produce the same or very similar designs without one copying the other, neither can sue the other. They both have copyright and could both sue real copyists. Also, copyright does not protect ideas. What is protected is the specific design – not the basic idea behind it. Nevertheless, slight changes in detail do not free a copyist from breach of copyright. He or she may still be seen as having copied 'in a material way'.

Copyright has the great advantage of being automatic and immediate – and there is no cost. The areas of design protected by copyright in the UK are:

1 'Printed material primarily of a literary or artistic character': this covers most items of graphic design from book jackets through to advertisements and trade forms.
2 'Artistic works' if industrially produced, e.g. decorative patterns on fabric and wallpaper or on three-dimensional products such as pottery and glass. The term 'artistic works' also includes sculpture and 'works of artistic craftsmanship'. Industrially produced products in these areas also qualify for copyright protection.
3 New designs for typefaces and the 'typographical arrangement of books'.

The full term of copyright protection (for works of literature, music and art) is for the life of the 'author' plus 50 years. The full term applies to the first category, graphic printed material. For artistic work industrially produced and for typefaces and books, it is 25 years from the date of first publishing or offering for sale.

Design right   This is a new form of design protection brought in by the 1988 Act and is unique to the UK. The problems that gave rise to design right were the long standing contentious issue of protection for purely functional designs and, more urgently, controversy about protection for the design of spare parts for motor vehicles, domestic equipment and industrial machinery.

Design right applies to both two- and three-dimensional products, and to purely functional products as well as those having appearance appeal. As with copyright, design right is automatic and immediate. There is no cost. The protection period is short: five years from the date of marketing plus a further five years subject to licences of right. (The phrase 'licences of right' means that if a would-be competitor applies for a licence to manufacture the product, it must be granted with the royalty payment determined by the Patent Office if the two parties cannot agree terms.) Even during the first five years of design right, if a monopoly situation damaging to the public interest is deemed to exist, it can be ordered that licences of right must be granted.

It will be seen, therefore, that while design right has value in allowing a manufacturer to recoup development costs over a short period, it does not give much long term benefit. There are other limitations that have particular relevance to spare parts.

Trade and service marks
These marks do not directly give protection to the design of products. However, the trading advantage of such marks does have a design protection benefit. If a product is sold under a name that is registered as a trade mark, the prestige of the mark may be the ruling factor when a purchaser is deciding between two similar products. The model name for a design may indeed be almost as valuable as the design itself. And if a firm is seen to be vigilant in protecting its mark it will indicate that a similar attitude will probably apply to protection for its designs.

Passing off
This activity is not dependent on particular statutes. It relates to deceitful business practices and to company goodwill. In other countries there is similar but stronger law, usually referred to as law against 'unfair competition'. The relevance of passing-off law to design has been in actions involving plagiarism in packaging and the design of containers. The possibility of the law being invoked in respect of the design of products as such may sometimes have been a deterrent to copying. Passing-off actions are usually difficult and expensive.

Managing Design Protection

If design is important in a firm's business it is essential to have good advice. Large companies usually have a special department. Small firms should have pre-arranged contact with a knowledgeable lawyer and a patent agent. If copying or other infringement occurs it is important to react confidently and quickly. If this is done on the basis of good advice, difficult entrenched situations are likely to be avoided. Supposed infringers may withdraw without resort to the upset and expense of court proceedings.

Ownership
The first need is to ensure proof of ownership of all designs produced and marketed. The rights to designs (and inventions) produced by staff in the course of their employment belong to the employer. Letters of engagement should refer to this and should also make it clear whether or not a staff designer may do freelance work for other companies in non-competing areas. (This may have value by giving the designer the opportunity to get greater experience and reputation.) With freelance or consultant designers, the employing company should require assignment of its rights when the final

fee for the design is paid, and this point should be covered in initial contracts commissioning designs. If designs (or inventions) are licensed to or from other companies, it is important to have good legal advice.

Record keeping

It is very important to be able to prove the origin and development of designs through to actual production and marketing. Rights to an invention or a registered design are established in the process of taking out a patent or a design registration. But if copyright is claimed on the basis of original graphic work or 'artistic work' industrially produced – usually a drawing or painting – the original work must be preserved. One reason for this is obvious: the copyright claimed is dependent on it. (This is not the case with design right because design right is not dependent on copyright in drawings.)

A second reason is that copying accusations may be the other way round. It may be necessary to appear as the defendant rather than the plaintiff – to show that the work produced was not a copy of something else. The evidence kept should include signed, dated and properly authenticated drawings, other documents and early samples. Authentication may take the form of a chronological record properly maintained or drawings may be deposited with a lawyer, bank or other organization and be date stamped.

Trade intelligence

A company in a competitive situation needs to be well informed about design trends – to know what competitors are doing. First, it is a myth to suppose that all good ideas can be generated from within. Other good ideas will also be around and, while there must be no suggestion of copying, ideas are free and can quite legitimately be developed independently. Secondly, as has been explained earlier, if other people copy it is important to know about it and to act quickly. Trade intelligence is normally best handled by a company's sales department – and sales staff feedback must be properly planned.

Trade precautions

Related to trade intelligence are precautions that can be taken within the trade to mitigate against copying. It should be a matter of policy to let it be known what the company's attitude is to design copying. This can be done in a friendly way, for example, by activity within the relevant trade association, by giving support to design events and by keeping contact with design colleges.

Product marking

This indicates that design protection is claimed. It is obviously desirable for trading reasons but is rarely done consistently. Where patents or design registration apply, the words 'patent', 'patent applied for' or 'registered design' should be given along with the number. If the protection claimed is by copyright, the sign © should be applied, supported by the copyright owner's name and the year of first marketing. (Copyright marking is not obligatory in the UK but is virtually so in the USA and some other countries.) Design right is new, so there is as yet no symbol and the words should be spelt out.

If marking cannot be applied to the product itself (and even if it can) it should certainly be applied to packaging and on any supporting literature. The markings for trade/service marks are various – M or TM or SM – usually applied in small letters after the mark, perhaps in a small circle.

Action against Infringers

If a UK competitor is thought to be infringing one of the company's designs (or infringing in some other way, e.g. by coming too close to a trade name or a trade mark) the first action to take may well be through the company's trade association. Many trade associations see it as one of their functions to handle disputes between members. Even if the supposed infringer is not a member it may still be useful to seek any help that can be given.

If such a move is not possible or proves to be unproductive, legal advice should be taken and an appropriate letter written to the supposed offender. It is unwise to make threats; the other firm may have a surprisingly good defence and there could be a counter legal action. However, a letter quoting the patent or registered design number or a plain statement about the apparent breach of copyright or design right may be enough to produce a solution. Ignorance or coincidence are always possible. Perhaps the offender will offer to withdraw the infringing product (or name or trade mark) or a satisfactory licensing arrangement may be made. The majority of intellectual property cases are settled out of court.

If a court action seems to be the only way, a calm assessment of the situation is to be strongly recommended. Moral indignation is certainly not enough. The real question is whether a costly court case (or a series of them) will be worth fighting, and the judgement must be a commercial one. If it is decided to fight, the first step may be an application for an interlocutory injunction in the High Court. If this is granted, the defendant will be prohibited from continuing the alleged infringement pending full trial of the case. Very often that will be the beginning of the end. The lawyers will work out a settlement covering costs, compensation and probably a public admission of infringement by the defendant.

Insurance For many small and medium firms the costs involved in bringing a design protection case to court are too great to be faced. Schemes have been devised, therefore, whereby legal costs can be met by insurance. The schemes cover protection for patents, registered designs, designs protected by copyright or design right and also trade and service marks.

There is a registration process for the patent, design or mark concerned. Actions are normally limited to plaintiff only and will only be started on independent advice that there is a better than 50 per cent chance of success. There is usually a registration fee and an annual premium for each patent, design or mark insured, and some part of the legal costs must be met by the insured. The great advantage of insurance is that its very existence gives credibility to a small firm facing up to an infringer who may be economically much stronger.

Plagiarism: the narrow dividing line

What has been said about legal problems in coping with plagiarism raises more fundamental issues. If a new invention results in a company's business being threatened, the directors of the company are quite entitled to look for ways of designing around it. However, a scheme for getting round a patent can come quite close to infringing it. There may be plenty of room for dispute.

With designs and possible copying, the scope for argument is even more obvious. Designers all live in the same world and the firms for which they work compete for the same business. It is not surprising that designers working for different companies often come up with very similar designs. Changes in fashion and technology can be acted on simultaneously and rarely emanate exclusively from one person. Most significant of all, a truly alternative design based on the same idea is within the law. But producing a 'truly alternative design' may not be easy; it may call for much ingenuity and creativity.

The best guide in assessing whether a new design is a copy or not is to look for functional or appearance differences that make it better than the old one. Judges are also concerned as to whether the possible copyist had the 'opportunity and motive' to copy. That the dividing line can be a narrow one is borne out by reference in some instances to 'flagrant copying' – cases where the act of copying was not in doubt.

Protection Overseas

Conventions

Legal protection for inventions and designs is to be found in all industrialized countries to a varying although gradually converging standard. International cooperation is based on three conventions: the Paris Convention for the Protection of Industrial Property, the Berne Copyright Convention and the Universal Copyright Convention.

The Paris and the Berne Conventions date back to the late nineteenth century. They provided in a very far-sighted way for a degree of cooperation between signatory nations. One of the provisions in both treaties was particularly important: that the same rights were to be granted to the nationals of other member countries as a country granted to its own. The Paris Convention also provided for a 'right of priority' whereby an applicant for a patent, a registered design or a trade mark was given a period in which to make a similar application in other member countries, the subsequent applications counting as though they had been made on the same day as the original application.

To keep them up to date with changing ideas, new texts of the Paris and the Berne Conventions are negotiated every ten to twenty years. The affairs of both conventions are managed by the World Intellectual Property Organization (WIPO) at Geneva, an agency of the United Nations. There are many treaties subordinate to the Paris Convention covering particular aspects of industrial property – patents, registered designs, trade and service marks, indications of the origin of goods.

The Universal Copyright Convention (UCC) is relatively recent (1956). It was negotiated because some countries were not willing to join the Berne Convention – notably the USA, the USSR and several South American states. The protection provided by the UCC is of a lower standard than that of the Berne Convention and signatory countries may require the use of the symbol © where copyright is claimed.

Many countries (including the UK) are signatories of all three conventions. A new copyright law made it possible for the USA to join the Berne Convention in 1989.

Europe   Although WIPO operates worldwide, much of the initiative for increasing international cooperation in intellectual property in recent years has come from Europe and the EEC. Most progress has been made with patents. The European Patent Convention covers most of the member states of the EEC and a few other European countries, with the European Patent Office established in Munich since 1978. Application may be made simultaneously in one language (English, French or German) for patents in several countries. There are procedural advantages and the cost, if more than three countries are covered, is less than if application had been made in each country separately. Application for European patents for UK residents must be made through the London Patent Office. A measure of the success of the scheme is that, despite gradually increasing patent activity, applications for national patents in the UK have fallen in the last decade from around 50,000 a year to about 30,000.

The EEC also has plans for a Community Patent – an even closer integration of patents and patenting. Discussions on this have been in progress for many years and still continue. The European Patent scheme is linked under the Patent Cooperation Treaty to an even wider scheme whereby patents are obtainable internationally using a similar formula – one language and modern procedural methods.

Things have gone more slowly in attempts to seek European harmonization of laws concerning design protection. In Europe there is a two-way split. The continental system, which is based on 'author's rights', emphasizes the continuing rights of the originator of a work, whether writer, artist or designer. In the UK, despite the introduction of 'moral rights', copyright law can still be said to be based to a greater extent on economic interests. Another difference is that design registration in the UK is given after search and examination; most other European countries only require formal application and deposit of a sample or a likeness. Most European countries are signatories to the Hague Agreement concerning design registration; the UK is not.

However, in the major industrialized countries of Europe things have moved closer together. The theories may vary greatly but design protection is generally available by registration and most countries limit registration to products with appearance value. There may be copyright protection for works of applied art. There is usually protection for a short period for purely functional designs (admittedly by very different legal means, e.g. petty patents in Germany, design right in the UK). And in many European countries protection by law against unfair competition fills in the gaps left by specific design protection law.

Likewise with trade and service marks, until recently, some countries (like the UK) only registered marks relative to goods. The move to include service mark registration is now almost complete. Most European countries (not the UK) are signatories to the Madrid Agreement on the international registration of marks. A Community Mark Convention is much nearer to realization than the comparable initiative in respect of patents.

United States of America

The inventiveness of US industry and the worldwide interests of huge American companies have long ensured for the USA a dominant position in the international patent scene. As they are fully entitled to do, US firms take out great numbers of European patents.

Design registration in the USA is more closely akin to the patent system. Copyright protection for designs is possible but registration is necessary before legal action can be taken. Where copyright is claimed, marking with the copyright symbol © is almost essential (but this may be further modified now that the USA is a party to the Berne Convention). There is an effective trade and service mark system.

Japan

The dominant position of the USA in industrial property matters remains true enough, but the Japanese have long since overtaken the Americans in sheer numbers of applications for patents and design and trade mark registrations. Again there is a correlation between intensity of inventiveness and product development coupled with worldwide commercial activity. Design registration covers both functional and aesthetic designs. Trade marks only relate to goods.

Other Far Eastern countries

Hong Kong, Singapore, Malaysia, Korea and Taiwan are all countries with considerable industrial potential and also an unenviable reputation for copying products at unbeatably low prices. In mitigation it must be said that pressures to improve their intellectual property laws have been heeded in many instances.

Action against infringements and copying from overseas

UK law protects manufacturers against patent infringements and product copying if the offending products are imported into Great Britain. Action must be taken against the importer. However, if an invention is not patented in their country, manufacturers abroad are entitled to use it (after its priority date is passed, assuming that the country concerned is a member of the Paris Convention). The same thing applies to registered designs – which tend to be more respected abroad than copyright.

If a design is protected in the UK by copyright (or design right), protection abroad varies greatly. In general, there is a possibility of some protection in countries with sophisticated intellectual property law. However, the prospects are poor in many countries of South East Asia. (The 1988 Copyright, Designs and Patent Act gives a special place to Europe – as it must – but includes reciprocity elsewhere as a factor in design right protection.)

The crunch comes with cost. To get patent protection virtually worldwide is extremely expensive (but may be very worthwhile if the invention justifies it). Again, the same thing applies to design registration. If legal actions have to be taken in other countries the costs will be heavy; in all probability there will be two sets of lawyers as well as language problems.

However, there are other measures that may help. Protests can be made directly to the offending firm (once identified). They can be backed up through trade and design channels, and also through government departments in the UK and the two embassies involved (in the UK and in the country concerned). Such action takes time but often succeeds. No country likes to have a reputation for plagiarism.

Conclusion

Maximum protection in the UK and abroad by patent and by design and trade mark registration will not always be affordable. However, essential action should be taken and many designs have automatic protection at no cost by copyright or design right. This involves good record keeping, including evidence of design ownership.

Well-planned design protection is the essential back-up to product innovation and development. Together they provide the basis for confident product marketing.

Further Reading

Chartered Society of Designers: *Protecting your Designs.* London, 1989.

Chartered Society of Designers: *A Designer's Guide to Design Protection within the EEC.* London, 1987.

Johnston, D.: *Design Protection.* London: Design Council, 1989.

Patent Office: *Applying for a Trade Mark.* London, 1985.

Patent Office: *Applying for Registration of a Service Mark.* London, 1988.

Patent Office: *Applying to Register a Design.* London, 1989.

Patent Office: *How to Prepare a UK Patent Application.* London, 1984.

Patent Office: *Introducing Design Registration.* London, 1989.

Patent Office: *Introducing Patents.* London, 1988.

39 Assignments and Licences

ROBERT AKROYD

Aston Business School, UK

Introduction

Companies may wish to exploit their own design successes by selling rights to their inventions or creations to others. Alternatively, they may need to secure permission to use the results of other companies' efforts. Similarly, contracts for the transfer of intellectual property rights over designs are often entered into as the foundation for a production and marketing joint venture package. As such, they frequently contain provisions which may fall within the regulation of such interstate and national competition as articles 85 and 86 of the Treaty of Rome and the UK Restrictive Trade Practices Act 1976.

Under English law, such contracts may be characterized as either assignments or licences. Broadly speaking the difference is that whereas an assignment is a permanent transfer of the whole right between an assignor and an assignee, a licence is a temporary transfer of a mere part of such right. Because both types of contract may be subject to extensive contractual qualifications, this distinction in principle may be difficult to identify in practice. Since they hold out the obvious advantage of retaining ultimate control, however, licences are more common than assignments. A contract of licence, therefore, may be defined as the grant of a proprietary power in return for payment of royalties.

Form of Assignments and Licences

The nature and obligations of assignments or licences are best outlined by describing some typical features of such contracts. Amongst other more general terms, intellectual property assignment or licensing contracts commonly include provision for choices of legal system to govern incidence of the intellectual property itself and matters appertaining to its assignment or licence, including dispute procedure, arbitration and litigation. The following clause is typical:

The construction validity and performance of this Agreement shall be governed by the Law of England. All disputes which may arise under out of or in connection with or in relation to this Agreement shall be submitted to the arbitration of the London Court of Arbitration under and in accordance with its Rules.

Definition and scope of the intellectual property to be assigned or licensed is normally part of the so-called 'recital' at the beginning of the agreement, along the following lines:

The Licensor is the registered proprietor of:

1 Letters Patent Number ... of the United Kingdom of Great Britain Northern Ireland and the Isle of Man (hereinafter called 'the British Patent') in respect of an invention (hereinafter called 'the Invention' relating to devices which may be used in connection with ... (as hereinafter defined).

2 Letters Patent corresponding to the British Patent in certain foreign countries short particulars whereof are set out in the first part of the First Schedule hereto (hereinafter called 'the Foreign Patents').

The Licensor is entitled to the benefit of pending applications for:

3 Letters Patent corresponding to the British Patent in certain other foreign countries short particulars whereof are set out in the second part of the First Schedule hereto (hereinafter called 'the Pending Applications').

The recital might then proceed:

The Licensee has requested the Licensor to grant it a licence under the British Patent and the Foreign Patents and under any patents granted in respect of the Pending Applications to carry out all or any one or more of the activities of importing manufacturing having manufactured using and selling throughout the Territory (as hereinafter defined) [products] embodying or making use of the Invention which the Licensor has agreed to do subject to the terms and conditions hereinafter set forth.

Having, as it were, cleared the ground, it is then necessary to state the fact of assignment of licence of the central proprietary rights, together with any ancillary rights; the exclusivity of a licence should be mentioned, as defined by reference to the product technology, its economic level or the market territory. It may also provide for any options on the results of further research and development by the assignor/licensor and perhaps by the assignee/licensee. The following is a simple example:

The Licensor hereby agrees for the consideration hereinafter mentioned to grant to the Licensee personal non-exclusive individual non-transferable Licences under the Patents to carry out all, or any one or more of the activities of [e.g. importing, manufacturing, etc.] in the licensed Territory under the Patents.

Pursuant to such Agreement the Licensor hereby undertakes at the Licensee's request to execute such formal Licence documents as the case may require in the form set forth in the Second Schedule hereto (or as near thereto as circumstances may admit) for the purpose (*inter alia*) of registration at the appropriate patent offices.

Extension or renewal of the licence and its termination may be provided for as follows:

The Licensee shall be entitled by giving not less than three months' notice to the Licensor expiring at the end of the ... Accounting Year to terminate this Agreement. If the Licensee shall exercise this right all Licences granted to it hereunder shall immediately and automatically terminate at the end of the said ... Accounting Year. If the Licensee shall not exercise the said right the minimum royalty of ... reserved by Clause ... hereof in respect of the ... Accounting Year shall continue to be applicable for the next following ... Accounting Years. Thereafter no further minimum royalties shall be payable.

There may be additional rights of termination for specific causes:

The Licensor shall have the right to terminate this Agreement and all Licences granted hereunder forthwith by notice in writing to the Licensee upon the happening of any of the following events:

1   ...

2   ...

[etc.]

Typically, these will include the licensee's default in payment of royalties due, the inaccuracy of statements or records and any events which tend to put the licensee's ability to pay royalties in jeopardy. It may also include matters relating to the security of competition-sensitive material. It is further likely to provide:

Any termination of this Agreement shall be without prejudice to any provisions of this Agreement which are expressed to apply thereafter and to any rights of either party against the other which may have accrued up to the date of such termination. If this Agreement shall be terminated by the Licensor pursuant to Clause ... hereof the Licensor shall also be entitled to recover damages from the Licensee as if this Agreement had been wrongfully terminated by the Licensee.

Other standard provisions deal with an assignee/licensee's recognition of the validity of assigned or licensed rights along the following lines:

The Licensee shall act loyally to the Licensor in all matters and in particular (but without limiting the generality of the foregoing) shall not at any time directly or indirectly during the Agreement term or thereafter dispute or question the validity of any of the Patents or oppose or contest any of the Patent Applications or in any of such cases encourage assist or instigate any other person to do so.

Consequential provisions include the allocation of responsibilities for protection against infringement by other parties and against breach of confidentiality, as well as applications for the legal extension of such rights. Important ancillary matters relating to licensed operations include quality control, technical assistance, know-how, etc.

Provisions for the calculation and variation of royalties, the keeping of accounts and records, methods of rendering payment, etc. will need to be drafted with utmost precision. The following is a fairly typical clause for the calculation of royalties:

In consideration of the Licences agreed to be granted pursuant to Clause ... hereof the Licensee shall pay royalties to the Licensor as follows:

1   On all [products] sold to or in [the licensed Territory] by or on behalf of the

Licensee either directly or indirectly irrespective of the country of manufacture royalties calculated in [a specified currency] at the following rates:

Royalty per [product] in [specified currency]	Number of [products] in each Accounting Year
e.g. 20	e.g. 1–50,000
15	50,001–300,000
10	over 300,000
etc.	etc.

2   On all [products] (a) manufactured in [the licensed Territory] by or for the Licensee and sold outside [the licensed Territory] and (b) sold by or on behalf of the Licensee from within [the licensed Territory] to any country outside [the licensed Territory] either directly or indirectly royalties calculated in [specified currency] at the following rates:

Royalty per [product] in [specified currency]	Number of [products] in each Accounting Year
e.g. 5	e.g. 1–50,000
3.75	50,001–300,000
2.5	over 300,000
etc.	etc.

A clause providing for the keeping of accounts is likely to be built up around some such basic provision, as follows:

The Licensee shall during the Agreement term keep at its principal place of business true and separate accounts and records in sufficient detail to enable the statements required under paragraph 1 of this Clause to be prepared and verified and such records shall be open for inspection by the Licensor through an independent Chartered Accountant or foreign equivalent appointed by the Licensor or through an accounting employee of the Licensor to whom the Licensee has no reasonable objection.

Other important clauses standard to business contracts provide for the definition of contractual terminology, the giving of notices between the parties, *force majeure*, including the insolvency of the parties, choice of law to govern the contract, actions on default and/or arbitration, and possibly estimated damages in the event of its breach.

Conclusions

It is essential for managers to be conscious of the complexities of design law as a major barrier to market penetration and to invest in good quality advice at an early stage in their strategic planning. The greatest danger is 'too little, too late'. Licences and assignments may be a key consideration for companies who wish either to disseminate their own technology or to benefit from the discoveries of others.

Having said that, it must be recognized that the value of legal protection for some product design may be very limited once the security of the initial development stage has passed. Far more important factors in these cases are likely to be the efficiency of integrated production/marketing systems geared to constant market research and feedback into an incremental innovation policy.

40 Sustaining Competitive Product Design

RICHARD HANDSCOMBE
UK

Introduction

Worldwide, individual designers, companies and national design campaigns are finding that customers are becoming more selective in the quality of products and services they are prepared to purchase. Many customers are participating in a shortening and more volatile buying cycle, parallel with – and in some cases driving – the shortening of product life cycles. Customers are willing to search widely for:

Cost effective basic products
Value-for-money, fun or experience products
Products which are part of an enhanced total package
Expensive image or status products with luxury added value.

As a result, they are demonstrating reduced product, brand and supplier loyalty. Designers, specialists and managers associated with the competitiveness of a product group therefore need to pay more attention to the productivity of individual products. Product productivity is concerned with value for money – the relationship between perceived and proven customer benefits and the relationship between the initial purchase price and anticipated lifetime costs. Inevitably, companies with poor product productivity find competition tough. They reach this position for a number of reasons:

1 Entering the market with new products with low productivity benefits to the company or customer, or at worst both.
2 Allowing the customer's value for money to slip by failing to introduce regular improvements in the basic product or product package. In more and more industries the product package includes product design, packaging design, financing facilities, before- and after-sales services, design compatibility, retail design, house style and so on.
3 Allowing the value for money to fall between one series of models and the next, perhaps as a result of infighting between functions such as design, manufacturing, marketing and financing. Also failing to achieve corporate-wide belief in the marketing concept and acceptance of a multi- as opposed to single-discipline approach to corporate design.

Many Japanese companies demonstrated these phenomena in the early 1970s but reversed the situation in the decade 1975–85. The reversal was characterized by speed, stealth, dedication, a thorough focus on the customer, an objective approach to design and the achievement of global competitiveness. However, these characteristics are not fundamentally Japanese but are associated with market leaders in all industries and all countries.

Competition and Design Quality

Good design gives customers more than they often expect: a product or service designed to give good performance or present an image that – in practice – goes beyond the customer's basic needs at a value-for-money price. Such an experience has significant commercial implications:

1 Customers are satisfied.
2 Customers are more likely to come back for a repeat or follow-on order, and to give overt or covert referrals to new prospective customers.
3 Payment for the product or service may be speedier, without reminders and without negotiation over any warranty retention.
4 Marketing and sales efforts may be reduced for the next order.
5 Customers' loyalty thresholds are raised, resulting in an extension of the buying cycle at a time when competitors' product life cycles may be fast reducing.

Inevitably, companies find themselves from time to time in markets where competing products are very similar and allow for little consumer differentiation. When this happens, customers tend to satisfy just basic needs and competing companies find a classic price war on their hands. This war will drive them to short term tactical marketing, suicide pricing, takeovers and, for some, corporate insolvency. In such situations, the designer's brief may be very basic: 'Design simple, no-frills, minimum fitness-for-purpose products with least cost materials, by least cost manufacturing methods or by low cost subcontracting and international sourcing.' At best, the brief may be: 'Get us out of this mass market and into a more attractive market segment.'

For the enterprising company such segments do exist. In many industrial and consumer markets, there are significant and attractive groups of buyers and end users who are interested in products which provide:

Economically attractive lifetime costs
Productivity/competitive potential
Enhanced corporate image
Enhanced personal experience, image and self-confidence.

Hunting out and responding to such customers provide significant design opportunities for companies as diverse as apparel makers, motor manufacturers, art galleries, transportation companies, agricultural genetic plant breeders and property developers. In their own way, the T-shirt, drought resistant maize and the Broadgate development in the City of London are design breakthroughs which have significant economic, social, cultural and design implications for the futures of their respective industries.

However, success requires changes in corporate behaviour for those companies interested in grasping follow-on or spin-off opportunities:

1   A recognition of the importance of design in developing and implementing corporate strategy.
2   The orientation of the total company – not just the design and marketing departments – to what the customer needs or can be stimulated to want.
3   Recognition across all functions in the company that products have 'nine lives' which comprise the product competitive mix listed in table 40.1.
4   A constant drive for improvement throughout product life cycles to achieve premium pricing and productivity improvement through the management of total design quality costs in times of good as well as poor market conditions.

Table 40.1 The product competitive mix: the table is used to analyse the situation in which you work or which you are evaluating

	How do you fare?				
	Badly	Fair	Average	Good	Excellent
1 *Product design is focused on fitness for the purpose* required by the purchaser and/or end user					
2 *Product quality and reliability* are built in from concept design through detailed design, component manufacture and assembly, and can be expected from each incremental design and series change					
3 *Product productivity potential* for (a) the supplier (b) the user organization in improving own competitiveness					
4 *Prompt delivery and reliability* for both the original and follow-on orders					
5 *Product package* in terms of total facilities, environment or experiences delivered integral with the core product					
6 *Product support* in commissioning and gaining confidence to use					
7 *Product servicing and spares*					
8 *Promotional follow-up and reality* in terms of delivering the customer at least what was anticipated					
9 *Price reductions* if 1 to 8 have been unsuccessful in establishing customer confidence and acceptance of a premium price					

Source: © Handscombe, R. S. *Chief Executive Magazine*, 1984

However, maintaining a time, cost and competitive balance of design activity throughout the product life cycle is not easy. Success requires a recognition of common problems and a search for practical solutions. The timing of the different aspects of design improvements is illustrated in table 40.2.

Table 40.2

Phases in product life cycle	Typical focus of design activity	
	Product design	Process design
Conception	Conceptual design	Conceptual review
Development	Detailed design/make process to achieve competitive balance	
Launch	Product troubleshooting	Process troubleshooting
Growth	Incremental product/package improvement	Incremental quality/productivity improvement
Maturity	Improvement in support services	Major productivity/cost drive
Old age	New product launch	Last ditch cost reductions, if viable

Typical Design Management Problems in Sustaining Competitive Product Improvements

The following are ten common problems faced by a wide range of consumer and industrial product companies. Few organizations avoid all the problems listed. Many readers will recognize from four to six of the problems and may have cause for concern.

1   *Conflicting objectives* over the need for, and timing of, product improvements. This situation is often stimulated and kept alive by professional rivalry between design, marketing, sales, production and finance over product priorities, the time for changes in technology, and the product improvement budget.
2   *Failure in interfunctional communication* resulting in the slow circulation of vital feedback on the success or failure of customer applications, new competitive products and so on. This situation is made more critical by the existence of: 'functional fortresses' which compartmentalize rather than integrate the concept, detail and manufacturing design processes; and middle-level designers who reject suggestions, particularly those that involve criticism from the shop floor, service engineers, etc.
3   *Failure to recognize design as a strategic issue* to be taken into consideration in analysing own and competitors' strengths and weaknesses and customer design needs in identifying new opportunities for product improvement.
4   *Vision gaps* between generations of designers. New graduate designers lose native creativity during their osmosis into the established design

department. The chance is lost for newcomers to be critical of existing designs before they themselves claim ownership and resist critique.

5 *Insularity* to outside ideas for product improvement – within the design function, between specialist design units, with other technologists and managers and, most seriously, with the customer.

6 *Poor design information base:* customers make products successful! Commencing and finalizing designs based on a shallow survey of market trends can be risky, especially in industries where the potential for technological leapfrog exists (figure 40.1).

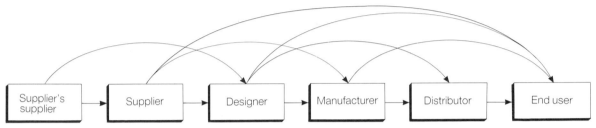

Figure 40.1 Technological leapfrog

7 *Insufficient recognition of the impact of a product on the environment:* in particular, failing to recognize customers' increased awareness of safety issues and the potential minefield of product liability laws.

8 *Passing the buck* in terms of accountability for resolving short term product problems and the initiation of effective medium term product improvement.

9 *Failure to learn* from the last design to the benefit of design improvement, new designs and new designers.

10 *Training of designers restricted* to specialist technical topics such as CAD or new materials, with too little attention paid to corporate topics such as strategic management, marketing, teamwork and project management.

Such problems require practical solutions.

Practical Approaches to Resolving the Problems

The approaches to resolving the above problems have been selected on the basis of the following criteria (see table 40.3):

Practicality
International relevance
Wide industrial application
Ease of implementation
Involvement of design in the strategic direction of the business.

Table 40.3

Practical approaches to problems	Impact of application on problems (numbers as in text)									
	1	2	3	4	5	6	7	8	9	10
Product group strategy session	●	●	●	●	●	●	●	●	●	●
Product technology workshops	●	●	●	●	●	●	●	●		●
Product project management	●	●	●	●	●	●	●	●		●
Combined sales/design teams	●	●	●		●		●		●	●
Product idea groups	●	●	●			●			●	●
Product problem prevention programmes	●	●	●		●			●	●	●
Product quality cost campaign	●	●	●		●			●	●	●
Product package audits	●	●	●		●			●	●	●
Product focused management development	●	●	●		●					●
Product management	●	●	●	●	●	●	●	●	●	●

Product group strategy session

Organized annually as an important feature of the corporate planning programme. The strategy should include consideration of the following questions:

1 What strategic opportunities could be opened up by more attention to the quality of design?
2 What opportunities exist for product improvement as well as new products?
3 What should be tomorrow's design strategy for the product group?
4 What design capability will be required to support the future strategy for the product group?
5 How will the capability be resourced with an open mind to subcontract design houses and joint ventures as well as increased in-house resources?
6 What will be the most productive link between corporate strategy and design strategy?

The participants required to debate such questions in an objective manner will include staff from head office, marketing, research and development, design, manufacturing, purchasing and sales.

Product technology workshops

Sponsored by the company to achieve a deeper insight into:

1 Potential changes in the pace and extent of technological developments impacting design.
2 The application of technologies by customers and competitors.
3 Potential product improvements.

Typically, the invited participants will be designers, researchers, market researchers and technical specialists from a cross-section of non-competing organizations, for example suppliers, research establishments, customers and other manufacturers. Each participating organization must be prepared to contribute to an open, cooperative sharing and challenging of facts and forecasts. The payback will be an enhanced vision of the state of the art and enriched ideas for product improvement.

Product project management   Using multidisciplined project teams to manage the product development programme from the initial concept stage to market launch. The project team should focus on achieving an integrated design/make process as outlined in figure 40.2 and should consider the following aspects:

Physical design including structure and styling
Cost effectiveness
Makeability
Reliability
Maintainability.

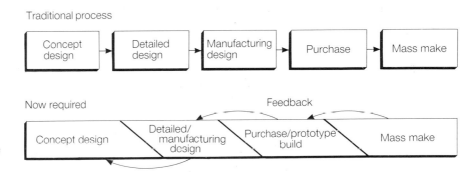

Figure 40.2 The design/make process

Combined sales/design teams   Used to tackle major contracts, particularly where:

1   The customer is technically aware.
2   The competition is intense.
3   The potential customer uses a multidisciplined buying team as in figure 40.3.

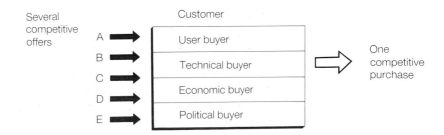

Figure 40.3 Multidisciplined competitive buying

Typically, the sales team will comprise an account executive or territorial sales manager, together with a senior product designer. The objective is to achieve a competitive edge in both tender and non-tender situations. The direct benefits are:

1   A clearer definition of customer needs.
2   A techno-commercial view of the business opportunity.
3   Improved design briefs.
4   An enhanced view of which product improvements would be attractive to the customer now and in one, two and three years' time.

Product idea groups   Taking advantage of product ideas from all levels in the organization, particularly those that make and service the product. In a typical situation, a multitrade group of twelve to sixteen blue collar employees plus four designers may come together for a day to debate openly the following brief: 'If you were the chief designer for a day, what would you change in the design of product X and the manufacturing process for product X?' A hundred or so ideas may be generated, to be followed up by appropriate subgroups each led by a designer. The cost benefits and degree of commitment obtained are often far in excess of those achieved by traditional quality circle approaches.

Product problem
prevention programmes   The establishment of a design management programme that requires a formal evaluation of design risks in the concept and manufacturing design phases, the objectives being to:

1   Ensure that problems experienced on previous designs are not repeated.
2   Reduce the need for expensive and time-consuming troubleshooting efforts to tackle production, commissioning and servicing problems.
3   Gain a reputation for first time right design.
4   Agree product improvements that will be introduced to the customer on a planned basis.

Product quality cost
campaign   A joint design/marketing-led identification and attack on the total corporate costs of:

Designing competitive products
Over- and under-design
Quality planning and control
Correction of poor quality.

Total quality costs in an industry can range from 6 to 30 per cent between the most and least competitive companies. What more challenge does a competitive design team require?

Product package audits   Carried out by a multidisciplined team with complementary skills to apply

Figure 40.4 Product package audit

value engineering concepts to the design of the total product package (figure 40.4). Typical audit questions will include:

1 How can the design of the basic product be made less complex to improve appearance, costs and reliability?
2 How can the before-sales package be best constructed and presented to the potential customer?
3 How can the after-sales package be best constructed and presented as a cost effective feature to the potential customer?
4 Which features of the total product package will offer greatest incremental value for money in the eyes of the customer?
5 Which aspects offer the greatest potential for low cost, high impact improvement?

Product focused management development Organized on an in-company basis for the team directly involved with the development, manufacture, marketing, sales, service and financing of a particular product or product group. The programme will focus on strategic management, marketing, team building, design management and project management, perhaps organized around a specific deal or a fictitious product. The objective is to achieve:

A dedicated, multidisciplined team
Interfunctional awareness and trust
Increased design literacy

Common customer and competition awareness
Cooperative practical action plans to follow up the development programme
A greater chance of product success.

Product management   Dedicated management is required for a specific product or product group, integrating product development, product marketing, and product sales and servicing on a regional or global basis. Product managers can achieve four significant benefits:

1   A more professional approach to gathering market intelligence about customers' needs and competitors' strategies.
2   Close integration between product design and development and market development efforts.
3   Strong links with customers through the international sales force, including feedback on own and competitive designs.
4   More timely product innovation and incremental improvements.

Initially product management was developed by consumer goods companies, but manufacturers of industrial equipment and related service companies are increasingly turning to this powerful form of organization structure.

Conclusion

Customer benefits and the productivity with which products are conceived, designed, manufactured, launched, marketed, sold and serviced in the marketplace are at the heart of competitive success. This success is achieved through a corporate strategy of continuous investment in the ongoing development of existing products and businesses rather than by acquisition. Designers have much to offer to this success provided they recognize their business and management roles. In the end, the only important perception of the design quality of a product improvement is that in the eyes of the customer.

PART VII

TOWARDS THE FUTURE

41 The Japanese Corporate Approach

BILL EVANS
Lunar Design, USA

Introduction

Most of us know of Japan's success through its products and massive trade surplus with the rest of the world. In accounting for the role of design in Japan's corporate success, it is also necessary to examine technical, cultural, economic and historical influences so that the Japanese approach to design management is seen in its proper perspective. Many factors are involved and it is Japan's thorough approach to some apparently peripheral factors, as well as its respect for design and designers, that often leads it to be so successful. What is it that enables Japan to manage design to rapidly meet changes in the market so cheaply, effectively and with such technological sophistication?

Even the most cursory viewer of current affairs will be aware of some of the contributory factors to Japan's success. The 'art of Japanese management' is often mentioned and has become something of a management fad; quality circles and the principles of lifetime employment are frequently cited as good examples. In fact only 35–40 per cent of Japanese employees enjoy the benefits of lifetime employment, the rest of the economy taking up the slack. And we also hear of the less acceptable side of Japanese success. Western commentators talk of workaholics and cramped social conditions. Japan is a high pressure society and much of its social welfare for the elderly and those not lucky enough to be in the lifetime employment system leaves a lot to be desired.

But to reduce discussion to a list of transferable factors and social problems is to misunderstand a complex social, cultural and industrial inter-relationship. We should not be so much interested in copying Japan but instead, by understanding its success, be searching for something more appropriate to our culture and expertise. The key lies in Japanese products themselves and the level of investment and confidence which supports them.

Economic Background

The economic build-up in Japan has happened primarily over the last 100 years but much of its present high technology success is based on plant less than seven years old. This investment, by all the major companies known as the *Zaibatsu* sector, is financed first by the banks that each group has under its wing and secondly by the very high level of personal savings. This high personal saving was initially at a cost to domestic consumption but it is now encouraged by tax incentives and a need to make personal provision for retirement. High investment is encouraged by a different attitude to borrowing by the companies, who enjoy low cost money from their links with the banks and hence look differently at long term investment. The Japanese stock market also tolerates much higher gearing ratios (the ratio of borrowing to capital) than most Western counterparts.

Design for world markets

These *Zaibatsu* companies – which rank among the largest in the world and are overtaking companies such as General Motors, Ford and IBM – have systematically worked their way through market sectors, first establishing products in their large domestic market (Japan has a population of 120 million) and then ruthlessly pursuing a proportion of the world market. They are often initially more concerned with market share than short term profitability. The oil shocks of the 1970s and the strengthening of the yen from Y260 per dollar to Y125 per dollar between 1985 and 1988 have brought home to the Japanese their delicate trading position, with its heavy reliance on imported raw materials and energy.

Japan has responded by shifting towards knowledge-intensive industries like electronics, telecommunications and robotics. This has also coincided with a period where its own R&D efforts have been bearing fruit and the country is now less dependent on imported technology. Investment in education, particularly technical education, has also helped; compared with the UK there are four times as many engineers per caput with an extra 70,000 graduating each year.

The *Zaibatsu* companies have always enjoyed a close relationship with their suppliers who do not provide lifetime employment schemes. It is these smaller companies which bear the brunt of the fluctuations necessary to cope with the inflexibility caused by the operation of the lifetime employment enjoyed only by an industrial elite. The industrial relations picture is very different from the West, with workers encouraged by their 'enterprise union' (single company unions rather than single craft unions) to identify strongly with the aims of the company. Bonus payments, based on corporate success rather than individual effort, are often 30–40 per cent of basic rates. A union's opposition to a company is usually only over the distribution of profits; both stand on common ground when it comes to the prosperity of the company upon which these profits are based.

Working life

Working life starts at recruitment when personnel begin to be moulded into the company way. The business community sees education as an almost subversive influence because it fosters individuality over the sense of group identity, commitment, obedience and discipline that the employment system demands. This means that recruits take an entrance examination which places greater emphasis on personal qualities than on technical ability. But there is a conflict between fostering a collective, almost subservient, attitude to work and the linear, logical analysis and dynamic entrepreneurial spirit which companies must display to survive in world markets. Nevertheless, the workers' willingness to subjugate themselves to the goals of the company is the first thing that is visible on any factory tour: the uniforms, innumerable notices encouraging less waste, zero defects, extra safety and general fillips have an almost 'McDonalds' quality. The pace of work leaves one breathless; workers literally run around their work positions with methodical but almost dangerous haste.

Japanese Corporate Design Strategy

Design is central to the corporate policy of the *Zaibatsu*; more specifically, design is part of product planning. Design is a service to management who see products essentially being pulled by the market. There are exceptions to this, where a strong design concept forges its own market – such as the Sony Walkman and Watchman. On the whole, a very careful and systematic product planning exercise is done on every new product, to the point where it becomes easy to see why many products look so similar. Market pull also has a slightly different definition in Japan, where the strength of many companies' marketing positions leaves a convenient blur between whether they pull the market around themselves or whether it pulls them. There is much talk of 'needs and seeds' but often customers have a need satisfied that they did not know even existed. There is a subtle distinction between giving people what they want and giving them what they get.

Many of the companies have slogans which emphasize their design approach. Sony says 'research makes the difference' whilst Nissan exclaims 'customer satisfaction above all else'. In the past Japan was renowned for exploiting foreign technology to the full, but it has now concentrated on controlling and developing all the fundamental building blocks of its industries. One year's leading edge research pilot plant is next year's essential production plant. Technological know-how and production expertise are acquired as essential to maintaining the competitive edge. The *Zaibatsu*, having mastered their own very large scale integrated (VLSI) semiconductor plants, move on to major new technical areas such as lasers and display technologies for new generation information technology (IT) where low cost compact colour displays will give marketing advantage. This attitude is helped considerably by a very pragmatic government approach to public spending on R&D. The government sees nothing contradictory about a vigorous private sector being made more vigorous by strategic public investment in new technologies. This is also true of basic research into materials, energy, aerospace and biotechnology.

To some extent a company's design strategy will depend on the product cycle time for both development and life. Where cycle times are short, for instance in consumer electronics, there is more emphasis on beating the competition to the marketplace and then staying one jump ahead. It is acknowledged by design managers that unscrupulous manufacturers can copy and bring to the market quite complex electronics products within six months. Innovation and speed are the key. If product cycles are longer, as with office automation (photocopiers take up to four years to develop), heavy patent protection and investment in process technology take over as the main strategy.

The human-ware age   Without exception, all the major companies are developing a sophisticated analysis of the future which is becoming increasingly user-conscious. They are looking to a period where advances in electronics are consolidating rather than rapidly advancing. As Ricoh puts it: 'We have moved from the hardware age (1965–1975), through the software age (1975–1985) towards the human-ware age.' By this they mean that users' requirements will take over as the major dictator of product capability. 'We have the technology' – so companies are now concentrating on allowing designers to input consumer lifestyle into the products, making the technology more intelligent and more flexible for users from different cultural backgrounds, and generally considering the social context. This is rather more than just rhetoric from an image-conscious Japan. Many companies say: 'You have to know people first before making any product.' Features are no longer seen as being good enough to differentiate products in the marketplace; image is taking over.

The trends in production technology also mate with the human-ware approach (figure 41.1). Flexible manufacturing systems (FMSs) are now common; these enable manufacturers to better serve niche markets. As systems become more truly flexible, smaller volumes and local market specials become viable.

Group technology   The availability of cheaper robotics for FMSs has enabled the rapid development of another major influence on design: group technology. Japanese electronic products have now become very systems and network oriented. Group technology as an approach to design and production is an extension of this. It is based around the idea that despite the many variations of items made by a company, many similarities exist. By concentrating on the similarities, the disadvantages of variety can be reduced. Products or components are divided into a limited number of groups which share qualities and are considered technologically similar. So at Sharp, for example, the divisions are: TV and video; audio; domestic appliances; solar systems; industrial instruments (mostly business machines); and finally electronic components. As much as possible of the plant producing any of these groups then becomes common or very similar.

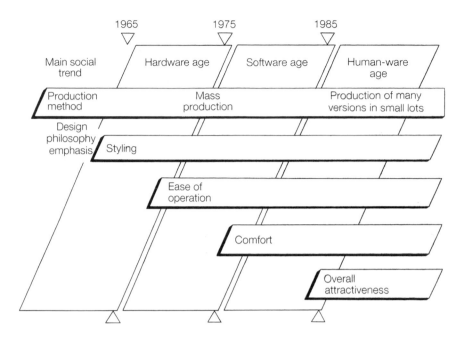

Figure 41.1 Evolution of design philosophy at Ricoh

FMSs have, of course, made this much easier with production lines becoming programmable rather than dedicated. Sharp now heralds these ideas as 'multiproduct mixed production systems' which, for instance, means a single line for producing both washing machines and refrigerators. Brother, whose business was founded on industrial sewing machines, has developed an FMS line for custom production. Consisting of 13 computer controlled machining centres and requiring only three operators, it can respond to an order of one machine whereas 150 was the previous minimum. Set-up time has been reduced from three hours to less than a minute.

Although some Western firms adopt a similar group technology approach, it is more vigorously pursued inside the *Zaibatsu* companies. This is because it is essentially a systems approach to designing and manufacturing which ideally suits the Japanese element-by-element analytical method. Such standardization has its design implications: traditional Western views of creativity and the creative atmosphere do not sit comfortably next to such a thorough, consultative approach. Implementation through to production is more rapid, with all departments sharing, understanding and working towards the group goal. The individual star designer has little place in such a framework.

Design Departments and the Product Planning Process

Product planning (PP) is really the heart of the modern Japanese company. Design and engineering may or may not be in this section. Product planning may be a department on its own or a function of corporate management (see figure 41.2).

Generally, the design department will be represented on the board but not necessarily by a designer. Sharp is believed to be the only large company where an industrial designer sits on the board; Ricoh has a graphic designer on the board, and Sony's head of PP reports directly to the chairman. Design staff tend to be well qualified and there is a high proportion of women (up to 30 per cent of industrial designers are women compared with less than 5 per cent in other design disciplines). Promotion through the company hierarchy is based on a combination of annual merit tests and the operation of the seniority system which reflects the length of service – although this approach is being undermined by an erosion of traditional values as Japan moves towards a meritocracy.

The design process   Basically, as figure 41.3 shows, Japanese companies adopt a very systematic linear flow of work from the initial concept, market research and planning to various stages of sketches, presentation models and pre-production prototypes. Design departments work closely with the engineering divisions to productionize designs. At all times, the latest state-of-the-art CAD systems facilitate the flow of work. Japanese companies are in the forefront of solids modelling and shading for use in industrial design visualizations. And one cannot overstate the importance of the design–marketing link; with near-constant formal meetings to assess progress, many iterative loops and final exhaustive testing of the pre-production units, not much is left to chance.

Sharp differs from Ricoh in that its design centre has all the disciplines in one division, making relations between engineering and design more fluid. Sony is a little looser in its definition of PP; it can be product planning, product proposal, product presentation or product promotion. However, these companies only really differ in where the ideas come from in the first place. Sony talks of industrial designers presenting 'engineers with themes for technological development'; at Sharp one suspects that the presence of a designer on the board gives added creditiblity to this department's innovations. In reality, ideas come from all departments, with technological breakthroughs sparking ideas from research departments, and designers perceiving the cultural importance of new fashion. Profit and/or turnover is still the objective. Marketing by whatever name is still pulling technology. This helps to explain why the Japanese are not renowned for the more risk-oriented seat-of-the-pants product development.

The rugby approach   Under the pressure of the relentless search for the competitive edge of being first into the marketplace, Japanese companies are constantly evolving new

Figure 41.2 Functional chart of the corporate design centre at Sharp

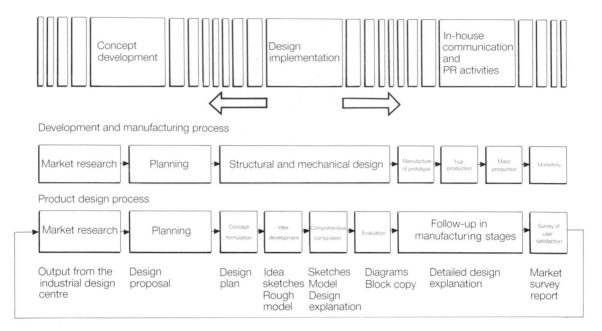

Development and manufacturing process

Product design process

| Output from the industrial design centre | Design proposal | Design plan | Idea sketches Rough model | Sketches Model Design explanation | Diagrams Block copy | Detailed design explanation | Market survey report |

Figure 41.3 Design process and work flow of industrial design centre at Ricoh

design management techniques to shorten the time from concept to market. Examples range from the linear stage-by-stage approach to the more maverick Sony PP centre with its integrated 'amoeba approach' simultaneously

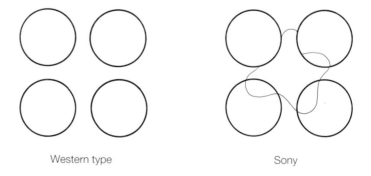

Figure 41.4 Amoeba-like job assignment at Sony

working on and passing between project stages (figure 41.4). Japanese researchers (Takeuchi and Nanaku 1986) are now referring to this more integrated and rapid method of new product development as the 'rugby approach'. The rugby analogy of the game where the ball is advanced by passing within the team, often backwards as a sacrifice to achieve the team goal, best describes the idea.

The approach is being adapted to further reduce development cycle times in companies such as FujiXerox, Canon, Honda and NEC. In essence the rugby approach consists of hand picked multidisciplinary teams who are set broad goals by their management and given full hands-off support. These teams may be quite radical in structure, perhaps breaking down traditional prejudices and specializations within the company. Generally, teams are encouraged to go through stages, gathering customer and marketing information by getting out into the field. Electronic engineers may have to concern themselves with marketing or production considerations and vice versa. The shared goals of the group are the most important.

The conventional development process can be considered as a relay race where each stage of the project is finished in turn and the baton is then passed on. The rugby method encourages an integrated style with problems being worked on simultaneously and passed around before being fully finished. For instance, a group of engineers may start to work on the product engineering aspects of the project before the results of the feasibility study are concluded. Decisions may be made that in the light of parallel work need to be reconsidered – but the overall effect is to move the project forward at greater speed.

Different companies exploit this technique with different amounts of interleaving between phases of the project. At FujiXerox, the *sashimi* approach – named after the Japanese culinary presentation of slices of raw fish overlapping each other – limits itself to adjacent phases of a project. Other companies adopt a more adventurous approach and allow several phases to overlap simultaneously – further shortening development times. This rather more daring approach to project management requires special skills from all concerned to enable it to flourish. The key is adaptability of project management techniques, leading to a product better adapted to its market. Design managers in different industries and markets have to adapt to the system that suits their goals and not blindly adopt a crude fast-track approach.

Design Case Studies

Sony The Sony Walkman is a famous design and marketing success. Since its initial launch in the early 1980s, it has undergone revolutionary design and marketing changes which are nearly as much a quantum leap as the initial product. Sony dominates the $1.3 billion world market for personal tape players with a 30 per cent share and still makes all its Walkman products in Japan. It decided to meet low cost competition head-on by designing and tooling up for a low price unit, retailing at $32 in the lucrative US market where about half its production is sold. The decision to compete at the low end forced Sony both to automate production and to integrate design. Production costs were reduced by 30 per cent through combining the playing mechanism with the printed circuit board.

Hence the design process had to shift its emphasis as the product matured in its market. Initially, it was sufficient for the designers to concentrate on creating an image for the product consistent with the company look. Extremes of size and cost reduction were not then factors but, as these became more of an issue, they had to work even more closely with production and marketing colleagues (figure 41.5) to balance the conflicting demands of low cost, further miniaturization, highly integrated parts and a quality image for market appeal. Sony was prepared to bring vast investment of resources to bear to maintain market share and now offers a full range of Walkman products from the $32 basic model to a $450 tape deck quality version.

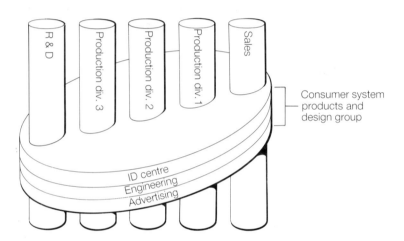

Figure 41.5 Relationship of industrial design centre to corporate structure at Sony

The Diskman, Sony's portable CD player, is also undergoing a similar design and marketing exercise but in a less mature marketplace. Morita, head of Sony, decided to peg the launch price at $200 even though the product initially cost more to make. He felt, rightly as it transpired, that this price would trigger the economies of scale that would finally make it profitable.

Sharp   Sharp has built an electronic house of the future with fully computer integrated appliances, heating, communications and entertainment systems. Although it looks rather like an updated 1950s home, there is a very serious side to the ultimate commercial exploitation of this inevitable level of domestic automation. Any company which has actually researched and built prototypes will be better placed to gain initial all-important market share.

As with some of their Western multinational counterparts, design studies of the future have become part of the design process within the *Zaibatsu*. Designers can apparently make real as yet technically impossible (or currently too costly) concepts to assist less product-aware corporate executives in their decision making. Increasingly the Japanese are proactive in marketplaces rather than reactive.

Minolta Minolta has brought together its expertise in optics, cameras, lasers and electronics to create a unified approach to document processing, storage and retrieval for the growing office automation market. Called the Minolta Integrated Information and Image Management System, it consists of a digital scanner for reading both conventional and microfilm originals into the system, an optical disc drive and computer for storage and control, and a variety of laser printers and microfilm creating output. This shows how large and diverse companies such as Minolta are able to group their technologies to create new markets or consolidate existing ones. Here the production and marketing expertise is the dominant motivator of new designs. These are permuted from the basic building blocks of group technologies to enhance market position.

The Human-ware Influence on Appearance

The initial flood of Japanese consumer products tended to mimic Western style, but with economic superiority has come a new design confidence. The Japanese have always had confidence in their technology and quality but now, in certain areas, they are actually path-finding with innovative visual design. This is not for its own sake but to differentiate their models in the vast spectrum of available products. They are now taking features out of their products to give a clean flowing look. The trend, particularly at Sharp and Sony, is away from the black and chrome look towards a simple, bright, form oriented design. Some designs echo a subtle 1950s curvaceous look; other fashion-oriented hi-fi reflects current post-modern flippant classical references, pioneered in Italy. And these trends are not confined to sound equipment: Sharp's new domestic appliance and home heating and cooling equipment has abandoned the American dominant teak and chrome look for a more simple European flavour.

This human-ware influence is important for two reasons. First, it marks the start of the final demise of the predominantly features-oriented approach to Japanese product design. Secondly, it heralds the marketing and packaging of lifestyle as the product; you buy the image first and the product second. Skilled in this crucial design–marketing management link, design departments are able to buy in design from around the globe and ensure that it enters production as both the designers and marketeers intended. This is already evident in Japan where a whole range of products from pink food mixers to cute blue cube TVs are sold in this fast moving image-conscious market. The human-ware approach has allowed designs to be greatly influenced by designers' perceptions of culture, subculture and lifestyle. It is fuelled by the Japanese consumer's insatiable appetite for the new and the different.

Conclusions

Japan, once famous for copying products and using innovative process and production techniques to drive costs down, is now a leader of design in many mass markets. Its attention to detail, sophisticated labour management and investment policies all show the importance of taking the longer term view. Design and design management are taken seriously in Japan, often with designers on company boards. The Japanese are not afraid to take a broad interdisciplinary view of design and creative teams. Well-developed links, both formal and informal, between marketing and design ensure the delivery of innovative products that customers actually want. Design is seen as a powerful visualizing and enabling tool to boost confidence in product ideas which will require massive corporate investment.

Many foreign observers visit Japan and look for transferable success factors to take home and implement. It is very difficult to install the necessary shift in attitudes and outlook in an organization surrounded by an inappropriate cultural sea. If looking for lessons from Japan, design managers must ask themselves such questions as: where are the subcontractors willing to adopt and cooperate with these methods? Does my education system foster the values I wish for and are they in keeping with my society's wider goals? Is the reality of my corporate structure compatible with rethinking a more consultative approach? It is precisely Japan's willingness to analyse itself and develop methodologies appropriate to this background that is its strength.

What is impressive from a designer's point of view is the emphasis put on investigating and meeting users' needs whilst pushing against constraints and ever increasing volumes of manufacture, coupled with an investment atmosphere which is prepared to listen to and respect the designer's view. More subtly, design sometimes takes on the role of *agent provocateur* to stir a reluctant product planning or marketing department into a new unfathomed area. *Soozoo* (creativity) is the new battle cry, as companies strive to gain or retain their share of the global market.

One has to look beyond structural differences in the Japanese economy to explain its economic dynamism. It is not possible to separate design management success from the culture in which it was born. Japanese management does have a different attitude and more respect for its workers but it is still part of a society with clear social divisions. To the Western observer, Japanese company paternalism looks pretty dominating. The lifetime employment and seniority system is under threat from a combination of new technology, an ageing work population and changing attitudes amongst the young about work. Referred to as *America-byo* (the American disease) it represents what has become known in Japan as flexible individualism. This is not the rugged American variety but a desire for self-expression in work, lifestyles and possession. Our social and economic histories put us at very different stages in our development.

Japanese government–industry links are also fundamental to success, enabling rapid progress in highly capital-intensive research. But perhaps it is Japan's reaction to the substantial appreciation of the yen that indicates strengths that will ensure its place as a major economic power well into the next century. Japan's response has been to stress added value, emphasizing quality and design over price and ploughing more finance back into R&D. If need be, companies return to the drawing board to further refine and integrate designs, bringing them more rapidly and cost effectively to markets so carefully identified.

References and Further Reading

Armstrong, L., Power, C. and Wallace, G.: Sony's challenge. *Businessweek USA*. 1 June 1987.

George, M. and Levie, H.: *Japanese Competition and the British Workplace*. London: CAITS, 1984.

NEDO: *Transferable Factors in Japan's Economic Success*. London: National Economic Development Office, 1982.

Smith, H.: *Inside Japan*. London: BBC, 1981.

Takeuchi, H. and Nanaku, I.: The new new product development game. *Harvard Business Review*. January–February 1986.

Wilkinson, E.: *Japan versus Europe*. London: Pelican, 1983.

42 The Future: Design and New Technology

TONY MEDLAND

Brunel University, UK

Introduction

The rapid growth of the CAD/CAM industry, and the continuous introduction of new techniques and processes, make it very difficult to predict the future with any degree of certainty. The size and power of a current personal computer system is far beyond that which was imagined for a complete central processor unit some 20 years ago when the industry was in its infancy. All the indications still point to these trends continuing for at least the next decade, with new hardware and communications systems completely changing the way such facilities are employed and managed within the design process. In an attempt to lay out the course of these developments it is first necessary to review the purpose and the growth of the design activity prior to the introduction of these computer-based techniques.

The Design Process

The process of design arises from the need to produce a real object that conforms to a set of functional requirements of an idea. The process thus passes through a number of phases: the specification of need; an idea that may achieve that need (the creative activity); a process of detailing to produce a physical interpretation of that idea; and finally the generation of the object by the application of skills and manufacturing processes. All design activities conform more or less to this pattern, be they the potter making a vase on a wheel through to the vast organization necessary to produce a modern fighter aircraft.

Such processes can be traced back to recognizable origins in the artisan workshop. Here the process of making fine arts was controlled and 'production engineered' for the first time. Processes needed to be formalized and communicated as exact descriptions. The master craftsman or artist distributed, instructed and oversaw all operations, contributing specialist skills where and when required. No longer was the idea and the execution performed by a single person.

The solitary craftsman can design, make and modify as a single integrated activity, letting the activity of moulding the clay to some extent dictate the form of the final design, whilst still keeping the purpose or intended form in mind. But once the activities of creating and implementing are separated, there arises a need to communicate. This must take the form of a two-way conversation, with needs and advice being communicated whilst the results or consequences of the chosen approach are reflected back. The form of the design process thus increases in complexity as the number of individuals increases and also with the level of technical communications that must pass between them.

In a small organization, the individuals are usually more aware of the overall aspects of the product that they are developing. This, together with the smaller number of people involved, allows control and communications to proceed in an informal manner with all members naturally attaining a similar level of understanding of the product and its development. Once the group size increases beyond the point at which individuals are uniquely responsible for a particular activity or skill, communications and control will break down if left on an informal basis.

Design Organization

The natural approach has, until now, been to group people together by skill or activities to perform as departments. These departments thus take on the identity of super-beings with the responsibilities previously resident in individuals. Large corporations thus perform their design processes by formal communications and controls being identified and executed between these departmental groups. Whilst this successfully identifies responsibilities and ensures that the correct activities are carried out, it does result in a system that is cumbersome, within which major changes are not easily implemented and the time to execute the complete process may be much extended. Many such highly regulated systems have been known to continue to produce and scrap items long after the original decision to abandon has been taken. Changes in new technologies are often resisted due to the difficulties of changing the responsibilities and skills of these large departmental groups.

This formal departmental structure fails to recognize two other major factors that have contributed to the success of the design plan to date. These are that the groupings were formed with the current 'local' nature of technology in mind, and that the organization also operates a second 'underground' level of communications.

This second level of communications is perhaps the easier of the two to recognize. Within any organization personal relationships are formed through daily contact within and, more importantly in this instance, between members of departments; these are reinforced by social activities and external contact (such as living in the same locality). Problems and information are thus exchanged via the grapevine so that everyone is kept informed of events before they are officially communicated between departments. Rumours and actual details of a test rig failure will be passed back to the design team long before the official report; this provides valuable time in which to prepare a case!

The other unrecognized factor controlling the departmental groups has been that, until now, all the technological developments have resulted in a local grouping of people and equipment in order to apply them effectively. The drawing boards need to be grouped in a single building to provide not only the clean conditions necessary but also the control and communications. The drawing office staff need access to information on mating components, company standards, bought-out items, etc. All of these are supplied as hard copy which has to be stored and maintained as a central facility available to all staff. What is being produced upon the boards must also be regulated and checked. The nature of drawing accurately upon a board makes it impractical to remove it for copying at frequent intervals. The checker must thus approach the drawing whilst it is still under construction on the board (in order to intercept problems as they occur) or wait until the drawing has been completed and attempt to resolve all problems at the end (usually by reworking the design). All such activities have contributed to the establishment of large departmental drawing offices in which the boards are laid out in rows 'as far as the eye can see'.

Similar groups centred on common need and localized technology exist throughout the organization. Whilst these often started by being grouped around manual records or card index systems, they are now considered wrongly to be simply groupings based on a particular computing facility. The department concentrating on stress calculations existed long before finite element techniques (in fact many of these departments, particularly in the aircraft industry, were instrumental in the original development and application of the techniques). It was also dependent on the availability of suitable equipment upon which the mathematical solutions could be performed. This being a computer (initially mechanical and later electronic), it was expensive and was only viable if made available as a central resource.

Similarly planning and control was, and in some cases is still, executed by card index systems. Here, in order to estimate delivery dates and shop floor loading, it is necessary to group the appropriate job cards in racks or ledgers that are maintained at a single location. The system breaks down as soon as any person establishes a personal, independent scheduling system.

Impact of New Technology

Such groupings based on technology and the control of the design process have until now dominated the structure of the organization. Current developments in technology will contribute to the dismembering of the departmental structure and its gradual replacement by product dependent groupings running vertically through the company. This will arise as the result of two distinct effects: an increase in the communication of relevant information and a declining workforce. This will return us to a situation like that of a small company where everybody is kept aware of the company-wide progress of the product, and not by informal communications but by an intelligent exchange of information.

The current state of technological development already makes it possible for departmental groupings to be broken down. On a CAD system, it is no longer necessary for the designer to be near other designers in order to share resources. Work stations can often be justified on an individual basis as was a drawing board previously; soon it will be unthinkable to share a work station. The designer/detailer does not need to be close to records of company standards and procedures; all can now be made available directly upon the system. Electronic copies can be taken at any time and communicated rapidly over large distances for checking and technical advice. The main reasons for grouping the design staff into a single building have now all but disappeared. Similarly, analysis programs can now be run on advanced work stations. These techniques have also advanced to a stage at which much of the expertise to run such programs can be built in. A single work station can either provide locally all the techniques and procedures that are necessary for the successful completion of a design exercise, or be arranged to access such activities over a network (to provide an illusion that the process is occurring locally).

The only major development still awaited is for an intelligent communications system. Such communications, however, depend not on any technological developments but more on an understanding of the decision making processes involved in design and their encoding into a language structure. Communications between CAD systems in a network are currently based upon the principles of requesting, sending and receiving files. This may be initiated to acquire a program in order to process some data locally, or to send the data in file form to be processed remotely.

Most activities in design, however, start as a vague definition of a need that has to be satisfied. Much of the creative part of the design process is centred on choosing arrangements that will satisfy an array of conflicting needs. Rarely will all the details of a design be uniquely defined. Often there are many ways to design a product. Its size, form, style and features are all arrived at by agreement based upon what is technically possible, what can be manufactured and what the market wants for a price. All of these constraints must be satisfied if a successful product is to be produced. Failure in any one means that there is no product to sell. The successful designer is one who is able to ensure that all the conditions are ultimately met and everyone is happy.

Thus it is not sufficient to generate and record a geometric description of a component without understanding the technical, production and commercial decisions that have been taken during its development. This group of constraints will thus limit the range of possible sizes or shapes for a particular feature (to ensure that it works, etc.). The drawing, however, will limit this to a single preferred value (with tolerances) in order that the complete design is uniquely defined and the requirements of inter-changeability and so on can be met. The engineering drawing only records the preferred values, leaving the designer to infer the constraints by deduction and experience.

In order that changes can be successfully made at a later date, it is necessary to store and communicate not only these selected geometric values but also the constraints upon which they are based. The main information traffic then becomes ideas and relationships, which are the true coinage of design, rather than highly detailed objects.

Such an approach brings with it a whole array of design management problems. If the chosen values fortuitously satisfy a new set of rules, then no changes are required (and it can be argued that all the rules need not have been sent to confuse the new designer). If on the other hand they create a non-truth or violation of some rules, changes have to be made. Further, if some rules or constraints are opposed so that it is impossible to satisfy all rules simultaneously, then a management decision has to be made. For example, if the rules for stress concentrations and avoiding an object compete, then can a higher stress be accepted (at the expense of a reduced component life) or must the clashing object be redesigned? The grouping of rules and their order of dominance in the actual design situation is as much a matter of management strategy as technical judgement. A new model for managing design is thus sought.

For an appropriate structure, we must return to the small company situation or even go back as far as the artisan workshop. Here the decisions were all made with the end objective and the skill of the workforce in mind. The workers were involved, their views were taken into account, but the overview was perceived by the master craftsman or artist who took the ultimate decisions.

Management Network

There is thus a need for a management network that links all computer-based work stations. It is through this that not only will information be exchanged, but the form and the level of that communication will be regulated.

There is little point in exchanging all the rules of the construction of an item if the recipient is not allowed to change any detail of the item. If the component is now in production and cannot be changed, then that is what the recipient should be told. If some changes can be made (such as only requiring extra material to be removed) then the management decision may be to include those in the communicated design information.

In order to meet deadlines it may be necessary for the management network to freeze or extract information from an incomplete design to allow preparatory work to be undertaken on the manufacturing side (such as ordering material, roughing out, or planning of machine loads). The management facility may also circulate pre-release information on a scheme to solicit views on likely problem areas in terms of its strength, cost, manufacturability or style. Such advice (or recommendations) would then be fed back to the designer.

The management network must thus be able to work on differing levels of interpretation of the product at differing times depending on the demands of the problem (and in order to achieve an answer in a reasonable time without getting bogged down in trivia). Information and rules must thus be associated through a hierarchical structure with an ability to perform relational searches to identify activities and designs that are similar. The proposed system is thus one based on problem bounding at the highest acceptable level within the structure. Upon the detection of an error or inconsistency, the network searches through progressively lower levels until the details and rules governing that violation are identified and appropriate action is taken to rectify the problem. If two mating components are being designed then the association of their individual spaces is held at the highest hierarchical level (a skeletal model) upon which function verification is performed. Each skeletal model can then be fleshed with material independently, applying its own internal set of design, manufacturing and management rules as necessary. Systematically, the two designs are reflected back to the assembly level to check their continuing ability to function. Violations caused by clash and non-assembly will be reflected back down to the component level and be resolved simultaneously in both designs. Thus the complexities of the interrelationship between separate designs are only faced when the need arises. Such cooperative design activity at the initial stages will greatly speed up the design process.

It is also envisaged that such a management network, whilst initially not product specific, will gradually acquire knowledge about the product under design. A successful set of rules can then be made available for application to a new set of designs. Similarly, solutions to a new set of rules can be sought by finding a design created from a closely related group of rules and modifying it to suit the new conditions (by applying the new constraint and relaxing those that are no longer necessary).

The future in design and its management is thus predicted to be highly dependent upon today's technology but to take on social forms quite unlike the present. The way of working and interaction of the designers will change to produce a more creative set of conditions. Designers are not paid for their ability to enter lines and circles into a CAD system; it is their ability to convey creative thoughts to the system that makes them invaluable. They will not disappear; the machine will not take over. They will through these proposed systems forge a true partnership. The designers will generate the needs and propose the solutions; the system will verify and cross-relate with its acquired knowledge and the requirements of other designers to seek out and resolve conflicts and failures. All personnel on the system will become 'designers' looking after differing aspects of the process. Some will operate on the functional needs, some on the material properties and strengths and some on the manufacturing requirements, whilst all will fall under the control of the management network.

Although these changes may appear to be far-reaching, they seem almost mundane when considered in an actual design situation. Let us imagine that such a networked system exists within an aircraft company and has been in use for some time, with the whole of the current aircraft range designed on it. So for this example let us consider not a new design but the modification of a previous one: a new, more powerful and more efficient set of engines is now available that should be incorporated if possible.

The redesign must thus commence at the highest design level. Here function, not structure, is under consideration. The first problem to be addressed is whether these changed engines will, in conjunction with the existing aerodynamic characteristics of the airframe, provide the performance required by the customer. This may range from that of a short takeoff requirement through to the overall operating efficiency. If it can be shown at this stage that these cannot be met then either the project should be abandoned or the specification should be renegotiated.

Such analysis will lead to the identification of those rules that govern the range of acceptable positions and attitudes of the engines. In parallel to those operations the structural design team will have been undertaking a study to ensure that the engine locations do not violate the structural integrity of the existing airframe system. If these two sets of rules can be satisfied (either by a single or by a number of different arrangements) then the detailed design work can commence. If, however, there is a conflict between the structural and performance requirements this will be identified and resolved at a management level, either by renegotiation of the specification or by instructions that a major structural redesign be undertaken (or possibly both).

As the design advances from the developing conceptual arrangement down through the assembly and the subassemblies to the detailed components, the results of decisions taken at each stage will be automatically reflected down to the lower levels. The definition that the engines will be hung by a bracket from a link to a strong point implies that a shear bolt is required to maintain that assembly. This will in its turn reflect into the design of both the bracket and the link. The relationship between all three items will be verified and maintained by an appropriate set of rules. A request that the shear bolt must be removable will result in clearance checking rules being evoked to ensure that violations do not occur during bolt withdrawal.

Details of the various components are chosen to provide various functions including strength and manufacturability. These are thus automatically verified when their forms become known. Changes and compromises are thus made at the earliest possible stages by the management process. All aspects of the design and manufacture are handled long before any metal is cut, ensuring that the design can be produced with relative ease.

Conclusions

Such a future may seem to be a long way off. The technology however already exists. The knowledge upon which such a system would be built is very advanced. All that is now needed to turn these dreams into tomorrow's designing tools is the commitment to make such radical changes to the way we handle design processes.

If it can be done, it is certain that it will be done by someone very soon. The potentials to be offered by such techniques suggest that those who decide not to pursue them could well regret their lack of courage as a whole new approach to design and its management emerges.

43 New Directions in Design and Management Education

WILLIAM CALLAWAY
Chelsea College of Art and Design, UK

Introduction

Greater understanding between those who design and those who manage industrial organizations can enhance both profitability and the quality of the product or service delivered. This chapter is concerned in general with the means by which that greater understanding may be promoted, and specifically with the role of education as one of those means. The main focus is on university-level training in design (both industrial and engineering) and in management and business subjects.

Higher education has enormous potential to be an agent of change and to alter practitioners' attitudes. Not all those who manage or design pass through higher education and this limit is acknowledged. Nonetheless, one recent educational project (CNAA 1984), which set out to improve design management competence in industry by identifying best practice and rendering it in the form of exemplars back to industry, has attracted wide interest in its outcomes.

The Problem

Higher education has perpetuated the lack of mutual understanding between those who emerge from the separate fields of design and management. The problem is twofold. First, there is a lack of information in each group about the technical knowledge and methods of the other; secondly, there is a gap between the collective cultures of the two groups. At its baldest, the problem is that in management courses the product is taken for granted or ignored. In design courses, the physical object itself is central but the context of its production and marketing tends to be ignored.

This simplification highlights the two issues to be tackled. The divide between management education and design education is both a result of different concerns – hence the absence of transfer of technical knowledge between the two fields – and more profoundly a manifestation of two very different cultures.

The origin of the division is readily seen. Design education is of some antiquity; the majority of design teaching institutions were established as a result of moves to improve manufactured goods in the second half of the nineteenth century through the foundation of art schools. 'Art' encompassed that industrial application which in the current century has come to be denoted by 'design'. Even where art schools have subsequently been absorbed into larger, multidiscipline institutions, the distinctive traditions have not been lost. The distinctiveness is confirmed by the persistence of specialist awards, design having been taken into the degree sector only since 1974 in the UK.

The quality of design education in Britain (and elsewhere) is often held to be a result of this separate tradition. The characteristics are essentially individual teaching by practitioners, a dominance of visual media over verbal and a pedagogy centred on the student's own experience of practical activity. Assessment involves artefacts as evidence of achievement far more than the written word, and examinations are rare. In contrast, management education is a comparatively recent phenomenon and, though innovative in the use of techniques such as gaming and simulation, it relies heavily on formal group methods of transmitting information, on study rather than practice and on assessment by written examination. Words, written and spoken, are the dominant medium.

In these circumstances, it is unsurprising that students in the two fields encounter little of the content of each other's courses and emerge with stereotyped attitudes towards each other. Ignorance of other fields is not unusual among graduates of specialist courses and is readily rectified in work situations where awareness of others' skills is demanded. For this very reason, the formation process for most professionals requires a post-qualification period of apprenticeship in which specialist skills are contextualized and their complementarity to those of other professionals established. The gap in knowledge between professional managers and professional designers, however, is exacerbated by the cultural gap engendered by the profoundly different premises on which the educations are rooted.

Management education is much concerned with those business functions associated with controlling – especially accounting, descriptive statistics and cost analysis – and with analysis of the market and presentation of the product to the customer. The education of designers is essentially concerned with creating. The divisive effect of two parallel, non-convergent educational tracks, one about controlling and one about creating, cannot be ignored.

Creativity, the ability to improve and innovate by intellectual leaps when faced with problems, is not the exclusive prerogative of designers. But socialization into the values of the design professions, plus the virtual hijack of the word 'design' by even the serious daily press to indicate that which is currently fashionable, plus non-designers' fear of the 'artistic', all conspire to suggest that creativity is a special quality reserved to those who are products of design education. This suggestion, moreover, can appeal both to non-designers and regrettably often to designers. Even that creativity which is the hallmark of new industrial enterprises can be all too easily overlooked.

It is not frivolous to note that this cultural division is manifest in the stereotypes fed by lack of knowledge of the other's territory. At worst, the stereotypes refer to external appearances. This has been expressed anecdotally to the writer by an industrial manager's suspicious reference to designers as the 'chaps in yellow trousers', and by the regret of the head of a design company that he was constrained to dress in a manner he considered more appropriate to a bank manager (see Callaway 1982).

The stereotype of mutual antipathy between yellow trousers and pinstriped suit would be merely amusing, did it not signify the ascription of relative status to the garments' occupants. The intensely practical training of designers tends to inculcate a 'technical expert' mentality. Little definitive evidence exists about the aspirations of design students, but it is regarded as a *sine qua non* among them and their teachers that they will pursue careers in design and occupy the traditional relationship to industry of consultant (either in that actual role or in-house as expert on tap). The first destinations of design graduates confirm this perception of aspirations; yet the fact that some design graduates eventually move to general management positions shows a capacity for transfer of their creative, problem solving minds. On the agenda for the future of design education must be that design graduates, like other graduates in technical subjects, should aspire to senior management positions as a normal outcome.

Towards a Solution

The most significant initiative in the UK at reconciling the stances of education in management and in design has been the Managing Design project (CNAA 1984). The project's approach to the business/design overlap has found much international support (Topalian 1985). The interposition of a new tier of experts between the design function and senior management was viewed as counter-productive; instead the model offered to education, and ultimately for the reshaping of industrial attitudes, was that of 'osmosis', a term adopted to describe the unique potential of having design schools operate within the same institutions as business schools and engineering schools. The experience of six teaching institutions in which business schools and design schools have collaborated as part of the project has been that this osmosis can indeed occur, the formation of an interdisciplinary teaching team being a microcosm of the breaking down of cultural barriers (Davisworth 1987).

If design management is not to be construed as a new, discrete discipline, it is necessary to offer an alternative theoretical framework for the relation of design issues to business practice and for appropriate education. The model offered here recasts the manager-as-controller, designer-as-creator tradition which has perpetuated the knowledge and culture gap, and replaces it with a model of associated disciplines.

Shifts in opinion have permitted a view of design as a multidisciplinary activity not confined to the specialist functions of design sections (for example, NEDO 1986). The process of design in industrial contexts from this new viewpoint is seen as involving almost all functional areas of business. Redefinition of the type of professionals needed has tended to centre on the design-aware manager, the focus of the Managing Design project and other initiatives to inject teaching about design into business schools. Equally, there is a case for reconsideration of the mainstream graduate from design education. The business-oriented designer, like the design-aware manager, is not entirely a theoretical type, but those who actually exist have most usually been formed by live experience rather than as a result of education. Design courses have the capability to contribute to producing the business-oriented designer if the essential attributes of the type are recognized and the learning experience is adjusted appropriately.

The role of the designer may be reconstructed as concerned not simply with creating but also with controlling, providing a balance which bridges the problematic cultural divide. The programme for design education follows from this reconstruction. A theme in recent discussions on the role of the designer in industry is that of designer as responsive to the needs of the client. A survey of design tutors' attitudes confirms the needs of customers to be the business input perceived as most important (Trustrum et al. 1987). Though few would dispute that designers must be informed about and responsive to the needs of the client, it should be recognized that such needs will, in practice, be defined through classic market analysis. However, students should not be trained to accept the current market as defining the parameters of their work. From a range of explorations, the Managing Design project has found marketing and design to be complementary functions and design not to be led by marketing considerations (Callaway 1987a). Examining a situation considered by many to be well in advance of the West, Lorenz (1986) describes the process by which a leading Japanese company developed ways to research and comprehend those changes in social attitudes and behaviour which would lead to opportunities for products as yet unknown. The nature of the needs to which designers are responding may thus be informed by analysis well beyond normal market research.

The bounds to designers' work, given a leap beyond the limits of conventional market research, are then the possibilities offered by technological developments and the future markets identified. A useful metaphor for the designer's role is that of the thermostat (Gorb 1986) controlling the flow of technologically driven innovation to the market; redefine the market as the future informed by social analysis, rather than the present, and a role emerges for the designer as controller as much as creator.

A common culture – or at least two not dissimilar cultures – may then be possible as an alternative to the traditional divide. In an education system which enables management students to understand the crucial role of design and enables design students both to see the potential role of the designer in corporate strategy and to aspire to management positions, the problem of sharing knowledge will cease to exist and cultural bridges can be built. The very categories 'design student' and 'management student' should not be regarded as immutable. The possibility should not be excluded of radical shifts in the current boundaries of the two disciplines.

Realizing this prescription demands careful experiment and identification of those elements which design students and management students need to experience and of the means by which that experience should be gained. It is in this field that current initiatives have particular value.

In management education the guidelines for courses are now well defined, though there remains much to be done in ensuring that the subject is allowed into the curriculum by those who control individual courses. The report from the Managing Design project (CNAA 1984) identified the crucial nine topics which all management students need:

The economic and business context of design
The nature of designers' work
Design and product strategies
Design policy making
Researching design and product requirements
Managing design projects
Elements of design work
Evaluating design results
Legal and quasi-legal aspects of design.

The subsequent pilot implementation of the recommendations from this project (CNAA 1987) has sharpened knowledge further on the practical aspects of introducing design to management students, as well as validating the project's basic proposition that design, to be effective in industry, requires commitment from the senior levels of the organization.

Less definitive evidence is available on the requirements for changes in design course curricula, although a number of valuable projects are under way. The writer has conducted a number of experiments and in the light of that experience offers an outline of the body of knowledge (Callaway 1987b) which all design students should encounter and of the experience they may be expected to gain.

Some objective guidance exists on design personnel needs as seen by UK industry. Technical skills are generally found adequate, but Hayes (1983) noted that employers saw adaptation to the corporate environment by industrial design graduates as a problem, and BTEC (1985) found that employers of young designers thought their business skills – people management and the understanding of market and consumer behaviour – could be better.

In a study of recent graduates' reflections on their educational experiences

(Brennan and McGregor 1987), a proportion of art and design graduates strikingly higher than in other subject fields said their courses had offered insufficient opportunities for developing written communication skills. Also, art and design was among the subject fields with the perceived worst record for the development of numeracy.

The business-oriented designer must, on this evidence, be equipped with better business skills, better communication skills and the ability to operate in a corporate environment. Technical skills, as might be expected from an education uniquely grounded in practical experience, do not need improvement; the shortfall is in some areas of knowledge and in transferable personal skills. Experimental projects confirm these as the areas needing attention, but the level of change needed is greater than the industrial/graduate evidence suggests. Remedial action to ensure skills for the current role of designer as technical expert will not bridge the cultural gap.

An agenda for the profound change in design education – bearing in mind the extensive change already set out for management education – demands a core of studies applicable to all graduates' future work situations, plus a preparation for at least a proportion of students which will raise aspirations and enable design education to produce future managers as well as technical professionals. Students' programmes, as with all degree programmes, need to equip them for subsequent further learning and transfer to areas of activity as yet not existing. Narrow responses to current manpower requirements – at best an inexact area – tend to limit the scope of learning.

The business-oriented designer will be formed by courses which enable the gaining of professional skills, the understanding of the structure of industry and commerce and the acquisition of knowledge of what is commercially viable. There are existing courses which already do much of this. Professional skills include students' own skills and attributes, transferable to successful practice in whatever work situation. Self-presentation and communication techniques are basic to this category. Some techniques may be apparently elementary, such as learning effective telephone use, a skill frequently mentioned by employers of design graduates as a weak area; others, such as interpersonal and negotiating skills, draw fully on the maturity of university-level students. Certain models, such as the presentation of work by students to critical peer groups, much used in courses within the architectural paradigm, are invaluable in fostering these skills.

Understanding the structure of industry and commerce demands that students be acquainted, albeit at an elementary level, with the nature of organizations and the theoretical categories into which organizations fall. Some introduction to the working of the economy is necessary, particularly the role of financial institutions in industry and commerce. At a more local level, the way that the functional areas of business interrelate is crucial, both in helping students to comprehend the conventional role of the designer in industry and in demonstrating the potential for change. The concept of corporate strategy and the actual and potential significance of design policy are important in building student confidence in the significance of their activity as more than just a technical specialism. The locus of design decision taking in an organization and the 'gatekeepers' of design need to be demonstrated. In these latter, though best practice needs always to be observed in case students are led to aspire to less than the best, it is salutary to demonstrate the nature of design hidden in organizations which ostensibly have no design function (Davies and Hird 1987).

Helping students to learn to be commercial, unlike the above two categories, requires concentration on the specific sector or sectors of industry and commerce to which a particular student's course tends to lead on graduation. The use, with discrimination as to their limitations, of the outcomes of market research and forecasting techniques; understanding what makes a product successful in the market; relating that success to technological innovation and the analysis of social change: these elements sharpen the commercial sense. Given the concentration of design education upon, in general, a single artefact per project, it is essential to introduce the notion of company profitability over a range of products.

These suggestions assume that the business-oriented designer can be created by a core of studies not specific to a particular sector. Design education has a strong tradition of specialism, often held to be the origin of the quality of its graduates. To suggest that there are common skills and approaches which equip graduates for more effective performance in industrial situations is not to suggest abandonment of the sector-specific study which forms the mass of a course. Rather it is to suggest that, over and above its unique properties (especially the pedagogic tradition of learning by experience), design education can confer on its graduates extra advantage by adding on those qualities which should be common to all who emerge from higher education, such as the ability to communicate cogently the approach which has been taken to a given problem.

Conclusions

Closing the gap in knowledge between those who pass through management education and those who pass through design education demands a shift in the content of both sets of courses, in order to heighten knowledge about the nature and value of design among senior managers and to enlighten prospective designers on the structure of industry and commerce and on their own potential as managers.

To close the knowledge gap might be counted a success; greater success would lie in bridging the cultural divide, but this is not achievable alone by teaching the basics. The two approaches towards enterprise – controlling and creating – have to achieve some balance inside the bodies of knowledge characterized as management theory and design theory, and the balance must be reflected in the philosophical stances underpinning educational programmes.

And more: individuals need to come together, since cultural exchange is dependent on familiarity. In educational contexts, experimentation with mixed-group projects (i.e. management and design students together) has revealed the practical difficulties of constructing cultural exchange but has not been without success (Heap and Slater 1987; Clipson 1986).

Similarly, the possibility that 'designer' and 'manager' can be combined in the same individual needs to be tackled. The postgraduate management sector is predominantly concerned with students who already have experience: the category 'management student' includes in practice many senior managers. It was for this reason that the Managing Design project was targeted at this category. Management students come from almost all industrial sectors and from public and private employment. Yet it remains a rarity for a design graduate to reappear as a management student, welding the two cultures into one individual.

Note

The Managing Design project is a collaboration between the Council for National Academic Awards, the Design Council and the Department of Trade and Industry. It has three stages: first, a major international research exercise to identify best practice and render it as curriculum topics for postgraduate management courses; secondly, a linked group of six pilot implementations at selected educational centres; and thirdly, a review by a distinguished US group funded jointly with the Fulbright Commission (see CNAA 1984; CNAA 1987). The editor of this book was research fellow for the first stage, and the writer of this chapter is the project's coordinator.

References

Brennan, J. and McGregor, P.: *Graduates at Work: Degree Courses and the Labour Market.* London: Jessica Kingsley, 1987.

BTEC: *The Design Needs of Industry.* London: Business and Technician Education Council, 1985.

Callaway, W.: *Supervised Work Experience in Educational Programmes: Sandwich Degree Courses in Design.* London: Council for National Academic Awards, 1982.

Callaway, W.: *Managing Design Reassessed.* In CNAA (1987), [1987a].

Callaway, W.: *Industrial and Commercial Studies for Design Students: a Handbook for Teachers in Higher Education.* London: Design Council, 1987b.

Clipson, C.: *First Things First: a Business and Design Educational Experiment.* Ann Arbor: University of Michigan Press, 1986.

CNAA: *Managing Design: an Initiative in Management Education.* London: Council for National Academic Awards, 1984.

CNAA: *Managing Design: an Update.* London: Council for National Academic Awards, 1987.

Davies, H. and Hird, M.: *Caravans, Designers, Lasers and Dustbins.* In CNAA (1987).

Davisworth, P.: *Managing Design, the Leicester Experience.* In CNAA (1987).

Gorb, P.: The business of design management. *Design Studies.* Vol. 7, no. 2, pp. 106–110, 1986.

Hayes, C.: *The Industrial Design Requirements of Industry.* London: Design Council, 1983.

Heap, J. and Slater, P.: *Managing Design, the Leeds Experience.* In CNAA (1987).

Lorenz, C.: *The Design Dimension.* Oxford and New York: Basil Blackwell, 1986.

NEDO: *Design Working Party Recommendations.* London: National Economic Development Office, 1986.

Topalian, A.: Waiting in the wings. *Designer.* p. 9, April 1985.

Trustrum, L., Grundy, R. and Bancroft, G.: *Professional and Business Awareness in Design Degrees, Summary of Preliminary Findings.* Stoke-on-Trent: North Staffordshire Polytechnic, 1987.

44 Objects in their Thousands: Progress or Evolution?

YVES DEFORGE
France

Introduction

If there is one concept which should be approached with caution, it is the concept of progress. Biased arguments are often used to illustrate it – arguments which ignore what should, for the sake of symmetry, be called counter-progress or even regression.

Take the motor car, which is to serve as our example in this chapter. It is commonly stated that progress from the first horseless carriages to the cars of today has consisted of improving all the factors that were merely in a rough, preliminary state early on. This is relatively true from car to car, although not everything can be considered an improvement. A car of 1900 had serious drawbacks but at least one advantage: it was spacious, like the carriages from which it had developed. If we go on to examine the car's relations with technical, economic and social systems, we see that the problems arising from its use – over-production, international competition, cumulative pollution, the exhaustion of energy resources and accidental death – all have to be set against progress in a generalized balance sheet showing costs and profits. These are not easy to quantify, and raise doubts about any simplistic reference to progress.

So we will leave this particular argument, which might be pursued to infinity and is likely to arouse passionate feelings and elicit opinions on points which are not really comparable, and instead simply state the obvious: during its century of existence the motor car has changed. The question we then have to ask is whether this change can be called evolution.

Like the animal kingdom or the vegetable kingdom, the mechanical kingdom can be considered from a naturalistic viewpoint. With the car, we even have the advantage of being able to study changes over a short period in a large number of individuals. This is an approach that may be applied to all technical objects produced by human industry, with the merit that it transfers the debate to the sphere of facts and avoids value judgements.

The Evolution of Technical Objects: Functions and Principles

When we consider technical objects, we have to take account of two basic characteristics:

1   The function or functions they provide in use.
2   The principle or principles of operation with which they are concerned.

In tracing their evolution, it is important to distinguish between objects which have the same form but not the same function, and objects which perform the same function but work on different principles. Objects which have the same function in use and operate on the same principle constitute a line of descent from the original formative act (bringing the principle into play) to extinction (either because the function no longer has to be fulfilled or because some other principle replaces the first). Looking at such a line of descent in chronological order, one generally notices a process of evolution towards various states (Deforge 1985).

Evolution towards autonomy and acclimatization

If a function is to be fulfilled in use, certain actions must take place. If the function of a motor car in use is to take the people inside it from A to B, it must be driven in accordance with certain procedures. The division of actions (or one might say of elementary functions) between user and machine indicates a kind of evolution which generally tends to make the machine take over actions (Giedion 1980) leading to autonomy of operation (the machine becomes independent of human operation). This autonomy is marked by evolutions:

Towards auto-regulation, an autonomous capacity to respond to internal and external disturbances and accidents. If evolution towards auto-regulation entails a reduction of the operational limits, then the environment becomes the determining factor.

Towards auto-correlation of the relations of various factors with each other. A simple example is the evolution towards auto-correlation of two parts in related motion correlated by lubrication. Originally a delicate manual operation, lubrication has evolved in the direction of 'permanent' lubrication, self-lubricating bearings, and the internalizing of lubrication by the use of porous and ceramic materials.

Towards auto-sufficiency, which may be regarded as a kind of final acclimatization, the supreme autonomy. The object maintains and repairs itself, as when circuits are doubled or even trebled; if one circuit fails, operation switches automatically to the other.

Towards automatic operation and cybernetics   In becoming independent of human operation, the object reaches a level of automatic operation which means that it can be associated in interconnecting units with objects on a similar level of automatic operation. The interconnection may be close (in the case of a multifunctional object or system) or distant (in the case of systems constituting networks).

Obviously, in this evolution towards autonomy, absolute functional complexity is increased by the fact that the object is taking over the elementary functions once performed by people.

Evolution towards the concrete and synergistic

When the object incorporates elementary functions, they are first juxtaposed and then enter into a relationship with each other, more or less merging. This is the phenomenon described by Simondon (1989) as the passage from the abstract to the concrete by synergy of functions.

Evolution towards smaller size: miniaturization

Synergy of functions causes structural complexity (the number of parts and pieces) to diminish. We may thus say that, all other things being equal, we are witnessing evolution in the sense of dimensional reduction reinforced by the complementary use of materials. The evolution may be countered when there is an ergonomic relationship to the user or to other less highly evolved factors. For instance, the functional complexity of the electronic wrist-watch has increased considerably in comparison with the mechanical watch and, thanks to the synergies which electronics allow, structural complexity has diminished a great deal. However, if the watch's function is to be fulfilled in use, its face, case and strap still need to be of suitable dimensions even if they themselves become synergistic (for example, if the case and strap are in the form of a bracelet made of a single piece of metal).

As the *principle* is that factor of the object which remains stable, it is not, by definition, affected by the evolution of a line of descent.

Examination of the origin of material responses to a function in use would cast light on the genesis of lines of descent, but since we are concerned with observing evolution in itself, we can simply note that in general the early period is marked by an anarchic abundance of attempts to make use of all the principles available. Next, constructive solutions based on certain wide principles emerge and become vigorous lines of descent. However, those lines which lead nowhere may be kept in reserve, so to speak, to re-emerge when new conditions are introduced, or when the object has evolved to saturation point on a certain principle (for instance, moving on from the piston engine to the turbocharger, an old principle tried out again in the 1970s). The picture we now have, then, is that of a family tree with:

1 Some lines of descent interrupted, some lines re-emerging, mutations caused by a change of principles.
2 Crossed lines when two separate lines marry, producing either a new or a hybrid object.
3 Dichotomies when a line divides into two branches.
4 Multiple divisions when the function is distributed between two different objects (e.g. the twin carburettor, the auxiliary engine, etc.). This mul-

tiplicity may be a response to functional overloading, although the phenomenon goes against the normal evolutionary process.

5   The ending of teratological lines, where perverse evolution can lead to ridiculous or monstrous developments (e.g. the Big Boy locomotive of 1936, which weighed 340 tonnes and had 24 wheels).

The Car as a Typical Example

Evolution towards autonomy

Figure 44.1 represents the general course of evolution from the chauffeur of the earliest cars – who used a crank to start the engine, operated numerous controls by means of handles, levers and knobs, and attended to repairs and lubrication – to the driver of a modern car with a vocal command system and route planner.

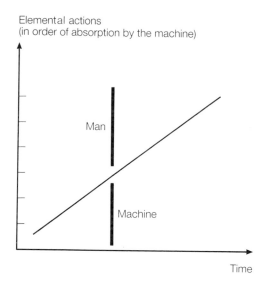

Figure 44.1 The motor car: evolution towards autonomy

It is possible to trace detailed parallel processes of evolution towards auto-sufficiency: for instance, active auto-maintenance (windscreen washing and defrosting) and passive auto-maintenance (resistant paint and plastic body-work). The same applies to automation of the controls: for instance, automatic control of the height of headlamp beams to suit the conditions, control of their direction by the movement of the steering wheel, automatic control of their brightness according to the environment, and so on.

Evolution towards the concrete and synergistic

Electronics have speeded up the car's evolution towards the concrete and the synergistic. It may be said, of mechanical as well as electronic parts, that absolute functional complexity (that is, the number of functions, not arranged in any hierarchy), seen in relation to the number of parts employed for the realization of those functions, provides a ratio applicable to each line of descent.

Normally, functional complexity is increased by the incorporation of functions and the number of parts is reduced by concretization, leading to a diminution of the ratio until we reach those asymptotic values which cannot be transcended except by a radical change of principle. Figure 44.2, showing the evolution of the ratio in cars between 1900 and 1980, is an approximation because of the variety of construction adopted by various makers and models in the same periods.

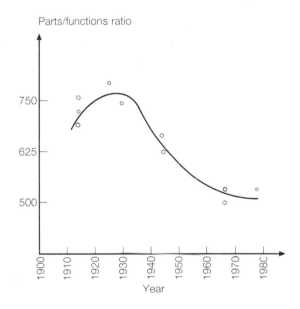

Figure 44.2 The motor car: evolution of the parts/functions ratio

Evolution towards the concrete can also be illustrated by the evolution of car bodies. Initially, a car was made of three (or four) clearly distinct parts: the engine, the driver's compartment, the passengers' compartment, and perhaps a luggage compartment. Progressively these parts merged into a whole, integrating the wings, the bonnet and the headlamps (now set into the wings, and sometimes with a protective shutter when not in use). The final stage of evolution towards the concrete was bodywork consisting of a single shell made of injected plastic or fibreglass.

Synergies of function appear when one part fulfils several functions initially covered by several parts. In the car, an example is lighting: the headlamp with its full beam and dipped beam, which were once separate, now consists of a single optical unit. Other functions, like lighting in fog, are still distinct, but are normally expected to be integrated with the optical unit.

Evolution towards smaller size

Because of synergy of functions, a reduction in the number of parts, the use of increasingly efficient materials and a limit on volume (particularly important in town cars), we may say that, all other things being equal, evolution moves towards a reduction of the dimensions of the object over a period of time.

This is illustrated by the parallel evolutions of power/weight ratio (watts per kilogram) and power/volume ratio (watts per cylinder cubic centimetre) in the car's internal combustion engine. Figure 44.3 represents the power/volume ratio (Robson 1977) and has the form of a classic life cycle curve: the ratio rises rapidly and then levels out. We can therefore formulate three hypotheses:

1 Evolution has reached the extreme limits of its potential and the object may be considered 'naturalized'.
2 Evolution may be repeated on the same principle, with the same object made out of other materials, such as ceramics.
3 Evolution will start again, based on another principle.

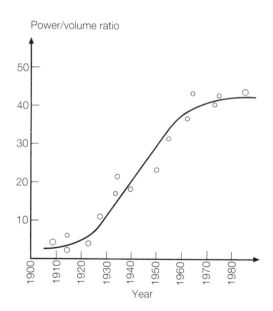

Figure 44.3 The motor car: power/volume ratio

Future Prospects

If the motor car was nothing but a means of transport, it might by now have become naturalized, or even standardized (as in the USSR, where until recently there were only one or two types of car per user category). The following is the argument to the contrary put forward by Mercedes-Benz in its advertising (1987):

If everything in the world were excellent, there would be no more excellence ... and the world would be much the poorer, a place of dismal uniformity. With nothing different and thus nothing desirable to offer. But such a theory has little chance of ever being put to the test: there are many areas in which excellence is not within everyone's reach. In the world of cars, it is the province of Mercedes-Benz. Its *raison d'être*. Perhaps that is what its famous star logo means.

However, if the car is a utilitarian technical and functional object, it is also everything that mythology, publicity and the deep-seated psychology of those who devise and use it want it to be. Hence, the car may be presented as not belonging to the (large) family of ordinary cars (as in the Mercedes publicity). Or else this ordinary car, of the three-door, four- to five-seat type (which at a distance much resembles all its rivals), claims to be different in some particular feature – to be honest, a minor feature – which is emphasized in its advertising: the car has three lateral stripes or a laminated windscreen, or has a few hundredths less of a drag coefficient.

These features are only symbolic variations on a single principle and a single process of evolution; they denote singularity, safety or speed. But because the motor car is not only a functional but also a cultural object, they are essential factors in relation to the logical – or technological – outcome which the modern car represents. A different prospect may be suggested to readers: an idea which is not entirely new but which may well be destined to make its mark in the medium term, not only with cars but with industrial objects in general.

One of the features of the classic system of industrial production is series production. With cars, this gives rise to a stream of identical vehicles, and if one car succeeds in standing out from its competitors it is only one among thousands. It has been said that as evolution leads to the naturalization of systems and the standardization of subsystems, the systems and subsystems may be regarded as constituting a stock of standard components, capable of combination, and which may subsequently be repackaged and recombined. This idea was put forward by engineers at the end of the nineteenth century (e.g. Reuleaux 1877); they postulated that mechanical systems constituted a finite and calculable whole, and concluded that all existing or potential machines could be only a combination of elementary machines that were already known. Their theory encountered a setback when it was pointed out that the calculations of the period, extensive as they were, were not exhaustive. For social, economic and technological reasons, matters are different today.

The difficulty in series production is having these elements in stock and putting them together at the right moment. However, the introduction of computer aided planning, data processing of production and flexible automated factories now makes it possible. So limited series may be proposed, with many different versions combining various factors and differing levels of equipment. We may even envisage the end of series production and the production instead of one-off items. Like the first motor cars, they would have originality, but with the added factors of reliability and functional capacity.

According to this hypothesis, the planning engineer who works on a model intended to be produced in its thousands (or millions) would become a designer engineer, setting out the range of possible combinations (including replicas of old models). The customer would choose a model (of a car or of some other object) from this range. Alternatively, employing the designer as practical adviser, the customer could devise the product to his or her own liking, constrained only by the limits of possible combinations. Then a code number describing the object (not as yet produced) would inform production of the requirement. Stocks held would be no longer of finished products, monotonously lined up to await delivery, but of components and/or information programs, allowing production to order the desired combination.

Conclusion

This is one possible prospect. It would allow the creation of something new through almost-infinite combinations of elements within a finite whole, and it would restore a certain power of creativity to the consumer. It is technically feasible and economically desirable; the rest is up to the imagination of planners and the education and culture of consumers.

References

Deforge, Y.: *Technologie et génétique de l'objet industriel.* Paris: Maloine, 1985.

Giedion, S.: *Mechanization Takes Command.* New York: Oxford University Press, 1948.

Giedion, S.: *La Mécanisation au pouvoir.* Paris: CCI, 1980.

Reuleaux, F.: *Cinématique (Principes fondamentaux d'une théorie générale des machines).* Paris: Sauvy, 1877.

Robson, G.: *Encyclopédie des voitures.* Paris and Brussels: Elsevier, 1980.

Simondon, G.: *Du mode d'existence des objets techniques.* Paris: Aubier, 1958.

Simondon, G. and Deforge, Y.: *Du mode d'existence* (rev. edn). Paris: Aubier, 1989.

45 Conclusion: Strategies for Design

MARK OAKLEY

Aston Business School, UK

Introduction

A wide range of issues and points of view has been presented in this book. It is clear that design management is still very much an evolving discipline and that it is by no means certain where its boundaries will eventually be drawn. At one level, some of our contributors show that there remains a strong allegiance to a traditionally segmented approach to the design process – with a resulting emphasis on 'engineering design' and a disregard for 'industrial design', or vice versa. Similarly, some contributors have projected a relaxed, almost intuitive approach to design and its management, while others have presented details of highly structured and rigorous methods.

Nevertheless, certain preoccupations do recur. In as much as it is possible to summarize any themes in a volume of this magnitude and scope, perhaps the editor may be allowed to conclude by drawing attention to aspects which he believes to be of special and generally applicable significance.

Design as a Neglected Contributor to Business Success

Despite the pressures of the modern marketplace, there are still managers around who are apparently oblivious to the effect that design (or, rather, lack of design) can have on the performance of their businesses. While competition on price, manufacturing ability and service levels is undeniably important, in more and more cases the roots of success now lie over-whelmingly in the design of the product or service which is being offered for sale.

Nowhere is this better illustrated than in Lorenz's case study of Ford of America (chapter 15). This shows not only that a major, marketing-aware corporation can lose sight of product design as the key ingredient of marketplace domination, but more importantly that substantial feats of managerial skill and commitment may be needed to bring about the necessary changes of attitude and policy within the organization.

Designers as Visionaries

There are dangers in allowing designers to be elevated to the status of idols or gurus; as with pop stars, actors or even evangelical management consultants, the true value of their contributions to society in general and clients in particular may be vastly over-stated. However, the point is made by Viti and Vidari, Palshoj, Woudhuysen (chapters 4, 5, 28) and others that designers can perform an important inspirational role and help to lead companies in vital new directions. They alone may provide creative clues to the possibilities which are available to a business; others in the firm may have highly developed analytical skills, well suited to understanding the past and the present but not capable of constructing or defining a vision of the future.

Need for Interfunctional Collaboration in Design Work

This is a strongly recurring theme throughout the book. For success in design work, it is essential that cooperation and commitment are achieved between all parts of the company and the design activity. This is especially so in the case of product design, where numerous contributors including Gunz, Holt and Maclean (chapters 18, 21 and 29) stress the importance of involving key participants, such as marketing, manufacturing and finance, *from the earliest stages of a design project*. The dangers of treating design as an isolated, out-of-the-mainstream activity cannot be over-stated; at best such an approach leads to disinterest and lack of commitment, and at worst it may give rise to total rejection of design results.

While few people in the firm may have the skills necessary to participate in the creative aspects of design work, many more will have some other kind of valid contribution to offer in the form of factual information or considered opinion. If the opportunity to contribute this knowledge is denied, the groups or individuals so affected may, quite rightly, demonstrate strongly negative feelings towards the results of the design exercise from which they have been excluded. This may well cause the new product to be a failure or to be abandoned even if, intrinsically, it is a good design.

Importance of both Project and Policy Aspects of Design

The distinction between project and policy issues (or operational and corporate activities) has been highlighted by Clipson, Oakley, Topalian (chapters 11, 12, 13) and others. There is some tendency for the project aspects of design management to receive the greatest attention and the policy considerations to be largely neglected. In many cases, this is a reversal of what ought to be happening. Too many managers focus their attention on the running of individual design projects and completely fail to consider why the projects have been set up in the first place, what they are expected to achieve and what are the measures by which a successful outcome will be judged.

Part of the problem that this reflects is the lack of interest, concern and knowledge which is a characteristic of the board-level management of a large number of companies. Policy making in general is a top management responsibility, so if design issues are not included in this process they are unlikely to be addressed elsewhere in the organization.

Amongst design policy issues, the most fundamental is the correct determination of a *design strategy* and its relevance to the overall business strategy. The editor's conviction is that this is the single most important issue raised in this volume and that it is therefore appropriate to conclude by presenting observations on the topic. In keeping with the practical intentions of the book, the reader is left with a checklist of questions which may help in arriving at an effective design strategy.

Business and Design Strategies

A strategy is a plan of action put into effect by a firm to achieve commercial advantage. Companies do not become market leaders just by improving accounting systems or re-equipping factories; it is their products or services which lead them to success.

Such companies achieve prosperity because they have formulated effective business strategies which make proper reference to design. They have identified their targets in terms of both the customers they want to reach and the qualities they know they must provide in order to satisfy those customers. They understand that the key to making a product or service succeed is design. Design and general business strategies are interdependent; spending money on design without attempting to understand customers' wants or trying to analyse competitors' actions will produce many failures and only occasional successes. Similarly, to research and define a business strategy, and then fail to use design to implement it, is equally unwise.

Some strategies work better than others

In practice, four types of business strategies may be encountered.

An *absent* strategy is like a black hole; it does not exist but has powerful effects. Typically, the company tries to sell what it can make; if customers do not buy then it is they who are wrong, not what they are being offered. Competitors are not acknowledged even when they capture 99 per cent of the market. Managers are preoccupied with many minor operational problems and design work is carried out only in response to extreme pressure.

A *misguided* strategy is more worthy of sympathy; at least an effort has been made to focus upon the role of the product or service in the affairs of the company. Unfortunately, perhaps because of poor information gathering or weak decision making, the strategy selected is the wrong one. An example would be where a firm decided to concentrate on the manufacture of a traditional mechanical product – not realizing that superior electronic devices were replacing it. Normally, the causes of weakness in strategies are more subtle. Lack of knowledge of trends in style and colour, ignorance of the effects of new assembly technologies, unawareness of demographic or social changes: these are just a few of the many factors which may render a strategy obsolete.

To be really effective, a strategy must be developed in the light of sound knowledge of company strengths and market opportunities. Depending on the nature of these, it may be appropriate to pursue either a reactive or a proactive strategy. A *reactive* strategy is one where market needs are correctly identified and the company is successfully able to respond with new or improved products or services. In contrast, a *proactive* strategy encourages the use of design to create new business opportunities by extending technology and styling and thus stimulating new market demands. The Japanese Sony company is an example; the cassette recorder is refined into the Walkman personal stereo and the television becomes a pocket TV or a high performance monitor.

Developing the right strategy

Put simply, a business strategy should be formulated to make the best use of company resources to build the strongest possible presence in the market. This involves three basic steps. The first is to understand the market: learning or confirming the preferences of customers, the activities of competing companies, the changes in technology and so on. The second is to appraise the real strengths of the company: what skills, knowledge and resources are really available (or can realistically be acquired)? Only as a result of these enquiries can the third step of defining a strategy be attempted with a high expectation of success.

Before exploring these three steps, it is worth pausing to underline the need for practicality in what is being proposed. In trying to order ideas, it is all too easy to produce a regimented prescription that might be rightly rejected by managers working in a far from ordered world. The guidelines are intended to form a basis for decision making; their use in practice will depend on the context in which they are applied. For example, in small companies, resources may not permit detailed information gathering and interpretation; managers have many pressing tasks and a sense of balance must be maintained. However, even a little time spent reviewing issues may help improve company performance; sometimes a few notes written about a problem may spark a new understanding or recall an old solution.

One further point is important: the three steps imply a rather tightly constrained environment. If a company settles on a strategy just on the basis of a review of itself and the market which it serves, it may be taken by surprise by some event in the wider environment. To minimize this risk, it is important to have an understanding of key national and international trends. Trade flows, levels of innovative activity or even political and social changes can all impinge on a company's plans and performance. Of particular concern are links between design and these trends; for example, if imports of goods increase at the expense of domestic production, is this connected with more distinctive styling and better quality?

Step 1: know the market

Most companies have an understanding of the market in which they operate, to the extent that they can measure sales levels, describe typical customers, acknowledge different segments of the market, etc. All these are important and should be taken into account when developing strategies for the future. However, there is often considerably less knowledge readily available about some equally important but rather more subtle factors.

The first of these concerns changes and trends in the market and in customers' behaviour. If customers' tastes are changing, it may be possible to plot the progression and even to estimate how tastes may develop in the future. In making such predictions it will be necessary to review the underlying causes, which may range from general factors such as increased leisure time to specific factors such as changed energy costs.

Secondly, the rate of change can be as important as the nature of the change itself. Most products and services have a finite life in the marketplace; when they lose favour with customers they must be replaced if sales income is to be maintained. The period that a particular design can survive unchanged is reflected by its life cycle – a marketing concept which reminds us that all successful products and services typically enjoy increasing, then stable, levels of sales before ultimately declining. In many sectors these cycles are shortening, either because purchasers are becoming bored more quickly or because greater competition encourages more frequent changes – or both.

At the same time that life cycles are getting shorter, customers' expectations of increasing sophistication often dictate design exercises of greater scope. Thus design projects have to start earlier (and more frequently) to satisfy market conditions more distant in the future. Inevitably this increases the risks of business, further underlining the crucial importance of deciding on the right strategy.

Accurate monitoring and predicting of developments achieved by competitors and others are essential parts of strategy formulation. There are many well-developed methods of forecasting which can help pin-point the technical features and performance that might replace today's norms. Correct interpretation of future conditions will enable one of the most important strategic parameters to be decided – the right balance between incremental refinement and radical change. If technology and/or customer tastes are developing steadily then a strategy of incremental design changes may be best. However, if significant discontinuities are expected then totally new design concepts may have to be created.

Step 2: assess company potential

A realistic strategy must take account not only of external influences but also of the company's own potential. Large companies have been known to overstretch themselves in attempting to meet a market challenge. Small companies may be extremely vulnerable; if they outstrip their resources, demise may be sudden and absolute. Some kind of audit is necessary to provide an unbiased assessment of the quality and quantity of resources available, together with a critical appraisal of whether previous design projects have been managed successfully and have achieved the intended results.

The review of resources should include all parts of the enterprise, with the aim of identifying not only those areas in which there is particular competence, but also those where there is weakness. Human resources are usually the single most important factor in design work; as well as people with specific design skills and knowledge, managers able to guide projects are vital to success. It is all too easy to make incorrect assumptions about the availability of skills, especially in large companies where the effects of changes in personnel and expertise may not be appreciated by senior managers until a disinterested assessment is made which confirms the true current profile of human resources. Likewise the state of technical resources may be poor; equipment and processes may be out of date and inadequate for the needs of the present, let alone the future.

However, in many cases the overriding concern will be availability of financial resources; often deficiencies in other resources can be put right if sufficient funds can be earmarked. Unfortunately, designing new products and services frequently consumes more money (and time) than was originally expected. Hence it is crucial to identify not only readily available funds but also the scope for securing contingency finance in case projects exceed original budgets.

Another important task is to evaluate previous design activities. Except in brand new companies there will be some history of designing, and this track record should be examined. The general question will be whether earlier projects were successful in terms of design excellence. Were results produced which gained enthusiastic market acceptance? If not, what were the causes and is it likely that history will be repeated, leading to further disappointing results? Were projects managed competently and did they stay within budgets? Were target dates met? If design expertise from outside the company was used, was it appropriate expertise, was value for money obtained, and was control maintained using competent project documentation?

Step 3: define the strategy

Analysing markets, competition and technology will have revealed the opportunities (and risks) that lie ahead; the hard look at the company will have confirmed its strengths and weaknesses. Putting these together will form the basis of a strategy to guide future actions. A design strategy for products or services should be a key part of the broader business strategy which sets out the general objectives for the enterprise – the corporate mission.

Usually the design strategy is so important that, as well as being a part of the business strategy, it is a very significant determinant of it. For example, a decision to design, produce and sell a world-standard business computer system by an electronics company might dictate corporate objectives that include mergers with or acquisitions of other manufacturers, upgrading of customer service systems, greater clarity of the corporate identity, better conditions of employment to attract highly skilled workers, promotion of a higher level of confidence amongst shareholders, and so on.

The detail in which a strategy is defined will depend on a number of factors, including the size of the company and the extent to which the new strategy departs significantly from existing practice. But in all cases, three key areas of information should be covered. First, product/service details should be set out to confirm where future design efforts will be needed. The explanation should be in terms of major design characteristics, especially those different from current models and competitors' expected developments. Characteristics might include those relating to technology, appearance, ergonomics, quality, price, reliability, etc.; they should be defined with as much precision as possible together with the levels of performance necessary to achieve objectives.

Next, market targets should be recorded – the sectors to be concentrated upon and the shares of sales required. Where a range of products or services is to be designed, then different parts of the range may be aimed at different markets or sectors and these should be coordinated. Finally, the management implications should be set out including the demands which will be made on financial and other resources. Similarly, the consequences for organizational structures in the company (and links with outside agencies) need to be addressed – particularly where changes in roles or responsibilities need to be agreed and understood. This last point should also take into account the importance of confirming the tasks assigned to the board and those to be tackled at the operating level; all major decision making should be guided by the strategy.

Checklist for Developing a Strategy

What trends in the wider world may have an impact on our products and business performance?

1   Are major social/economic/political trends affecting the markets for our products? How have customers' preferences, habits and tastes changed during the last five, ten and fifteen years? Will they continue to change?
2   Have new competitors appeared, especially overseas? Have others disappeared? (Why?) What is the current performance of competitors in terms of indicators such as productivity, investment, return on sales and market shares? How innovative are competitors as measured by patents and R&D expenditure?
3   Is our business threatened by shortages of raw materials, by new technology (perhaps from outside the industry), by lack of skills, etc.?

How good is our understanding of our markets and customers?

4   Can we describe typical market and customer profiles? Who are our customers? What are their incomes, ages, preferences, etc.? Are there several distinct types of customers; do they fall into a number of market segments which we can identify and measure?

5   Do we try to cater for all these market segments? If so, is this wise? Would we do better by concentrating on just one or two? Would we do more business if the product was redesigned to suit more precisely the needs of existing or new customers?

6   Do we really know why customers buy our product? Is it mainly because of price? Or is it because of other factors relating to design and quality, such as:
Technical   Durability, reliability, flexibility, performance, ergonomics, maintenance, etc.
Aesthetic   General appearance, styling, colour, finish, texture, noise, comfort, etc.
Economic   Selling price, depreciation, running cost, service/repair cost, etc.?

7   What are our competitors doing? Are they aiming for the same market segments? What is their performance on price, quality and design? How often do they revise their products? What technical and other trends are they pursuing?

Do we have an accurate understanding of the resources and potential of our company?

8   Would it be useful to conduct an audit of skills and strengths in the company? What kind of design skills do we have (or can easily obtain)? Are they of the right kind and quality? What about management skills to set up and guide design projects? And marketing skills to help draw up design specifications?

9   What is the relationship between design and other departments in the company? Are there obstacles and barriers to new ideas? Are designers physically isolated from other functions? What is the reason for this?

10 How does the design department relate to customers? Is there direct contact or is information filtered by managers in other departments?

11 Do design staff receive adequate motivation, support and encouragement? Are they given the opportunity to learn about new developments? Is their contribution understood and acknowledged?

In short, does management generally understand the nature of design work and how to organize it?

12 What about time and cost considerations? Does management know the cost of design projects? Is there an effective cost recording system? Is there a good understanding of time deadlines and the design results that can be realistically expected within them?

13 Who is responsible for evaluating design results? What reference standards are used? Are clear design objectives set and are they generally achieved?

14 What is the company's design track record?

What strategy will give the best chance of success?

15 What does the market really want? What can our company provide? What is the best way to use design to match company strengths with market opportunities?

16 Should we start with changes or improvements to existing products (and/or processes)? Can we reduce costs and improve the attractiveness of our current ranges? What are the priorities (based on maximum potential benefits)?

17 Do existing products need to be replaced by new ideas or new technology? If so, what is the source of these: customers, suppliers, ourselves, competitors?

18 What are the time constraints? Do we have to aim for a particular window in the market (an opportunity we have predicted or an important event such as a trade show)? Have we taken account of design lead times, product life cycles and the need to avoid design gaps?

19 Have we calculated how much money will be needed? Can we afford it? Do we know when it will be required? Is there a contingency fund? Is the right manpower available? Are other resources available?

20 Are we properly organized for design work? Who is responsible? What targets have to be achieved, and against what standards? Who will judge results?

Index